T0202825

Lecture Notes in Artificial Intelligence 13588

Subseries of Lecture Notes in Computer Science

More information about this subseries at https://link.springer.com/bookseries/1244

Leszek Rutkowski · Rafał Scherer ·
Marcin Korytkowski · Witold Pedrycz ·
Ryszard Tadeusiewicz ·
Jacek M. Zurada (Eds.)

Artificial Intelligence and Soft Computing

21st International Conference, ICAISC 2022
Zakopane, Poland, June 19–23, 2022
Proceedings, Part I

 Springer

Editors
Leszek Rutkowski (iD)
Systems Research Institute of the Polish
Academy of Sciences
Warsaw, Poland

The Institute of Computer Science
AGH University of Science and Technology
Kraków, Poland

Marcin Korytkowski (iD)
Częstochowa University of Technology
Częstochowa, Poland

Ryszard Tadeusiewicz (iD)
AGH University of Science and Technology
Kraków, Poland

Rafał Scherer (iD)
Częstochowa University of Technology
Częstochowa, Poland

Witold Pedrycz (iD)
University of Alberta
Edmonton, AB, Canada

Jacek M. Zurada (iD)
University of Louisville
Louisville, KY, USA

ISSN 0302-9743   ISSN 1611-3349 (electronic)
Lecture Notes in Artificial Intelligence
ISBN 978-3-031-23491-0   ISBN 978-3-031-23492-7 (eBook)
https://doi.org/10.1007/978-3-031-23492-7

LNCS Sublibrary: SL7 – Artificial Intelligence

This Springer imprint is published by the registered company Springer Nature Switzerland AG
The registered company address is: Gewerbestrasse 11, 6330 Cham, Switzerland

Preface

This volume constitutes the proceedings of the 21st International Conference on Artificial Intelligence and Soft Computing ICAISC 2022, held in Zakopane, Poland, on June 19–23, 2022. The conference was organized by the Polish Neural Network Society in cooperation with the Department of Intelligent Computer Systems at the Częstochowa University of Technology, the University of Social Sciences in Łódź, and the IEEE Computational Intelligence Society, Poland Chapter. The conference was held under the auspices of the Committee on Informatics of the Polish Academy of Sciences.

Previous conferences took place in Kule (1994), Szczyrk (1996), Kule (1997) and Zakopane (1999, 2000, 2002, 2004, 2006, 2008, 2010, 2012, 2013, 2014, 2015, 2016, 2017, 2018, 2019, 2020 and 2021) and attracted a large number of papers and internationally recognized speakers: Lotfi A. Zadeh, Hojjat Adeli, Rafal Angryk, Igor Aizenberg, Cesare Alippi, Shun-ichi Amari, Daniel Amit, Plamen Angelov, Sanghamitra Bandyopadhyay, Albert Bifet, Piero P. Bonissone, Jim Bezdek, Zdzisław Bubnicki, Jan Chorowski, Andrzej Cichocki, Swagatam Das, Ewa Dudek-Dyduch, Włodzisław Duch, Adel S. Elmaghraby, Pablo A. Estévez, João Gama, Erol Gelenbe, Jerzy Grzymala-Busse, Martin Hagan, Yoichi Hayashi, Akira Hirose, Kaoru Hirota, Adrian Horzyk, Tingwen Huang, Eyke Hüllermeier, Hisao Ishibuchi, Er Meng Joo, Janusz Kacprzyk, Nikola Kasabov, Jim Keller, Laszlo T. Koczy, Tomasz Kopacz, Jacek Koronacki, Zdzislaw Kowalczuk, Adam Krzyzak, Rudolf Kruse, James Tin-Yau Kwok, Soo-Young Lee, Derong Liu, Robert Marks, Ujjwal Maulik, Zbigniew Michalewicz, Evangelia Micheli-Tzanakou, Kaisa Miettinen, Krystian Mikołajczyk, Henning Müller, Christian Napoli, Ngoc Thanh Nguyen, Andrzej Obuchowicz, Erkki Oja, Nikhil R. Pal, Witold Pedrycz, Marios M. Polycarpou, José C. Príncipe, Jagath C. Rajapakse, Šarunas Raudys, Enrique Ruspini, Roman Senkerik, Jörg Siekmann, Andrzej Skowron, Roman Słowiński, Igor Spiridonov, Boris Stilman, Ponnuthurai Nagaratnam Suganthan, Ryszard Tadeusiewicz, Ah-Hwee Tan, Dacheng Tao, Shiro Usui, Thomas Villmann, Fei-Yue Wang, Jun Wang, Bogdan M. Wilamowski, Ronald Y. Yager, Xin Yao, Syozo Yasui, Gary Yen, Ivan Zelinka and Jacek Zurada.

The aim of this conference is to build a bridge between traditional artificial intelligence techniques and so-called soft computing techniques. It was pointed out by Lotfi A. Zadeh that "soft computing (SC) is a coalition of methodologies which are oriented toward the conception and design of information/intelligent systems. The principal members of the coalition are: fuzzy logic (FL), neurocomputing (NC), evolutionary computing (EC), probabilistic computing (PC), chaotic computing (CC), and machine learning (ML). The constituent methodologies of SC are, for the most part, complementary and synergistic rather than competitive".

These proceedings present both traditional artificial intelligence methods and soft computing techniques. Our goal is to bring together scientists representing both areas of research. This volume is divided into five parts:

- Neural Networks and Their Applications
- Fuzzy Systems and Their Applications
- Evolutionary Algorithms and Their Applications
- Pattern Classification
- Artificial Intelligence in Modeling and Simulation

I would like to thank our participants, invited speakers and reviewers of the papers for their scientific and personal contribution to the conference. The advice and constant support of the Honorary Chair of the conference Prof. Ryszard Tadeusiewicz is acknowledged with many thanks. Finally, I thank my co-workers Łukasz Bartczuk, Piotr Dziwiński, Marcin Gabryel, Rafał Grycuk, Marcin Korytkowski and Rafał Scherer, for their enormous efforts to make the conference a very successful event. Moreover, I would like to acknowledge the work of Marcin Korytkowski who was responsible for the Internet submission system.

June 2022   Leszek Rutkowski

Organization

ICAISC 2022 was organized by the Polish Neural Network Society in cooperation with the University of Social Sciences in Łódź and the Institute of Computational Intelligence at Częstochowa University of Technology.

ICAISC Chairpersons

General Chair

Leszek Rutkowski, Poland

Co-chair

Rafał Scherer, Poland

Technical Chair

Marcin Korytkowski, Poland

Financial Chair

Marcin Gabryel, Poland

Area Chairs

Fuzzy Systems

Witold Pedrycz, Canada

Evolutionary Algorithms

Zbigniew Michalewicz, Australia

Neural Networks

Jinde Cao, China

Computer Vision

Dacheng Taom, Australia

Machine Learning

Nikhil R. Pal, India

Artificial Intelligence with Applications

Janusz Kacprzyk, Poland

International Liaison

Jacek Żurada, USA

ICAISC Program Committee

Hojjat Adeli, USA
Cesare Alippi, Italy
Shun-ichi Amari, Japan
Rafal A. Angryk, USA
Robert Babuska, Netherlands
James C. Bezdek, Australia
Piero P. Bonissone, USA
Bernadette Bouchon-Meunier, France
Jinde Cao, China
Juan Luis Castro, Spain
Yen-Wei Chen, Japan
Andrzej Cichocki, Japan
Krzysztof Cios, USA
Ian Cloete, Germany
Oscar Cordón, Spain
Bernard De Baets, Belgium
Włodzisław Duch, Poland
Meng Joo Er, Singapore
Pablo Estevez, Chile
David B. Fogel, USA
Tom Gedeon, Australia
Erol Gelenbe, UK
Jerzy W. Grzymala-Busse, USA
Hani Hagras, UK
Saman Halgamuge, Australia
Yoichi Hayashi, Japan
Tim Hendtlass, Australia
Francisco Herrera, Spain
Kaoru Hirota, Japan
Tingwen Huang, USA
Hisao Ishibuchi, Japan
Mo Jamshidi, USA

Robert John, UK
Janusz Kacprzyk, Poland
Nikola Kasabov, New Zealand
Okyay Kaynak, Turkey
Vojislav Kecman, USA
James M. Keller, USA
Etienne Kerre, Belgium
Frank Klawonn, Germany
Robert Kozma, USA
László Kóczy, Hungary
Józef Korbicz, Poland
Rudolf Kruse, German
Adam Krzyzak, Canada
Věra Kůrková, Czech Republic
Soo-Young Lee, Korea
Simon M. Lucas, UK
Luis Magdalena, Spain
Jerry M. Mendel, USA
Radko Mesiar, Slovakia
Zbigniew Michalewicz, Australia
Javier Montero, Spain
Eduard Montseny, Spain
Kazumi Nakamatsu, Japan
Detlef D. Nauck, Germany
Ngoc Thanh Nguyen, Poland
Erkki Oja, Finland
Nikhil R. Pal, India
Witold Pedrycz, Canada
Leonid Perlovsky, USA
Marios M. Polycarpou, Cyprus
Danil Prokhorov, USA
Vincenzo Piuri, Italy

Sarunas Raudys, Lithuania
Olga Rebrova, Russia
Vladimir Red'ko, Russia
Raúl Rojas, Germany
Imre J. Rudas, Hungary
Norihide Sano, Japan
Rudy Setiono, Singapore
Jennie Si, USA
Peter Sincak, Slovakia
Andrzej Skowron, Poland
Roman Słowiński, Poland
Pilar Sobrevilla, Spain
Janusz Starzyk, USA
Jerzy Stefanowski, Poland
Vitomir Štruc, Slovenia
Ron Sun, USA
Johan Suykens, Belgium
Ryszard Tadeusiewicz, Poland

Hideyuki Takagi, Japan
Dacheng Tao, Australia
Vicenç Torra, Spain
Burhan Turksen, Canada
Shiro Usui, Japan
DeLiang Wang, USA
Jun Wang, Hong Kong
Lipo Wang, Singapore
Paul Werbos, USA
Bernard Widrow, USA
Kay C. Wiese, Canada
Bogdan M. Wilamowski, USA
Donald C. Wunsch, USA
Ronald R. Yager, USA
Xin-She Yang, UK
Gary Yen, USA
Sławomir Zadrożny, Poland
Jacek Zurada, USA

ICAISC Organizing Committee

Rafał Scherer
Łukasz Bartczuk
Piotr Dziwiński
Marcin Gabryel (Finance Chair)
Rafał Grycuk
Marcin Korytkowski (Databases and Internet Submissions)

Contents – Part I

Fuzzy Systems and Their Applications

Evolutionary Algorithms and Their Applications

Pattern Classification

Artificial Intelligence in Modeling and Simulation

Contents – Part II

Various Problems of Artificial Intelligence

Bioinformatics, Biometrics and Medical Applications

Neural Networks and Their Applications

A Fast Learning Algorithm
for the Multi-layer Neural Network

Jarosław Bilski$^{(\boxtimes)}$ ⓘ and Bartosz Kowalczyk ⓘ

Department of Computational Intelligence, Częstochowa University of Technology,
Częstochowa, Poland
{Jaroslaw.Bilski,Bartosz.Kowalczyk}@pcz.pl

Abstract. In this paper, the computational improvement for the scaled Givens rotation-based training algorithms is presented. Application of the scaled rotations boosts the algorithm significantly due to the elimination of the computation of the square root. In a classic variant scaled rotations utilize so-called scale factors — χ. It turns out that the scale factors can be omitted during the computation which boosts the overall algorithm performance even further. This paper gives a mathematical explanation of how to apply the proposed improvement to the scaled variants of the training algorithms. The last section of the paper contains several benchmarks which prove the proposed method to be superior to the classic approach.

Keywords: Neural network training algorithm · QR decomposition · Scaled givens rotations · Optimization · Approximation · Classification

1 Introduction

The artificial neural networks are flexible mathematical abstractions capable of solving many problems, whereas the classic mathematical apparatus fails to succeed. To date, neural networks are the very common utilities in the scope of science and industry. Recently many researches revolving around various types of neural networks were released eg. in [4,26,33,34]. Many branches of artificial intelligence are discussed and applied in various sectors of industry such as biology, medicine [1,20,24,28] and others [13,15,16,21,25–27,29,31].

Most modern neural network training algorithms originate from the classic Backpropagation algorithm that was initially proposed in [30]. As the artificial intelligence evolved many training methods superior to the classic Backpropagation were made [14,23,32]. These algorithms are close to their archetype making use of the error function's gradient and the learning rate occasionally adding some additional hyperparameters such as momentum. The other group of training algorithms is more complex. Their cores originate from various mathematical

This work has been supported by the Polish National Science Center under Grant 2017/27/B/ST6/02852.

L. Rutkowski et al. (Eds.): ICAISC 2022, LNAI 13588, pp. 3–15, 2023.
https://doi.org/10.1007/978-3-031-23492-7_1

principles such as Hessian matrices (eg. Levenberg-Marquardt [19]) or RLS [5]. Each training method has its advantages but also some drawbacks. In most cases more complex algorithms struggle with huge datasets due to computational complexity.

The SGQR training algorithm for feedforward neural networks was initially proposed in [12] to boost the classic GQR method [11]. In this paper, the method to increase the original SGQR algorithm performance is described. This can be achieved by removing the so-called *scale factors* from the equations which leads to the overall boost in performance while maintaining similar training metrics in terms of success ratio and average epoch count per training. The core of the paper contains a detailed mathematical background of the proposed improvement. The equations are followed by a set of benchmarks and conclusions.

2   Givens Rotations Basics

The Givens rotation is one of the well-known orthogonal transformations. It is used in many engineering aspects especially, in image processing. A rotation can be multidimensional but most often it is limited to two dimensions defined by vectors $span\{e_p, e_q\}(1 \leq p < q \leq n)$. Such rotation is described by an orthogonal matrix of the following structure [18,22]:

$$\mathbf{G}_{pq} = \begin{bmatrix} 1 & & & \cdots & & & 0 \\ & \ddots & & & & & \\ & & c & \cdots & s & & \\ & \vdots & \vdots & \ddots & \vdots & \vdots & \\ & & -s & \cdots & c & & \\ & & & & & \ddots & \\ 0 & & & \cdots & & & 1 \end{bmatrix} \begin{matrix} \\ \\ p \\ \\ q \\ \\ \\ \end{matrix} \tag{1}$$

The $\mathbf{G}_{pq}$ matrix given by (1) from now on will be referred to as *rotation matrix* or simply *rotation*. Based on the rotation structure given by (1) it is easy to see that in comparison to the Identity matrix it differs only in terms of four elements $g_{pp} = g_{qq} = c$ and $g_{pq} = -g_{qp} = s$, where

$$c^2 + s^2 = 1 \tag{2}$$

Equation (2) reveals that $\mathbf{G}_{pq}^T \mathbf{G}_{pq} = \mathbf{I}$ which is the proof for matrix $\mathbf{G}_{pq}$ being an orthogonal matrix. Assume that $a \in \mathbb{R}^n$. The following transformation is achieved by a single rotation

$$\mathbf{a} \rightarrow \bar{\mathbf{a}} = \mathbf{G}_{pq}\mathbf{a} \tag{3}$$

Knowing the structure of the rotation matrix (1) it is easy to notice that only two elements of vector **a** will be changed during this transformation. Due to this property, it is possible to compute the values of c and s, so the a_q will be replaced with 0 after being rotated. Let us consider

$$\bar{a}_q = -sa_p + ca_q = 0 \tag{4}$$

To achieve that the parameters c and s of rotation matrix $\mathbf{G}_{pq}$ are calculated as follows

$$c = \frac{a_p}{\rho}, \quad s = \frac{a_q}{\rho}, \quad \text{where} \quad \rho = \sqrt{a_p^2 + a_q^2} \tag{5}$$

3 The Scaled Givens Rotation

For vector $\mathbf{a} \in \mathbb{R}^n$, consider transformation given by (3) [17,22]. Matrix $\mathbf{G}_{pq}$ has to meet the condition (4). The scaled Givens rotation is obtained by using scaled multipliers $\mathbf{K}^2$ and $\bar{\mathbf{K}}^2$:

$$\begin{aligned}
\mathbf{a} &= \mathbf{K}\mathbf{d}, \quad \text{where} \quad \mathbf{K} = diag\left(\sqrt{\chi_l}\right) \\
\bar{\mathbf{a}} &= \bar{\mathbf{K}}\bar{\mathbf{d}}, \quad \text{where} \quad \bar{\mathbf{K}} = diag\left(\sqrt{\bar{\chi}_l}\right)
\end{aligned} \tag{6}$$

where $\chi_l, \bar{\chi}_l > 0\,(l = 1,\ldots,n)$. Also matrix $\mathbf{G}_{pq}$ will be presented in a scalable form

$$\mathbf{G}_{pq} = \mathbf{K}\mathbf{F}_{pq}\mathbf{K}^{-1} \tag{7}$$

where $\mathbf{F}_{pq}$ is:

$$\mathbf{F}_{pq} = \begin{bmatrix} 1 & & \cdots & & 0 \\ & \ddots & & & \\ & & \alpha & \cdots & \beta & & p \\ \vdots & & \vdots & \ddots & \vdots & & \vdots \\ & & -\gamma & \cdots & \delta & & q \\ & & & & & \ddots & \\ 0 & & \cdots & & & & 1 \end{bmatrix} \begin{matrix} p & q \end{matrix} \tag{8}$$

Equation (3) takes the form

$$\begin{aligned}
\mathbf{K}^2 &\to \bar{\mathbf{K}}^2 \\
\mathbf{d} &\to \bar{\mathbf{d}} = \mathbf{F}_{pq}\mathbf{d}
\end{aligned} \tag{9}$$

and Eq. (4) becomes the following

$$\bar{d}_q = -\gamma d_p + \delta d_q = 0 \tag{10}$$

From (7) the following is obtained

$$\bar{\chi}_l = \chi_l \quad \text{for} \quad (l \neq p, q; l = 1,\ldots,n) \tag{11}$$

$$c = \alpha\sqrt{\frac{\bar{\chi}_p}{\chi_p}} = \delta\sqrt{\frac{\bar{\chi}_q}{\chi_q}}, \quad s = \beta\sqrt{\frac{\bar{\chi}_p}{\chi_q}} = \gamma\sqrt{\frac{\bar{\chi}_q}{\chi_p}} \tag{12}$$

Equation (2) must also be satisfied.

Because there are six variables $\alpha, \beta, \delta, \gamma, \bar{\chi}_p, \bar{\chi}_q$ and only four Eqs. (10), (12) and (2), two cases have to be treated as parameters. Two variants are possible, see the important parts of the $\mathbf{F}_{pq}$ matrix

$$
\begin{bmatrix} 1 & \beta \\ -\gamma & 1 \end{bmatrix} \quad \text{and} \quad \begin{bmatrix} \alpha & 1 \\ -1 & \delta \end{bmatrix} \tag{13}
$$

From (5) and (6) the following is obtained

$$
c^2 = \frac{a_p^2}{a_p^2 + a_q^2} = \frac{\chi_p d_p^2}{\chi_p d_p^2 + \chi_q d_q^2}, \quad s^2 = \frac{a_q^2}{a_p^2 + a_q^2} = \frac{\chi_q d_q^2}{\chi_p d_p^2 + \chi_q d_q^2} \tag{14}
$$

There are two computational cases:

Case 1: $c \neq 0$ i.e. $d_p \neq 0$. The two parameters are set as follows

$$
\alpha = \delta = 1 \tag{15}
$$

from (10), (12) and (14) the following is obtained

$$
\gamma = \frac{d_q}{d_p}, \quad \beta = \frac{\gamma \chi_q}{\chi_p} = \frac{\gamma \bar{\chi}_q}{\bar{\chi}_p}. \tag{16}
$$

From (12) is $\bar{\chi}_i = \chi_i c^2$ for $i = p, q$. Taking into account equation

$$
\frac{1}{c^2} = 1 + \beta\gamma \overset{def}{=} \tau \tag{17}
$$

and (12), the following values are obtained

$$
\bar{\chi}_p = \frac{\chi_p}{\tau}, \quad \bar{\chi}_q = \frac{\chi_q}{\tau}, \quad \bar{d}_p = d_p\tau. \tag{18}
$$

Case 2: $s \neq 0$ i.e. $d_q \neq 0$. The two parameters are set as follows

$$
\beta = \gamma = 1 \tag{19}
$$

from (10) we obtain

$$
\delta = \frac{d_p}{d_q}, \quad \alpha = \frac{\delta \chi_p}{\chi_q} = \frac{\delta \bar{\chi}_q}{\bar{\chi}_p}. \tag{20}
$$

From (12) is $\bar{\chi}_p = \chi_q s^2$ and $\bar{\chi}_q = \chi_p s^2$. Taking into account equation

$$
\frac{1}{s^2} = 1 + \alpha\delta \overset{def}{=} \tau \tag{21}
$$

and (12) the obtained values are

$$
\bar{\chi}_p = \frac{\chi_q}{\tau}, \quad \bar{\chi}_q = \frac{\chi_p}{\tau}, \quad \bar{d}_p = d_q\tau. \tag{22}
$$

Equations (11, 15–22) allow to determine parameters $\alpha, \beta, \gamma, \delta$ of matrix $\mathbf{F}_{pq}$ and scaling multipliers $\bar{\chi}_i$.

4   Weights Update in the FSGQR Algorithm

The FSGQR algorithm is designed for any multi-layered neural network with any differentiable activation function. The weight update is computed based on the error measure given as

$$J\left(n\right)=\sum_{t=1}^{n}\lambda^{n-t}\sum_{j=1}^{N_L}\varepsilon_j^{(L)2}\left(t\right)=\sum_{t=1}^{n}\lambda^{n-t}\sum_{j=1}^{N_L}\left[d_j^{(L)}\left(t\right)-f\left(\mathbf{x}^{(L)T}\left(t\right)\mathbf{w}_j^{(L)}\left(n\right)\right)\right]^2$$
(23)

Finding the minimum of function (23) is a primary target for the SGQR algorithm. It starts with the classic error backpropagation phase followed by linearisation of the activation function, which yields

$$\sum_{t=1}^{n}\lambda^{n-t}f'^2\left(s_i^{(l)}\left(t\right)\right)\left[b_i^{(l)}\left(t\right)-\mathbf{x}^{(l)T}\left(t\right)\mathbf{w}_i^{(l)}\left(n\right)\right]\mathbf{x}^{(l)T}\left(t\right)=\mathbf{0}$$
(24)

The SGQR algorithm is using rotation matrices, hence Eq. (24) needs to be presented in the matrix notation as follows

$$\mathbf{A}_i^{(l)}\left(n\right)\mathbf{w}_i^{(l)}\left(n\right)=\mathbf{h}_i^{(l)}\left(n\right)$$
(25)

where

$$\mathbf{A}_i^{(l)}\left(n\right)=\sum_{t=1}^{n}\lambda^{n-t}\mathbf{z}_i^{(l)}\left(t\right)\mathbf{z}_i^{(l)T}\left(t\right)$$
(26)

$$\mathbf{h}_i^{(l)}\left(n\right)=\sum_{t=1}^{n}\lambda^{n-t}f'\left(s_i^{(l)}\left(t\right)\right)b_i^{(l)}\left(t\right)\mathbf{z}_i^{(l)}\left(t\right)$$
(27)

and

$$\mathbf{z}_i^{(l)}\left(t\right)=f'\left(s_i^{(l)}\left(t\right)\right)\mathbf{x}^{(l)}\left(t\right)$$
(28)

$$b_i^{(l)}\left(n\right)=\begin{cases}f^{-1}\left(d_i^{(l)}\left(n\right)\right)&\text{for}\quad l=L\\s_i^{(l)}\left(n\right)+e_i^{(l)}\left(n\right)&\text{for}\quad l=1\ldots L-1\end{cases}$$
(29)

$$e_i^{(k)}\left(n\right)=\sum_{j=1}^{N_{k+1}}f'\left(s_i^{(k)}\left(n\right)\right)w_{ji}^{(k+1)}\left(n\right)e_j^{(k+1)}\left(n\right)\quad\text{for}\quad k=1\ldots L-1$$
(30)

To solve Eq. (25) we use QR decomposition with the Givens rotations for $k=0\ldots ni-1$ and for $q=k+1\ldots ni-1$ $(p<q)$ as follows

$$\mathbf{G}_{pq}^{(k)}\mathbf{A}_i^{(l)}\left(n\right)\mathbf{w}_i^{(l)}\left(n\right)=\mathbf{G}_{pq}^{(k)}\mathbf{h}_i^{(l)}\left(n\right)$$
(31)

or in the scaled form

$$\mathbf{KF}_{pq}^{(k)}\mathbf{K}^{-1}\mathbf{KE}_i^{(l)}\left(n\right)\mathbf{w}_i^{(l)}\left(n\right)=\mathbf{KF}_{pq}^{(k)}\mathbf{K}^{-1}\mathbf{Kd}_i^{(l)}\left(n\right)$$
(32)

where $\mathbf{h}_i^{(l)}(n) = \mathbf{K}\mathbf{d}_i^{(l)}(n)$, and $\mathbf{A}_i^{(l)}(n) = \mathbf{K}\mathbf{E}_i^{(l)}(n)$. After multiply the Eq. (32) by $\mathbf{K}^{-1}$ it is obtained

$$\mathbf{IF}_{pq}^{(k)}\mathbf{E}_i^{(l)}(n)\,\mathbf{w}_i^{(l)}(n) = \mathbf{IF}_{pq}^{(k)}\mathbf{d}_i^{(l)}(n) \tag{33}$$

or simply

$$\mathbf{F}_{pq}^{(k)}\mathbf{E}_i^{(l)}(n)\,\mathbf{w}_i^{(l)}(n) = \mathbf{F}_{pq}^{(k)}\mathbf{d}_i^{(l)}(n) \tag{34}$$

Now it is easy to see the all

$$\chi_i = \bar{\chi}_i = 1. \tag{35}$$

This leads to a Fast Scaled GQR algorithm. The two computational cases take the form:
Case 1: $c \neq 0$ i.e. $d_p \neq 0$. The two parameters are set as follows

$$\alpha = \delta = 1 \tag{36}$$

the remaining parameters are calculated

$$\gamma = \frac{d_q}{d_p}, \quad \beta = \gamma. \tag{37}$$

$$\tau = 1 + \gamma^2 \tag{38}$$

$$\bar{d}_p = d_p\tau. \tag{39}$$

Case 2: $s \neq 0$ i.e. $d_q \neq 0$. The two parameters are set as follows

$$\beta = \gamma = 1 \tag{40}$$

other parameters are calculated

$$\delta = \frac{d_p}{d_q}, \quad \alpha = \delta. \tag{41}$$

$$\tau = 1 + \alpha^2 \tag{42}$$

$$\bar{d}_p = d_q\tau. \tag{43}$$

Equations (35–43) allow to calculate parameters $\alpha, \beta, \gamma, \delta$ of matrix $\mathbf{F}_{pq}$. The scaling factors $\bar{\chi}_i = 1$ and should not be calculated.

In the FSGQR algorithm the linear response $(s_i^{(l)})$ is calculated for each neuron. That forces the Eq. (25) to be solved as many times as there are neurons in the network. This is achieved by the Givens QR decomposition that utilizes a fast variant of the scale rotations as described in the previous section. The intermediary rotation matrix $\mathbf{Q}^T$ does not need to be stored in the memory since, only the rotation parameters α, β, γ, and δ are needed to accomplish the decomposition process.

$$\mathbf{Q}_i^{(l)T}(n)\,\mathbf{A}_i^{(l)}(n)\,\mathbf{w}_i^{(l)}(n) = \mathbf{Q}_i^{(l)T}(n)\,\mathbf{h}_i^{(l)}(n) \tag{44}$$

$$\mathbf{R}_i^{(l)}(n)\,\mathbf{w}_i^{(l)}(n) = \mathbf{Q}_i^{(l)T}(n)\,\mathbf{h}_i^{(l)}(n) \tag{45}$$

Once Eq. (45) is calculated, the matrix $\mathbf{A}_i^{(l)}(n)$ is fully transformed to its upper-triangle form depicted as $\mathbf{R}_i^{(l)}(n)$. Due to the upper triangle matrices properties inversion of $\mathbf{R}_i^{(l)}(n)$ is not burdened with high computational complexity. Finally, the weight update formula is

$$\hat{\mathbf{w}}_i^{(l)}(n) = \mathbf{R}_i^{(l)-1}(n)\,\mathbf{Q}_i^{(l)T}(n)\,\mathbf{h}_i^{(l)}(n) \tag{46}$$

$$\mathbf{w}_i^{(l)}(n) = (1-\eta)\,\mathbf{w}_i^{(l)}(n-1) + \eta\,\hat{\mathbf{w}}_i^{(l)}(n) \tag{47}$$

5 Experimental Results

To test the performance of the proposed FSGQR algorithm it has been compared to its predecessors — SGQR and GQR. All three algorithms share the same core — the QR decomposition, but they differ in terms of rotation types. The algorithms have trained several neural networks of various topologies. The main effort was put on the classical multilayered perceptrons (MLP) and fully connected multilayered perceptrons (FCMLP). The FCMLP networks are characterized by additional connections between layers and inputs. That means each layer is connected to the network's input and outputs of all preceding layers.

The conducted experiment yields two measures — Success Ratio (SR), and average training Time (T) in milliseconds. The FSGQR algorithm utilizes two hyperparameters, the learning rate (η) and the forgetting factor (λ). During the experiment, the best combinations of η and λ have been found and used to generate the statistics presented in this paper. For gathering the most valuable data, each experiment was conducted 100 times. The criterion of best matching hyperparameters is called the performance factor and is given by the following equation

$$\xi = \frac{SR}{Ep \cdot T} \tag{48}$$

where Ep is the average epoch count that was required to meet the training target.

5.1 Logistic Function Approximation

The logistic function approximation benchmark assumes training the network in order to resolve the problem given by the following equation

$$y = f(x) = 4x(1-x) \tag{49}$$

The training set for this benchmark contains 11 samples where $x \in [0,1]$. The training target is to drop the average error measure below 0.001.

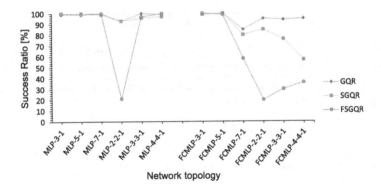

Fig. 1. The logistic function success ratio.

Figure 1 presents the outcome in terms of success ratio which is similar to other GQR variants. Lower values of the SR can be observed for the bigger FCMLP networks.

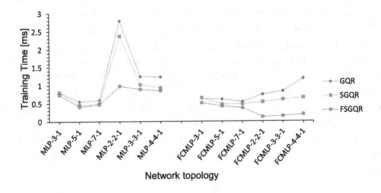

Fig. 2. The logistic function average time.

Figure 2 presents the average training time. The FSGQR algorithm establishes convergence about 50% faster than the classic GQR method. It is worth noticing that the time difference is bigger as the network size grows.

5.2 Hang Function Approximation

In the Hang function benchmark, the network is expected to be able to find a solution for a non-linear two-dimensional function given as

$$y = f(x_1, x_2) = \left(1 + x_1^{-2} + \sqrt{x_2^{-3}}\right)^2 \tag{50}$$

The training set for Hang benchmark contains 50 samples where $x_1, x_2 \in [1, 5]$. The training target is to achieve average error lower or equal to 0.001.

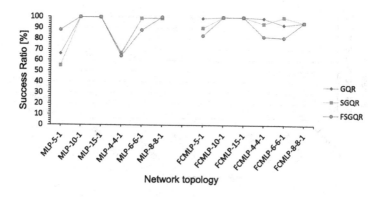

Fig. 3. The Hang success ratio.

As presented in Fig. 3, all three algorithms manifest similar performance in terms of success ratio.

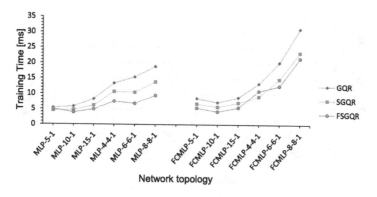

Fig. 4. The Hang average time.

The Hang benchmark average training time is shown in Fig. 4. The overall time boost of the FSGQR over the classic GQR is about 38%. Again, the bigger the network is, the training time difference grows.

5.3 Two Spirals Classification

The Two Spirals benchmark originates from the classification problems domain. The neural network needs to group the input set into two categories — the lower and the upper spiral. The training set for this benchmark consists of 96 samples. The training target is to drop the average epoch error below 0.05 threshold.

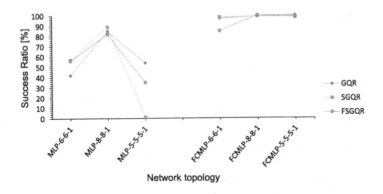

Fig. 5. The Two Spirals success ratio.

The overall success ratio (Fig. 5) of the FSGQR algorithm is similar to it's predecessors.

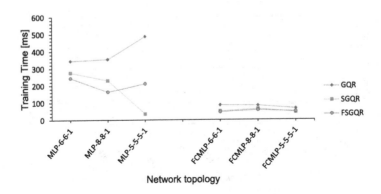

Fig. 6. The Two Spirals average time.

The training time for the Two Spirals benchmark, as presented in Fig. 6, is similar for FSGQR and SGQR algorithms. The FSGQR method manifests about 32% time boost compared to the classic GQR variant.

6   Conclusion

This paper presents the computational optimization for the scaled rotations that are utilized in the SGQR algorithm. The proposed modification is called FSGQR

(Fast Scaled Givens QR decomposition). The core of this paper contains a mathematical explanation of eliminating the redundant scaled factors (χ) calculation which brings a significant time boost for the algorithm. The experimental results that are shown in Sect. 5 confirm that the proposed FSGQR algorithm is superior to its predecessors in terms of the average training time. As expected, the FSGQR algorithm retains a similar (very high) success ratio as other variants of the GQR algorithm. The FSGQR algorithm is eligible to train any feedforward neural network. It is also flexible enough to seek for further improvements and modifications in terms of momentum or parallel computation as initially presented in [2,3,6–10].

References

1. Alsaadi, F.E., et al.: On knowledge discovery and representations of molecular structures using topological indices. J. Artif. Intell. Soft Comput. Res. **11**(1), 21–35 (2021)

2. Bilski, J.: Momentum modification of the RLS algorithms. In: Rutkowski, L., Siekmann, J.H., Tadeusiewicz, R., Zadeh, L.A. (eds.) ICAISC 2004. LNCS (LNAI), vol. 3070, pp. 151–157. Springer, Heidelberg (2004). https://doi.org/10.1007/978-3-540-24844-6_18

3. Bilski, J.: Struktury równoległe dla jednokierunkowych i dynamicznych sieci neuronowych. Akademicka Oficyna Wydawnicza EXIT (2013)

4. Bilski, J., Kowalczyk, B., Marchlewska, A., Żurada, J.M.: Local Levenberg-Marquardt algorithm for learning feedforwad neural networks. J. Artif. Intell. Soft Comput. Res. **10**(4), 299–316 (2020)

5. Bilski, J., Rutkowski, L.: A fast training algorithm for neural networks. IEEE Trans. Circ. Syst. Part II **45**(6), 749–753 (1998)

6. Bilski, J., Smoląg, J.: Parallel Realisation of the recurrent multi layer perceptron learning. In: Rutkowski, L., Korytkowski, M., Scherer, R., Tadeusiewicz, R., Zadeh, L.A., Zurada, J.M. (eds.) ICAISC 2012. LNCS (LNAI), vol. 7267, pp. 12–20. Springer, Heidelberg (2012). https://doi.org/10.1007/978-3-642-29347-4_2

7. Bilski, J., Smoląg, J.: Parallel approach to learning of the recurrent Jordan neural network. In: Rutkowski, L., Korytkowski, M., Scherer, R., Tadeusiewicz, R., Zadeh, L.A., Zurada, J.M. (eds.) ICAISC 2013. LNCS (LNAI), vol. 7894, pp. 32–40. Springer, Heidelberg (2013). https://doi.org/10.1007/978-3-642-38658-9_3

8. Bilski, J., Smoląg, J.: Parallel architectures for learning the RTRN and Elman dynamic neural network. IEEE Trans. Parall. Distrib. Syst. **26**(9), 2561–2570 (2015)

9. Bilski, J., Smoląg, J., Galushkin, A.I.: The parallel approach to the conjugate gradient learning algorithm for the feedforward neural networks. In: Rutkowski, L., Korytkowski, M., Scherer, R., Tadeusiewicz, R., Zadeh, L.A., Zurada, J.M. (eds.) ICAISC 2014. LNCS (LNAI), vol. 8467, pp. 12–21. Springer, Cham (2014). https://doi.org/10.1007/978-3-319-07173-2_2

10. Bilski, J., Smoląg, J., Żurada, J.M.: Parallel approach to the levenberg-marquardt learning algorithm for feedforward neural networks. In: Rutkowski, L., Korytkowski, M., Scherer, R., Tadeusiewicz, R., Zadeh, L.A., Zurada, J.M. (eds.) ICAISC 2015. LNCS (LNAI), vol. 9119, pp. 3–14. Springer, Cham (2015). https://doi.org/10.1007/978-3-319-19324-3_1

11. Bilski, J., Kowalczyk, B., Marjański, A., Gandor, M., Zurada, J.: A novel fast feedforward neural networks training algorithm. J. Artif. Intell. Soft Comput. Res. **11**(4), 287–306 (2021)

12. Bilski, J., Kowalczyk, B.: A new variant of the GQR algorithm for feedforward neural networks training. In: Rutkowski, L., Scherer, R., Korytkowski, M., Pedrycz, W., Tadeusiewicz, R., Zurada, J.M. (eds.) ICAISC 2021. LNCS (LNAI), vol. 12854, pp. 41–53. Springer, Cham (2021). https://doi.org/10.1007/978-3-030-87986-0_4

13. Bougueroua, N., Mazouzi, S., Belaoued, M., Seddari, N., Derhab, A., Bouras, A.: A survey on multi-agent based collaborative intrusion detection systems. J. Artif. Intell. Soft Comput. Res. **11**(2), 111–142 (2021)

14. Duchi, J., Hazan, E., Singer, Y.: Adaptive subgradient methods for online learning and stochastic optimization. J. Mach. Learn. Res. **12**, 2121–2159 (2011)

15. Gabryel, M.: The bag-of-words method with different types of image features and dictionary analysis. J. Univers. Comput. Sci. **24**(4), 357–371 (2018)

16. Gabryel, M., Grzanek, K., Hayashi, Y.: Browser fingerprint coding methods increasing the effectiveness of user identification in the web traffic. J. Artif. Intell. Soft Comput. Res. **10**(4), 243–253 (2020)

17. Gentleman, M.W.: Least squares computations by givens transformations without square roots. IMA J. Appl. Math. **12**(3), 329–336 (1973)

18. Givens, W.: Computation of plain unitary rotations transforming a general matrix to triangular form. J. Soc. Ind. Appl. Math. **6**, 26–50 (1958)

19. Hagan, M.T., Menhaj, M.B.: Training feedforward networks with the marquardt algorithm. IEEE Trans. Neuralnetworks **5**, 989–993 (1994)

20. Izonin, I., Tkachenko, R., Dronyuk, I., Tkachenko, P., Gregus, M., Rashkevych, M.: Predictive modeling based on small data in clinical medicine: Rbf-based additive input-doubling method. Math. Biosci. Eng. **18**(3), 2599–2613 (2021)

21. Łapa K., Cpałka K., Galushkin A.I.: A new interpretability criteria for neuro-fuzzy systems for nonlinear classification. In: International Conference on Artificial Intelligence and Soft Computing, vol. 9119, pp. 448–468 (2015)

22. Kiełbasiński, A., Schwetlick, H.: Numeryczna Algebra Liniowa: Wprowadzenie do Obliczeń Zautomatyzowanych. Wydawnictwa Naukowo-Techniczne, Warszawa (1992)

23. Diederik, P.: Kingma and Jimmy Ba. A method for stochastic optimization, Adam (2014)

24. Larsen, B.M., et al.: A pan-cancer organoid platform for precision medicine. Cell Rep. **36**(4), 109429 (2021)

25. Napoli, C., Pappalardo, G., Tramontana, E., Nowicki, R.K., Starczewski, J.T., Woźniak, M.: Toward work groups classification based on probabilistic neural network approach. In: Rutkowski, L., Korytkowski, M., Scherer, R., Tadeusiewicz, R., Zadeh, L.A., Zurada, J.M. (eds.) ICAISC 2015. LNCS (LNAI), vol. 9119, pp. 79–89. Springer, Cham (2015). https://doi.org/10.1007/978-3-319-19324-3_8

26. Niksa-Rynkiewicz, T., Szewczuk-Krypa, N., Witkowska, A., Cpałka, K., Zalasinski, M., Cader, A.: Monitoring regenerative heat exchanger in steam power plant by making use of the recurrent neural network. J. Artif. Intell. Soft Comput. Res. **11**(2), 143–155 (2021)

27. Nowicki, R.K., Starczewski, J.T.: A new method for classification of imprecise data using fuzzy rough fuzzification. Inf. Sci. **414**, 33–52 (2017)

28. Rahman, J.S., Gedeon, T., Caldwell, S., Jones, R., Jin, Z.: Towards effective music therapy for mental health care using machine learning tools: human affective reasoning and music genres. J. Artif. Intell. Soft Comput. Res. **11**(1), 5–20 (2021)

29. Szczypta, J., Przybył, A., Cpałka, K.: Some aspects of evolutionary designing optimal controllers. In: Rutkowski, L., Korytkowski, M., Scherer, R., Tadeusiewicz,
R., Zadeh, L.A., Zurada, J.M. (eds.) ICAISC 2013. LNCS (LNAI), vol. 7895, pp.
91–100. Springer, Heidelberg (2013). https://doi.org/10.1007/978-3-642-38610-7_9
30. Werbos, J.: Beyond Regression: New Tools for Prediction and Analysis in the
Behavioral Sciences. Harvard University, Cambridge (1974)
31. Zalasiński, M., Cpałka, K., Er, M.J.: A new method for the dynamic signature
verification based on the stable partitions of the signature. In: Rutkowski, L.,
Korytkowski, M., Scherer, R., Tadeusiewicz, R., Zadeh, L.A., Zurada, J.M. (eds.)
ICAISC 2015. LNCS (LNAI), vol. 9120, pp. 161–174. Springer, Cham (2015).
https://doi.org/10.1007/978-3-319-19369-4_16
32. Matthew, D.: Zeiler. An adaptive learning rate method, Adadelta (2012)
33. Zhao, X., Song, M., Liu, A., Wang, Y., Wang, T., Cao, J.: Data-driven temporal-
spatial model for the prediction of AQI in Nanjing. J. Artif. Intell. Soft Comput.
Res. 10(4), 255–270 (2020)
34. El Zini, J., Rizk, Y., Awad, M.: An optimized parallel implementation of non-
iteratively trained recurrent neural networks. J. Artif. Intell. Soft Comput. Res.
11(1), 33–50 (2021)

A New Computational Approach to the Levenberg-Marquardt Learning Algorithm

Jarosław Bilski$^{(\boxtimes)}$ ⓘ, Barosz Kowalczyk ⓘ, and Jacek Smoląg ⓘ

Institute of Computational Intelligence, Częstochowa University of Technology,
Częstochowa, Poland
{Jaroslaw.Bilski,Barosz.Kowalczyk,Jacek.Smolag}@pcz.pl

Abstract. A new parallel computational approach to the Levenberg-Marquardt learning algorithm is presented. The proposed solution is based on the AVX instructions to effectively reduce the high computational load of this algorithm. Detailed parallel neural network computations are explicitly discussed. Additionally obtained acceleration is shown based on a few test problems.

Keywords: Neural network learning algorithm · Levenberg-marquardt learning algorithm · Vector computations · Approximation · Classification

1 Introduction

Artificial feedforward neural networks have been studied by many scientists e.g. [2,12,14,27,28,31,43,45]. One of the most frequently used methods for training feedforward neural networks are gradient methods, see e.g. [18,29,44]. Most of the simulations of neural networks learning algorithms, like other learning algorithms [19,20,30,33,34,36,40,41], work on a serial computer. The computational complexity of many learning algorithms is very high. This makes serial implementation very time consuming and slow. The Levenberg Marquart (LM) algorithm [21,26] is one of the most effective learning algorithms, unfortunately, it requires a lot of calculations. But, for very large networks the computational load of the LM algorithm makes it impractical. A suitable solution to this problem is the use of high performance dedicated parallel structures, see eg. [3,5–13,38,39,48]. This paper shows a new parallel computational approach to the LM algorithm based on vector instruction. The results of the study of a new parallel approach to the LM algorithm is shown in the last part of the paper.

A sample structure of the feedforward neural network is shown in Fig. 1. This sample network has L layers, N_l neurons in each $l-th$ layer, and N_L outputs.

This work has been supported by the Polish National Science Center under Grant 2017/27/B/ST6/02852.

L. Rutkowski et al. (Eds.): ICAISC 2022, LNAI 13588, pp. 16–26, 2023.
https://doi.org/10.1007/978-3-031-23492-7_2

The input vector contains N_0 input values. The Eq. (1) describes the recall phase of the network

$$s_i^{(l)}(t) = \sum_{j=0}^{N_{l-1}} w_{ij}^{(l)}(t) x_i^{(l)}(t), y_i^{(l)}(t) = f(s_i^{(l)}(t)). \qquad (1)$$

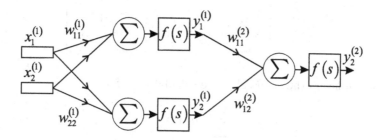

Fig. 1. Sample feedforward neural network.

The Levenberg-Marquard method [21,26] is used to train the feedforward neural network. The following loss function is minimized

$$E\left(\mathbf{w}\left(n\right)\right) = \frac{1}{2}\sum_{t=1}^{Q}\sum_{r=1}^{N_L}\varepsilon_r^{(L)^2}(t) = \frac{1}{2}\sum_{t=1}^{Q}\sum_{r=1}^{N_L}\left(y_r^{(L)}(t) - d_r^{(L)}(t)\right)^2 \qquad (2)$$

where $\varepsilon_i^{(L)}$ is defined as

$$\varepsilon_r^{(L)}(t) = \varepsilon_r^{(Lr)}(t) = y_r^{(L)}(t) - d_r^{(L)}(t) \qquad (3)$$

and $d_r^{(L)}(t)$ is the $r-th$ desired output in the $t-th$ probe.

The LM algorithm is a modification of the Newton method and is based on the first three elements of the Taylor series expansion of the loss function. A change of weights is given by

$$\Delta\left(\mathbf{w}(n)\right) = -\left[\nabla^2\mathbf{E}\left(\mathbf{w}(n)\right)\right]^{-1}\nabla\mathbf{E}\left(\mathbf{w}(n)\right) \qquad (4)$$

this requires knowledge of the gradient vector

$$\nabla\mathbf{E}\left(\mathbf{w}(n)\right) = \mathbf{J}^T\left(\mathbf{w}(n)\right)\varepsilon\left(\mathbf{w}(n)\right) \qquad (5)$$

and the Hessian matrix

$$\nabla^2\mathbf{E}\left(\mathbf{w}(n)\right) = \mathbf{J}^T\left(\mathbf{w}(n)\right)\mathbf{J}\left(\mathbf{w}(n)\right) + \mathbf{S}\left(\mathbf{w}(n)\right) \qquad (6)$$

where $\mathbf{J}\left(\mathbf{w(n)}\right)$ in (5) and (6) is the Jacobian matrix

$$\mathbf{J}(\mathbf{w}\left(n\right)) = \begin{bmatrix} \dfrac{\partial \varepsilon_1^{(L)}(1)}{\partial w_{10}^{(1)}} & \cdots & \dfrac{\partial \varepsilon_1^{(L)}(1)}{\partial w_{ij}^{(k)}} & \cdots & \dfrac{\partial \varepsilon_1^{(L)}(1)}{\partial w_{N_L N_{L-1}}^{(L)}} \\ \vdots & \cdots & \vdots & \cdots & \vdots \\ \dfrac{\partial \varepsilon_{N_L}^{(L)}(1)}{\partial w_{10}^{(1)}} & \cdots & \dfrac{\partial \varepsilon_{N_L}^{(L)}(1)}{\partial w_{ij}^{(k)}} & \cdots & \dfrac{\partial \varepsilon_{N_L}^{(L)}(1)}{\partial w_{N_L N_{L-1}}^{(L)}} \\ \vdots & \cdots & \vdots & \cdots & \vdots \\ \dfrac{\partial \varepsilon_{N_L}^{(L)}(Q)}{\partial w_{10}^{(1)}} & \cdots & \dfrac{\partial \varepsilon_{N_L}^{(L)}(Q)}{\partial w_{ij}^{(k)}} & \cdots & \dfrac{\partial \varepsilon_{N_L}^{(L)}(Q)}{\partial w_{N_L N_{L-1}}^{(L)}} \end{bmatrix}. \tag{7}$$

In the hidden layers the errors $\varepsilon_i^{(lr)}$ are calculated as follows

$$\varepsilon_i^{(lr)}(t) \overset{\wedge}{=} \sum_{m=1}^{N_{l+1}} \delta_i^{(l+1,r)}(t)\, w_{mi}^{(l+1)}, \tag{8}$$

$$\delta_i^{(lr)}(t) = \varepsilon_i^{(lr)}(t)\, f'\left(s_i^{(lr)}(t)\right). \tag{9}$$

Based on this, the elements of the Jacobian matrix for each weight can be computed

$$\frac{\partial \varepsilon_r^{(L)}(t)}{w_{ij}^{(l)}} = \delta_i^{(lr)}(t)\, x_j^{(l)}(t). \tag{10}$$

It should be noted that derivatives (10) are computed in a similar way it is done in the classical backpropagation method, except that each time there is only one error given to the output. In this algorithm, the weights of the entire network are treated as a single vector and their derivatives form the Jacobian matrix $\mathbf{J}$.

The $\mathbf{S}\left(\mathbf{w}(n)\right)$ component (6) is given by the formula

$$\mathbf{S}\left(\mathbf{w}(n)\right) = \sum_{t=1}^{Q} \sum_{r=1}^{N_L} \varepsilon_r^{(L)}(t)\nabla^2 \varepsilon_r^{(L)}(t). \tag{11}$$

In the Gauss-Newton method it is assumed that $\mathbf{S}\left(\mathbf{w}(n)\right) \approx 0$ and that equation (4) takes the form

$$\Delta\left(\mathbf{w}(n)\right) = -\left[\mathbf{J}^T\left(\mathbf{w}(n)\right)\mathbf{J}\left(\mathbf{w}(n)\right)\right]^{-1}\mathbf{J}^T\left(\mathbf{w}(n)\right)\varepsilon\left(\mathbf{w}(n)\right). \tag{12}$$

In the Levenberg-Marquardt method is is assumed that $\mathbf{S}\left(\mathbf{w}(n)\right) = \mu\mathbf{I}$ and that equation (4) takes the form

$$\Delta\left(\mathbf{w}(n)\right) = -\left[\mathbf{J}^T\left(\mathbf{w}(n)\right)\mathbf{J}\left(\mathbf{w}(n)\right) + \mu\mathbf{I}\right]^{-1}\mathbf{J}^T\left(\mathbf{w}(n)\right)\varepsilon\left(\mathbf{w}(n)\right). \tag{13}$$

By defining

$$\begin{aligned} \mathbf{A}\left(n\right) &= -\left[\mathbf{J}^T\left(\mathbf{w}(n)\right)\mathbf{J}\left(\mathbf{w}(n)\right) + \mu\mathbf{I}\right] \\ \mathbf{h}\left(n\right) &= \mathbf{J}^T\left(\mathbf{w}(n)\right)\varepsilon\left(\mathbf{w}(n)\right) \end{aligned} \tag{14}$$

the Eq. (13) is as follows

$$\Delta\left(\mathbf{w}(n)\right) = \mathbf{A}(n)^{-1}\mathbf{h}\left(n\right).\tag{15}$$

The Eq. (15) can be solved using the QR factorization

$$\mathbf{Q}^{T}\left(n\right)\mathbf{A}\left(n\right)\Delta\left(\mathbf{w}(n)\right) = \mathbf{Q}^{T}\left(n\right)\mathbf{h}\left(n\right),\tag{16}$$

$$\mathbf{R}\left(n\right)\Delta\left(\mathbf{w}(n)\right) = \mathbf{Q}^{T}\left(n\right)\mathbf{h}\left(n\right).\tag{17}$$

This paper used the Givens rotations for the QR factorization. The operation, in 5 steps, of the LM algorithm is described below:

1. The calculation of the network outputs for all input data, errors, and the loss function.
2. The calculation of the Jacobian matrix, using the backpropagation method for each error individually.
3. The calculation of weight changes $\Delta\left(\mathbf{w}(n)\right)$ using the QR factorization.
4. The recalculation of the loss function (2) for new weights $\mathbf{w}(n) + \Delta\left(\mathbf{w}(n)\right)$. If the loss function is less than the one calculated earlier in step 1, then μ should be reduced β times, the new weight vector is saved and the algorithm returns to Step 1. Otherwise, the μ value is increased β times and the algorithm repeats step 3.
5. The algorithm stops running when the loss function falls below a preset value or the gradient falls below a preset value.

2   Vector Solution for Levenberg-Marquardt Algorithm

The Levenberg-Marquardt algorithm needs high computing power. Each epoch starts with steps 1 and 2, and next steps 3 and 4 can be repeated a few times. Figure 2 shows a single epoch of the LM algorithm, showing the first two steps and repeating steps 3 and 4. It is worth noting that the next pairs of steps 3 and 4 are independent of each other and can be performed at the same time. They only differ in the μ parameter value and have the same starting point. Thus, they can be run parallel on separate processor cores. However, the solution proposed in this article uses processor vector instructions. Vector instructions allow 4, 8, and even 16 operations to be performed in parallel. This approach enables simultaneous determination of new 4, 8, or 16 points in the weight space using only one processor core, see Fig. 3. Figure 3a shows the epoch of the LM algorithm with the use of four-element vectors. After completing the first two

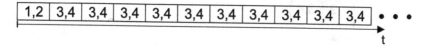

Fig. 2. Sample illustration for computational steps in LM algorithm.

steps, the algorithm calculates steps 3 and 4 for the next 4 parameters μ at one time. Thus, the three consecutive computations of steps 3 and 4 are performed earlier and therefore do not take computational time. The rectangles with the line in the middle symbolize steps 3 and 4, which are used in the standard calculation method and are omitted in calculations using vector instructions. Figure 3b shows the version with eight-element vectors.

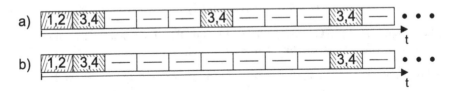

Fig. 3. Sample illustration for calculating method with vector instructions. a) the 4-elements vector, b) the 8-elements vector

Figure 4 shows an example of the learning process using the LM algorithm. In the following epochs, you can see a different number of steps 3 and 4 repetitions. There are epochs where the repetition does not occur and there are those with a

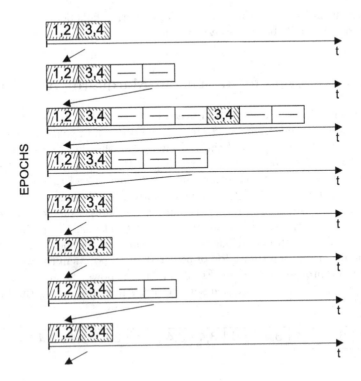

Fig. 4. Sample illustration for training process with vector instructions.

large number of repetitions, in this case, vector instructions can be used, which makes it possible to calculate up to four pairs of steps 3 and 4 at the same time and consequently shortening the learning time. Of course, eight- or sixteen-element vectors can be used instead of using four-element vectors. This increases the parallelism and speed of the proposed calculation method.

3   Experimental Results

The proposed solution was tested against the classical variant of the Levenberg-Marquardt learning algorithm on several test problems. Two types of forward-coupled artificial neural networks were tested in the experiment: MLP — Multilayer Perceptron, FCMLP — Fully Connected Multilayer Perceptron. The performance of the presented calculation method was measured in average training time in milliseconds. The presented results are compiled according to the best combination of training parameters. In all cases, the initial weights were randomly selected from the range [−0.5,0.5]. The number of epochs has been limited to 1,000. Each training session was repeated 100 times.

3.1   Logistic Function Approximation

The logistic function is a unary function given by the formula

$$y = f(x) = 4x(1 - x) \tag{18}$$

The teaching sequence contains 11 samples where $x \in [0,1]$. The average accepted error threshold has been set to 0.001. Table 1 shows the simulation results for two kinds of neural networks MLP and FCMLP. Both networks have five neurons in the hidden layer. The symbols LM, LMP 4, LMP 8, and LMP 16 represent the average network training time using the LM algorithm and its vector versions for 4, 8, and 16 element vectors, respectively. The speed factor means how many percent the vector version is faster then the classical one and is given by the formula

$$SF = \left(1 - \frac{LMPx}{LM}\right) * 100\% \tag{19}$$

Table 1. Training results for the LOG function.

Network	LM [ms]	LMP4 [ms]	SF [%]	LMP8 [ms]	SF [%]	LMP16 [ms]	SF [%]
MLP-1-5-1	0.880	0.440	50	0.434	50	0.433	50
FCMLP-1-5-1	0.588	0.311	47	0.306	48	0.305	48

3.2 Hang Function Approximation

The Hang function is a nonlinear two-argument x_1 and x_2 function with the following formula

$$y = f(x_1, x_2) = \left(1 + x_1^{-2} + \sqrt{x_2^{-3}}\right)^2 \tag{20}$$

The Hang teaching sequence contains 50 samples that cover arguments in the range of $x_1, x_2 \in [1, 5]$. The target error threshold was set to 0.001 as the epoch average. The results of simulations for the Hang function are shown in Table 2. Both tested networks have 15 neurons in the hidden layer.

Table 2. Training results for the HANG function.

Network	LM [ms]	LMP4 [ms]	SF [%]	LMP8 [ms]	SF [%]	LMP16 [ms]	SF [%]
MLP-2-15-1	27.235	13.191	51	12.553	53	12.462	54
FCCMLP-2-15-1	34.237	16.691	51	16.165	52	16.111	52

3.3 IRIS Function Classification

The iris dataset contains 150 instances describing three species of iris flowers. The flowers are identified with 4 numerical attributes describing the lengths and widths of the petals of the flower. The target error has been set to 0.05. Table 3 shows the simulation results.

Table 3. Training results for the IRIS function.

Network	LM [ms]	LMP4 [ms]	SF [%]	LMP8 [ms]	SF [%]	LMP16 [ms]	SF [%]
MLP-4-6-6-3	528.183	242.789	54	229.337	56	223.374	57
FCCMLP-4-6-6-3	1851.720	870.468	52	842.894	54	831.464	55

3.4 The Two Spirals Classification

Two spirals is a well-known classification problem where a neural network has to identify one of the two helices based on two-dimensional coordinates. The training set for this problem contains 96 samples. The target error has been set to 0.05. Table 4 shows the simulation results.

Table 4. Training results for the TS function.

Network	LM [ms]	LMP4 [ms]	SF [%]	LMP8 [ms]	SF [%]	LMP16 [ms]	SF [%]
MLP-2-5-5-5-1	166.819	77.954	53	76.139	54	75.555	54
FCMLP-2-5-5-5-1	349.704	165.037	52	161.613	53	161.192	53

4 Conclusion

In this paper, the new computational approach to the Levenberg-Marquardt learning algorithm for a feedforward neural network is proposed. Two types of feedforward neural networks were used in the experiments: multilayer perceptron and fully interconnected multilayer perceptron. The networks were trained with different training sets: Logistic function, Hang, Iris, and Two Spirals. We can compare the computational performance of the proposed solution, based on vector instructions of the Levenberg-Marquardt learning algorithm, with a classical solution. The conducted experiments showed a significant reduction of the real learning time. For all training sets, calculation times have been reduced by an average of 50%. It has been observed that the performance of the proposed solution is promising.

A vector approach can be used for other advanced learning algorithms of feedforward neural networks, see eg. [2,8]. In the future research, we plan to design parallel realization of learning of other structures including probabilistic neural networks [32] and various fuzzy [1,15,20,22,24,37,42,46,47], and neuro-fuzzy structures [16,17,23,25,35].

References

1. Bartczuk, Ł, Przybył, A., Cpałka, K.: A new approach to nonlinear modelling of dynamic systems based on fuzzy rules. Int. J. Appl. Math. Comput. Sci. (AMCS) **26**(3), 603–621 (2016)
2. Bilski, J.: The UD RLS algorithm for training the feedforward neural networks. Int. J. Appl. Math. Comput. Sci. **15**(1), 101–109 (2005)
3. Bilski, J., Litwiński, S., Smoląg, J.: Parallel realisation of QR algorithm for neural networks learning. In: Rutkowski, L., Siekmann, J.H., Tadeusiewicz, R., Zadeh, L.A. (eds.) ICAISC 2004. LNCS (LNAI), vol. 3070, pp. 158–165. Springer, Heidelberg (2004). https://doi.org/10.1007/978-3-540-24844-6_19
4. Bilski, J., Smoląg, J.: Parallel realisation of the recurrent RTRN neural network learning. In: Rutkowski, L., Tadeusiewicz, R., Zadeh, L.A., Zurada, J.M. (eds.) ICAISC 2008. LNCS (LNAI), vol. 5097, pp. 11–16. Springer, Heidelberg (2008). https://doi.org/10.1007/978-3-540-69731-2_2
5. Bilski, J., Smoląg, J.: Parallel realisation of the recurrent elman neural network learning. In: Rutkowski, L., Scherer, R., Tadeusiewicz, R., Zadeh, L.A., Zurada, J.M. (eds.) ICAISC 2010. LNCS (LNAI), vol. 6114, pp. 19–25. Springer, Heidelberg (2010). https://doi.org/10.1007/978-3-642-13232-2_3

6. Bilski, J., Smoląg, J.: Parallel realisation of the recurrent multi layer perceptron learning. In: Rutkowski, L., Korytkowski, M., Scherer, R., Tadeusiewicz, R., Zadeh, L.A., Zurada, J.M. (eds.) ICAISC 2012. LNCS (LNAI), vol. 7267, pp. 12–20. Springer, Heidelberg (2012). https://doi.org/10.1007/978-3-642-29347-4_2

7. Bilski, J., Smoląg, J.: Parallel approach to learning of the recurrent Jordan neural network. In: Rutkowski, L., Korytkowski, M., Scherer, R., Tadeusiewicz, R., Zadeh, L.A., Zurada, J.M. (eds.) ICAISC 2013. LNCS (LNAI), vol. 7894, pp. 32–40. Springer, Heidelberg (2013). https://doi.org/10.1007/978-3-642-38658-9_3

8. Bilski, J.: Parallel structures for feedforward and dynamical neural networks (in Polish). AOW EXIT (2013)

9. Bilski, J., Smoląg, J., Galushkin, A.I.: The parallel approach to the conjugate gradient learning algorithm for the feedforward neural networks. In: Rutkowski, L., Korytkowski, M., Scherer, R., Tadeusiewicz, R., Zadeh, L.A., Zurada, J.M. (eds.) ICAISC 2014. LNCS (LNAI), vol. 8467, pp. 12–21. Springer, Cham (2014). https://doi.org/10.1007/978-3-319-07173-2_2

10. Bilski, J., Smoląg, J.: Parallel architectures for learning the RTRN and elman dynamic neural networks. IEEE Trans. Parallel Distrib. Syst. PP(99), (2014). https://doi.org/10.1109/TPDS.2014.2357019

11. Bilski, J., Kowalczyk, B., Marchlewska A., Żurada J.M.: Local levenberg-marquardt algorithm for learning feedforwad neural networks. J. Artif. Intell. Soft Comput. Res. 10(4), 299–316 (2020). https://doi.org/10.2478/jaiscr-2020-0020

12. Bilski, J., Kowalczyk, B., Marjański, A., Gandor, M., Żurada, J.M.: A novel fast feedforward neural networks training algorithm. J. Artif. Intell. Soft Comput. Res. 11(4), 287–306 (2021). https://doi.org/10.2478/jaiscr-2021-0017

13. Bilski J., Rutkowski L., Smoląg J., Tao D., A novel method for speed training acceleration of recurrent neural networks. Inf. Sci. 553, 266–279 (2021). https://doi.org/10.1016/j.ins.2020.10.025

14. Chu, J.L., Krzyzak, A.: The recognition of partially occluded objects with support vector machines, convolutional neural networks and deep belief networks. J. Artif. Intell. Soft Comput. Res. 4(1), 5–19 (2014)

15. Cpałka, K., Rutkowski, L.: Flexible takagi-sugeno fuzzy systems. In: Proceedings of the International Joint Conference on Neural Networks, Montreal, pp. 1764–1769 (2005)

16. Cpałka, K., Łapa, K., Przybył, A., Zalasiński, M.: A new method for designing neuro-fuzzy systems for nonlinear modelling with interpretability aspects. Neurocomputing 135, 203–217 (2014)

17. Cpałka, K., Rebrova, O., Nowicki, R. et al.: On design of flexible neuro-fuzzy systems for nonlinear modelling. Int. J. Gener. Syst. 42(6), Special Issue: SI, 706–720 (2013)

18. Fahlman, S.: Faster learning variations on backpropagation: an empirical study. In: Proceedings of Connectionist Models Summer School, Los Atos (1988)

19. Gabryel, M., Przybyszewski, K.: Methods of searching for similar device fingerprints using changes in unstable parameters. In: Rutkowski, L., Scherer, R., Korytkowski, M., Pedrycz, W., Tadeusiewicz, R., Zurada, J.M. (eds.) ICAISC 2020. LNCS (LNAI), vol. 12416, pp. 325–335. Springer, Cham (2020). https://doi.org/10.1007/978-3-030-61534-5_29

20. Gabryel, M., Scherer, M.M., Sułkowski, Ł, Damaševičius, R.: Decision making support system for managing advertisers by ad fraud detection. J. Artif. Intell. Soft Comput. Res. 11, 331–339 (2021)

21. Hagan, M.T., Menhaj, M.B.: Training feedforward networks with the Marquardt algorithm. IEEE Trans. Neuralnetworks 5(6), 989–993 (1994)

22. Korytkowski, M., Rutkowski, L., Scherer, R.: From ensemble of fuzzy classifiers to single fuzzy rule base classifier. In: Rutkowski, L., Tadeusiewicz, R., Zadeh, L.A., Zurada, J.M. (eds.) ICAISC 2008. LNCS (LNAI), vol. 5097, pp. 265–272. Springer, Heidelberg (2008). https://doi.org/10.1007/978-3-540-69731-2_26

23. Korytkowski, M., Scherer, R.: Negative correlation learning of neuro-fuzzy system. LNAI **6113**, 114–119 (2010)

24. Łapa, K., Przybył, A., Cpałka, K.: A new approach to designing interpretable models of dynamic systems. In: Rutkowski, L., Korytkowski, M., Scherer, R., Tadeusiewicz, R., Zadeh, L.A., Zurada, J.M. (eds.) ICAISC 2013. LNCS (LNAI), vol. 7895, pp. 523–534. Springer, Heidelberg (2013). https://doi.org/10.1007/978-3-642-38610-7_48

25. Łapa, K., Zalasiński, M., Cpałka, K.: A new method for designing and complexity reduction of neuro-fuzzy systems for nonlinear modelling. In: Rutkowski, L., Korytkowski, M., Scherer, R., Tadeusiewicz, R., Zadeh, L.A., Zurada, J.M. (eds.) ICAISC 2013. LNCS (LNAI), vol. 7894, pp. 329–344. Springer, Heidelberg (2013). https://doi.org/10.1007/978-3-642-38658-9_30

26. Marqardt, D.: An algorithm for last-sqares estimation of nonlinear paeameters. J. Soc. Ind. Appl. Math. 431–441 (1963)

27. Niksa-Rynkiewicz, T., Szewczuk-Krypa, N., Witkowska, A., Cpałka, K., Zalasiński, M., Cader, A.: Monitoring regenerative heat exchanger in steam power plant by making use of the recurrent neural network. J. Artif. Intell. Soft Comput. Res. **11**(2), 143–155 (2021). https://doi.org/10.2478/jaiscr-2021-0009

28. Patan, K., Patan, M.: Optimal training strategies for locally recurrent neural networks. J. Artif. Intell. Soft Comput. Res. **1**(2), 103–114 (2011)

29. Riedmiller, M., Braun, H.: A direct method for faster backpropagation learning: the RPROP Algorithm. In: IEEE International Conference on Neural Networks, San Francisco (1993)

30. Romaszewski, M., Gawron, P., Opozda, S.: Dimensionality reduction of dynamic msh animations using HO-SVD. J. Artif. Intell. Soft Comput. Res. **3**(3), 277–289 (2013)

31. Rumelhart, D.E., Hinton, G.E., Williams, R.J.: Learning internal representations by error propagation. Parallel Distributed Processing, vol. 1, ch. 8, Rumelhart, D.E., McCelland, J. (red.). The MIT Press, Cambridge, Massachusetts (1986)

32. Rutkowski, L.: Multiple Fourier series procedures for extraction of nonlinear regressions from noisy data. IEEE Trans. Sig. Process. **41**(10), 3062–3065 (1993)

33. Rutkowski, L.: Identification of MISO nonlinear regressions in the presence of a wide class of disturbances. IEEE Trans. Inf. Theor. **37**(1), 214–216 (1991)

34. Rutkowski, L., Jaworski, M., Pietruczuk, L., Duda, P.: Decision trees for mining data streams based on the gaussian approximation. IEEE Trans. Knowl. Data Eng. **26**(1), 108–119 (2014)

35. Rutkowski, L., Przybył, A., Cpałka, K., Er, M.J.: Online speed profile generation for industrial machine tool based on neuro-fuzzy approach. In: Rutkowski, L., Scherer, R., Tadeusiewicz, R., Zadeh, L.A., Zurada, J.M. (eds.) ICAISC 2010. LNCS (LNAI), vol. 6114, pp. 645–650. Springer, Heidelberg (2010). https://doi.org/10.1007/978-3-642-13232-2_79

36. Rutkowski, L., Rafajlowicz, E.: On optimal global rate of convergence of some nonparametric identification procedures. IEEE Trans. Autom. Control **34**(10), 1089–1091 (1989)

37. Rutkowski, T., Łapa, K., Jaworski, M., Nielek, R., Rutkowska, D.: On explainable flexible fuzzy recommender and its performance evaluation using the akaike

information criterion. In: Gedeon, T., Wong, K.W., Lee, M. (eds.) ICONIP 2019. CCIS, vol. 1142, pp. 717–724. Springer, Cham (2019). https://doi.org/10.1007/978-3-030-36808-1_78

38. Smoląg, J., Bilski, J.: A systolic array for fast learning of neural networks. In: Proceedings of V Conference Neural Networks and Soft Computing, Zakopane, pp. 754–758 (2000)

39. Smoląg, J., Rutkowski, L., Bilski, J.: Systolic array for neural networks. In: Proceedings of IV Conference Neural Networks and Their Applications, Zakopane, pp. 487–497 (1999)

40. Starczewski, A.: A clustering method based on the modified RS validity index. In: Rutkowski, L., Korytkowski, M., Scherer, R., Tadeusiewicz, R., Zadeh, L.A., Zurada, J.M. (eds.) ICAISC 2013. LNCS (LNAI), vol. 7895, pp. 242–250. Springer, Heidelberg (2013). https://doi.org/10.1007/978-3-642-38610-7_23

41. Starczewski J.T. Advanced Concepts in Fuzzy Logic and Systems with Membership Uncertainty, volume 284 of Studies in Fuzziness and Soft Computing. Springer, Berlin (2013). https://doi.org/10.1007/978-3-642-29520-1

42. Starczewski, J.T., Goetzen, P., Napoli, Ch.: Triangular fuzzy-rough set based fuzzification of fuzzy rule-based systems. J. Artif. Intell. Soft Comput. Res. 10, 271–285 (2020)

43. Tadeusiewicz, R.: Neural Networks (in Polish). AOW RM (1993)

44. Werbos, J.: Backpropagation through time: what it does and how to do it. In: Proceedings of the IEEE, vol. 78, no. 10 (1990)

45. Wilamowski, B.M., Yo, H.: Neural network learning without backpropagation. IEEE Trans. Neural Network. 21(11), 1793–1803 (2010)

46. Zalasiński, M., Cpałka, K.: New approach for the on-line signature verification based on method of horizontal partitioning. In: Rutkowski, L., Korytkowski, M., Scherer, R., Tadeusiewicz, R., Zadeh, L.A., Zurada, J.M. (eds.) ICAISC 2013. LNCS (LNAI), vol. 7895, pp. 342–350. Springer, Heidelberg (2013). https://doi.org/10.1007/978-3-642-38610-7_32

47. Zalasiński, M., Łapa, K., Cpałka, K.: Prediction of values of the dynamic signature features. Expert Syst. Appl. 104, 86–96 (2018)

48. El Zini J., Rizk Y., Awad M.: An optimized parallel implementation of non-iteratively trained recurrent neural networks. Journal of Artif. Intell. Soft Comput. Res. 11(1), 33–50 (2021). https://doi.org/10.2478/jaiscr-2021-0003

Training Subjective Perception Biased Images of Vehicle Ambient Lights with Deep Belief Networks Using Backpropagation- and Enforcing-Rules Supervised

Gregor Braun[1], Michel Brokamp[1], and Christina Klüver[2](✉)

[1] University of Applied Sciences, Düsseldorf, Germany
gregor.braun@hs-duesseldorf.de
[2] University of Duisburg-Essen, Essen, Germany
christina.kluever@uni-due.de

Abstract. Quality measurement of vehicle ambient lighting during series production can be influenced by subjective perceptions of light homogeneity. In consequence, the labels correspond to the decisions whether the lights appear homogeneous or not. In this article we demonstrate how images of ambient lighting were trained by Deep Belief Networks using the learning rules "backpropagation" (BP) and "enforcing-rule supervised" (ERS). In addition, the effect of the contrastive divergence pre-training is analyzed on the accuracy of the trained networks. The results are promising for decision support in the production process to minimize the influence of subjectivity by human evaluators.

Keywords: Deep belief networks · Enforcing-rule supervised · Subjective perception · Vehicle ambient lighting

1 Introduction

Ambient lights have been used in luxury cars for years to enhance the mood, orientation, and comfort of the driver [1–3]. Several experiments were conducted to find out the individual perceptions, emotions, or the suited positions in the vehicles (e.g., [4–8]).

Because the concept and technology of ambient lighting have been expanded over the years, the effects of automotive lights were examined under further aspects, e.g., positive psycho-physiological reactions on driver mood or carsickness [9,10]. Likewise, the negative effects are studied in relation to cognitive stress [11,12], accidents caused by distraction or load because of diverse contrast at night [1,13,14]. In addition, the implications for future automatic driving are considered [15–17].

Supported by MENTOR GmbH & Co. Präzisions-Bauteile KG.

On the other hand, few analyses refer to the evaluation of ambient lighting or light guides from the point of view of homogeneity distribution [12,18]. Minor defects can hardly be noticed [12] but have importance for the perceived quality of the interior [18]. This is a major challenge for inspection during production.

Since perception and human evaluation of ambient lighting is a matter of subjectivity and depends on personal condition, decision support is desirable. For the production process, a small amount of training data is required on the part of the company, as well as quick adaptation. For this task we propose deep belief networks (DBN) trained with images of light guides.

The challenge is that there are no open data sets available and no comparative studies. Neural networks are proposed for e.g., evaluation of fiber optic connectors [19] or LED chips [20], and in general for surface inspection, but there are no findings for the evaluation of homogeneity distribution of ambient lighting for vehicles.

The proposed solution using DBN with the learning rules "backpropagation" and "enforcing-rule supervised" is quite promising, as will be shown in the following sections.

This contribution is structured as follows: in the next section the problem and the data sets are described in more details. In Sect. 3 the architecture of the Deep Belief Networks, the learning rules and the experimental design are presented. The results and implications are discussed in Sect. 4.

2   Aesthetic Ambient Lighting in Vehicles

As already mentioned, the light emission in the interior of a vehicle should be pleasant and perceived as uniform for the driver. In addition to aesthetic properties, ambient lighting should help with orientation inside the car, but not distract while driving.

During the production process of the ambient lighting, the light guides undergo a manual quality check by human evaluators. Labeling the light guides into categories of "homogenous" or "not homogenous" as well as "pleasant" or "not pleasant" can be subjective and depended on the evaluator's experience, opinion, and current state, e.g., the degree of fatigue or environmental influences.

As it was pointed out in [[12], p. 662] the problem is to detect an "invisible" inhomogeneity. Figure 1 shows two such ambient lights, where it is difficult for laypersons to decide which light guide produces a homogeneous light and which one does not.

Fig. 1. Light guides for car interior lighting. Flawless light guide (upper one) and faulty light guide (lower one). (Images by Mentor GmbH & Co. Präzisions-Bauteile KG)

To verify whether decision support can be provided through image processing and neural networks, the project was supported by the company Mentor GmbH & Co. Präzisions-Bauteile KG. Because most of the light guides are of course flawless in the production process, the dataset was specially created, including manually scratched or manipulated light guides.

Dataset. Several light guides have been clamped into a specially designed system with a camera to ensure that the light guide will always be in the same position within the image. At a later stage, the images of the light guides were evaluated and labeled.

A total of 201 images of light guides with a size of 4112×188 pixels are available: 82 images are labeled as flawless and 119 images of possible production errors like scratches, faulty spots, too dark light emission, too yellow light emission and combinations of it. These images were divided into different classes and serve as a training and test dataset for the Deep Belief Network (DBN), which will be discussed in the next section.

3 Deep Belief Networks and Learning Rules

DBN is a generative model and successfully used for monitoring in manufacturing due to its feature extraction capability [21–23]. For the given task DBN have been trained besides fully connected neural networks. DBN have been used as a pre-training mechanism to find a good initial weight matrix for the network.

For pre-training, two layers of the network at a time form a Restricted Boltzmann Machine (RBM), where the hidden layer of the first RBM is used as the visible layer of the second RBM. This way all created RBMs are stacked as described by [24] and shown in Fig. 2. Just the last layer for classification is left out since it can only be trained supervised.

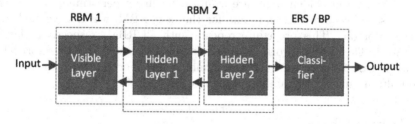

Fig. 2. The RBM architecture.

In the unsupervised pre-training, the RBMs learn a probabilistic model of the given input vectors. More details on the unsupervised learning are given below.

3.1   Unsupervised Pre-training

The weight matrix W and the bias vectors $\vec{b_v}$ (visible units) and $\vec{b_h}$ (hidden units) for a single RBM are updated for each training vector $\vec{x}$ by performing the following steps:

1 $\vec{p_v^0} = \vec{x}$ Use normalized input vector as probabilities for visible units

2 $\vec{p_h^0} = \sigma\left(\vec{b_h} + W \cdot \vec{p_v^0}\right)$ Calculate probabilities units for hidden units and sample binary states
   $\vec{h^0} = sample\left(\vec{p_h^0}\right))$

3 $\vec{p_v^1} = \sigma\left(\vec{b_v} + W^{tr} \cdot \vec{h_v^0}\right)$ Calculate probabilities for visible units (reconstruction)

4. Repeat steps 2 and 3 (alternating Gibbs Sampling) n–1 times.

5 $\vec{p_h^n} = \sigma\left(\vec{b_h} + W \cdot \vec{p_v^n}\right)$ Calculate probabilities for hidden units the last time

6 $\Delta W = \eta \cdot \left(\vec{p_h^0} \cdot \left(\vec{p_v^0}\right)^{tr} - \vec{p_h^n} \cdot \left(\vec{p_v^n}\right)^{tr}\right)$ Calculate weight matrix changes and change weights: $W' = W + \Delta W$

7 $\Delta \vec{b_v} = \eta \cdot \left(\vec{p_v^0} - \vec{p_v^n}\right)$ Calculate bias vector changes and change biases: $\vec{b_v} = \vec{b_v} + \Delta\vec{b_v}; \vec{b_h} = \vec{b_h} + \Delta\vec{b_h}$
   $\Delta \vec{b_h} = \eta \cdot \left(\vec{p_h^0} - \vec{p_h^n}\right)$

Note for step 2, 3, and 5 (1): The functions σ and *sample* calculate the following output for each vector component:

$$\sigma(x) := \frac{1}{1 + e^{-x}}; \quad sample(p) := \begin{cases} 1 \text{ if } p > random([0|1]) \\ 0 \text{ else} \end{cases} \quad (1)$$

After the changes for the first RBM are calculated from the input vector (steps 1 to 7), the hidden units are sampled again by performing the second step. These values are used as inputs for the next RBM, which is trained just like the prior one. This procedure repeats until all RBMs respectively layers are trained. Then the network continues learning the next input vector. This practice is called Greedy Layer Wise Training [23]. For the experiments n=1 step of alternating Gibbs Sampling was used.

3.2   Learning Rules

The used learning rules are the well-known "backpropagation" (BP) and the "enforcing-rule supervised" (ERS). The learning rule introduced by [25] is a simplified alternative to BP, that is not based on the gradient descent method. Moreover, the weighting values are allowed only in an interval between $[-1, 1]$.

ERS is based on the following Eq. (2), where c is the equivalent of the learning rate:

$$\Delta w_{i,j} = c \cdot |(1 - |w_{i,j}|)| \cdot \delta_j \cdot sgn\,(o_i) \tag{2}$$

The factor $sgn\,(o_i)$ is used only as the sign of the whole product.

ERS2 is a version in which the value of the sending neuron o_i is multiplied to allow better comparability with BP (3).

$$\Delta w_{i,j} = c \cdot |(1 - |w_{i,j}|)| \cdot \delta_j \cdot o_i \tag{3}$$

The error is computed as follows (4):

$$\delta_j = \begin{cases} (t_j - o_j) & \text{if j is an output unit} \\ \sum_k \delta_k \cdot w_{j,k} & \text{if j is a hidden unit} \end{cases} \tag{4}$$

The similarity with BP is obvious, but the function is much simplified by omitting the derivatives.

To cope with deep architectures and many neurons, the error function was modified to (5):

$$\delta'_j = 2d \cdot \delta_j / n_i \tag{5}$$

where d is the number of layers and n_i the number of neurons in the layer directly above the layer of neuron j.

The ERS and ERS2 variants are called ERS-DL and ERS2-DL for deep networks, respectively.

These five learning rules, i.e., BP and the four variants of ERS, are used for the experiments.

3.3 Experimental Design

To reduce computational load, while keeping the RGB values, the images are scaled to $180 \times 35 \times 3$, resulting in 18.900 input neurons.

The input values for the RBM pre-training then are converted into binary values either 0 and 1 or bi-polar values -1 and 1. For the subsequently supervised training the input values are normalized to the corresponding interval with real values between either 0 and 1 or -1 and 1.

The labels are coded as a vector with one component per class with the values 0 (wrong class) and 1 (correct class) for networks that use the logistic activation function and -1 and 1 for networks that use the hyperbolic tangent activation function.

The images were split into sets of 55% for training and 45% for testing to meet the requirement to use less training data than usual.

Since the learning rules each work best with their own hyper parameters (learning rate, activation function, number of layers, neurons per layer), various architectures have been tested against each other. To discuss the results, only the best architectures will be pointed out for each learning rule, even thou every learning rule was tested with that particular architecture.

The output layer contains 2 or 4 neurons corresponding to the number of classes. In the first experiments, the images were only classified as "flawless" and "faulty". As a further task, the images were sorted into four classes (flawless, light guides with a yellow cast, light guides that are too dark, other defects such as scratches).

The hidden layers vary between one and three layers. Experiments have shown that accuracy decreases when more than three layers are used.

The number of neurons per hidden layer varied from 10 neurons to a maximum of 1.500 neurons. It has been shown that the use of more than 1.500 neurons does not lead to any improvement or even to a degradation of the performance of the network. Before training, the matrices were initialized in three different ways: Gaussian distributed small random values (SRV), evenly distributed random values between -0.5 and +0.5 (ESRV) as well as evenly distributed random values between −1.0 and +1.0 (EBRV).

In addition, the contrastive divergence (CD) algorithm introduced by [26] was alternately used or switched off for the same network configurations and initial matrices. In this way, it is possible to validate whether there is a positive effect on training results and whether different learning rules respond differently to CD.

In the experiments, after initializing the weighting matrices, the next training step was either to apply CD with a learning rate of 0.1 and one iteration, and then apply the learning rules, or to proceed directly to supervised learning without CD.

All experiments have been executed with learning rates 0.1, 0.05 and 0.01.

4   Experimental Results

In the experiments, over 1600 networks with different architectures and hyperparameters were tested to find the networks with the highest accuracy. The training data set, i.e., 111 images for training and 90 images for testing, was divided into two classes, "flawless" and "faulty". An accuracy of 97.78% could be achieved with ERS2-DL. The best configurations per learning rule are shown in Table 1.

The corresponding confusion matrices are shown only for the best ERS rule in comparison to BP (Table 2).

To differentiate possible defects, the images were assigned as specified in Table 3.

The results of the best performing networks per rule can be seen in Table 4 and 5.

Summary of the Results. The most promising candidates are ERS2-DL and ERS2, with ERS2-DL being a good candidate for deeper networks.

Table 1. Classification results of the test data

Rule	ERS2-DL	ERS2	ERS-DL	ERS	BP
Accuracy (%)	**97.78**	96.67	95.56	91.11	90.00
Architecture	18900-500-500-2	18900-500-500-2	18900-500-500-2	18900-500-2	18900-500-2
Initialization	ESRV	ESRV	ESRV	ESRV	ESRV
CD	No	Yes	No	Yes	No
Learning Rate	0.01	0.1	0.01	0.01	0.01
Batch Size	2	2	2	2	2
Activation Function	Logistic	Logistic	Tanh	Tanh	Logistic
Target vector encoding	[0,1]	[0,1]	[−1,1]	[−1,1]	[−1,1]

Table 2. Confusion matrices for two classes

predicted	**ERS2-DL**	**Flawless**	**Faulty**	
	flawless	34	0	100.00%
	Faulty	2	54	96.43%
		94.44%	100.00%	97.78%
predicted	**BP**	**Flawless**	**Faulty**	
	flawless	36	9	80.00%
	faulty	0	45	100.00%
		100.00%	83.33%	90.00%

Table 3. Image assignment to four classes

	Flawless	Yellow	Faulty	Dark
Training data set	46	9	47	9
Test data set	36	8	39	7

Table 4. Classification in four classes

Rule	ERS2	ERS2-DL	ERS-DL	BP	ERS
Accuracy(%)	**88.89**	87.78	87.78	85.56	81.11
Architecture	18900-500-500-4	18900-100-100-4	18900-1500-500-4	18900-100-100-4	18900-100-100-4
Initialization	ESRV	ESRV	ESRV	SRV	ESRV
CD	Yes	No	No	No	Yes
Learning Rate	0.01	0.01	0.1	0.01	0.01
Batch Size	4	4	4	4	4
Activation Function	Logistic	Tanh	Tanh	Logistic	Tanh
Target Vector Encoding	[0,1]	[−1,1]	[−1,1]	[0,1]	[−1,1]

Table 5. Confusion matrices for 4 classes

Predicted ERS2	Flawless	Yellow	Faulty	Dark	
Flawless	36	0	5	1	89.47%
Yellow	0	6	0	0	100.00%
Faulty	0	1	32	0	96.97%
Dark	0	1	2	6	66.67%
	100.00%	75.00%	82.05%	85.71%	88.89%
Predicted BP	Flawless	Yellow	Faulty	Dark	
flawless	36	0	6	1	83.72%
yellow	0	6	1	1	75.00%
faulty	0	2	32	2	88.89%
dark	0	0	0	3	100.00%
	100.00%	75.00%	82.05%	42.86%	85.56%

The results also show that the benefits of CD can be especially used together with ERS learning rules.

In addition, the classification of faulty light guides was the most successful. This is true for both two-class and four-class classifications. When the BP learning rule was applied, the fault-free light guides were detected, but not the faulty ones, and the dark ones in the case of the four classes.

The Problem of Subjectivity and Labeling of Data. Since the labels are the result of subjective influences, it is likely that some of them should be reviewed.

When subjective influences play a role, 100% accuracy of results cannot be expected. The studies e.g., by [19, 27] on different subjects came to similar results as the ones presented here.

Jahani [27] has studied the forest landscape quality and received an accuracy of R2 = 0.871 with a multilayer feed-forward network and R2 = 0.782 with multiple regression. Fernandez et al. [19] deal with the classification of fiber optic connectors through image processing and neural networks. The results are a training accuracy of 97% and 92% for the test data. In this case, three classes were defined for "clean", "dirty" and "very dirty", with most problems occurring in the classification of "dirty".

Hassib et al. [9] have analyzed the influence of the emotions on drivers' capabilities with camera-based methods, psycho-physiological sensors, and different ambient light colors. Using the subject-dependent random forest classifier, an average accuracy of 78.9% was achieved for valence classification and 68.7% for arousal.

In a previous work of our study, the same data set was trained with a Convolutional Neural Network (CNN) with similar results as ERS2-DL [28]. However, the assignment of faulty light guides as correct ones was twice that of ERS2-DL.

For the company, it is very important that as few faulty light guides as possible are classified as fault-free.

As the question of interest in our case is why no network reached 100% in the learning process, the images that were incorrectly assigned according to the label were examined more closely.

It turned out that the images were assigned twice with uncertainty: once with a "yellow tint" and the same image as "dark" (Fig. 3).

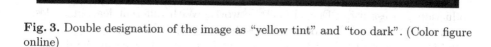

Fig. 3. Double designation of the image as "yellow tint" and "too dark". (Color figure online)

Due to this double classification, of course, no network can learn the images unambiguously or to classify them correctly. This is also the case when these are split by random selection once in the training data set and once in the test data set.

Another problem occurs when light guides are labeled as e.g., "yellow tint" while also having other flaws that belong into other classes like scratches that are classified as "faulty". The image then can only be classified as either "yellow" or "faulty" when in reality both classes can apply.

Other cases will be shown in Fig. 4 and 5.

Fig. 4. Image is labeled as "flawless", but classified as "faulty".

Fig. 5. Image is labeled as "dark", but classified as "yellow". (Color figure online)

Most interesting is the classification of the image shown in Fig. 6: each network, including the CNN, has classified the image as "flawless".

A human being will certainly classify this image of a light guide as faulty. This phenomenon needs to be investigated further.

Fig. 6. Image classified as "flawless".

5   Conclusion and Future Work

We proposed a solution to support human evaluators of ambient lights in the production process using Deep Belief Networks. With different learning rules, satisfactory results could be obtained despite subjective influences.

Subjective perception led to some images being listed twice in the dataset with different labels. These could be identified by the classification results through the networks.

The ERS learning rule introduced here has proven best for this problem. The most suitable variation needs to be investigated in the production process. In particular, it should be examined whether the use of both learning rules, BP and ERS, is an option to identify light guides that cannot be unambiguously classified. Only these then need to be evaluated by experts.

An automated support system offers the advantage that fatigue, one's own emotional state or subjective perceptions do not play a role. In this respect, the number of ambient guides to be manually checked by human evaluators can be reduced, as shown in this contribution.

References

1. Schellinger, S., Franzke, D., Klinger, K., Lemmer, U.: Advantages of ambient interior lighting for drivers contrast vision. In: Proceedings of SPIE 6198, Photonics in the Automobile II, 61980J (2006)
2. Flannagan, M.J., Devonshire, J.M.: Effects of automotive interior lighting on driver vision. Leukos **9**(1), 9–23 (2012)
3. Winklbauer, M., Bayersdorfer, B., Lang, J.: Evocative lighting design for premium interiors. ATZ Worldw **117**, 32–35 (2015)
4. Caberletti, L., Elfmann, K., Kummel, M., Schierz, C.: Influence of ambient lighting in a vehicle interior on the driver's perceptions. Lighting Res. Technol. **42**(3), 297–311 (2010)
5. Luo, W., Luo, X.: User experience research on automotive interior lighting design. In: Ahram, T., Falcão, C. (eds.) AHFE 2017. AISC, vol. 607, pp. 240–246. Springer, Cham (2018). https://doi.org/10.1007/978-3-319-60492-3_23
6. Nandyala, S., Gayathri, K., Sharath, D.H., Manalikandy, M.: Human Emotion Based Interior Lighting Control. No. 2018-01-1042, SAE Technical Paper (2018)
7. Weirich, C., Lin, Y., Khanh, T.Q.: Evidence for human-centric in-vehicle lighting: part 1. Appl. Sci. **12**, 552 (2022)
8. Kim, T., Kim, Y., Jeon, H., Choi, C.-S., Suk, H.-J.: Emotional response to in-car dynamic lighting. Int. J. Autom. Technol. **22**(4), 1035–1043 (2021). https://doi.org/10.1007/s12239-021-0093-4

9. Hassib, M., Braun, M., Pfleging, B., Alt, F.: Detecting and influencing driver emotions using psycho-physiological sensors and ambient light. In: Lamas, D., Loizides, F., Nacke, L., Petrie, H., Winckler, M., Zaphiris, P. (eds.) INTERACT 2019. LNCS, vol. 11746, pp. 721–742. Springer, Cham (2019). https://doi.org/10.1007/978-3-030-29381-9_43

10. Bohrmann, D., Bruder, A., Bengler, K.: Effects of dynamic visual stimuli on the development of carsickness in real driving. IEEE Trans. Intell. Transp. Syst. **23**(5), 4833–4842 (2022)

11. Shelton, B., Nesbitt, K., Thorpe, A., Eidels, A.: Assessing the cognitive load associated with ambient displays. Pers. Ubiquit. Comput. **26**(1), 185–204 (2022)

12. Blankenbach, K., Hertlein, F., Hoffmann, S.: Advances in automotive interior lighting concerning new LED approach and optical performance. J. Soc. Inf. Display. **28**, 655–667 (2020)

13. Fotios, S., Robbins, C.J., Uttley, J.: A comparison of approaches for investigating the impact of ambient light on road traffic collisions. Lighting Res. Technol. **53**(3), 249–261 (2020)

14. Fotios, S., Robbins, C.J.: Effect of ambient light on the number of motorized vehicles, cyclists, and pedestrians. Transp. Res. Record 03611981211044469 (2021)

15. Blankenbach, K., Brezing, L., Reichel, S.: Evaluation of luminance vs. brightness for automotive RGB LED light guides in autonomous cars. In: Proceedings of SPIE 11874, Illumination Optics VI, p. 1187406 (2021)

16. FakhrHosseini, S., Ko, S., Alvarez, I., Jeon, M.: Driver emotions in automated vehicles. In: Riener, A., Jeon, M., Alvarez, I. (eds.) User Experience Design in the Era of Automated Driving. SCI, vol. 980, pp. 85–97. Springer, Cham (2022). https://doi.org/10.1007/978-3-030-77726-5_4

17. Mangla, A., Gulati, D., Jhamb, N., Vashist, D.: Design analysis of dimmer light for autonomous vehicles. In: Khosla, A., Aggarwal, M. (eds.) Smart Structures in Energy Infrastructure. SIC, pp. 145–152. Springer, Singapore (2022). https://doi.org/10.1007/978-981-16-4744-4_15

18. Stylidis, K., Woxlin, A., Siljefalk, L., Heimersson, E., Söderberg, R.: Understanding light. A study on the perceived quality of car exterior lighting and interior illumination. Procedia CIRP **93**, 1340–1345 (2020)

19. Fernandez, V., Chavez, J., Kemper, G.: Device to evaluate cleanliness of fiber optic connectors using image processing and neural networks. Int. J. Electr. Comput. Eng. (IJECE) **11**(4), 3093–3105 (2021)

20. Lin, H., Li, B., Wang, X., Shu, Y., Niu, S.: Automated defect inspection of LED chip using deep convolutional neural network. J. Intell. Manuf. **30**(6), 2525–2534 (2019)

21. Fu, Y., Downey, A.R.J., Yuan, L., Zhang, T., Pratt, A., Balogun, Y.: Machine learning algorithms for defect detection in metal laser-based additive manufacturing: A review. J. Manuf. Process. **75**, 693–710 (2022)

22. Liu, Y., Zhou, H., Tsung, F., Zhang, S.: Real-time quality monitoring and diagnosis for manufacturing process profiles based on deep belief networks. Comput. Ind. Eng. **136**, 494–503 (2019)

23. Huang, X., Zhang, X., Xiong, Y., Liu, H., Zhang, Y.: A novel intelligent fault diagnosis approach for early cracks of turbine blades via improved deep belief network using three-dimensional blade tip clearance. IEEE Access **9**, 13039–13051 (2021)

24. Bengio, Y., Lamblin, P., Popovici, D., Larochelle. H.: Greedy layer-wise training of deep networks, Technical Report 1282 (2006)

25. Klüver, C., Klüver, J.: New learning rules for three-layered feed-forward neural networks based on a general learning schema. In: Madani K. (ed.) Proceedings of ANNIIP: International Workshop on Artificial Neural Networks and Intelligent Information Processing. Portugal: Scitepress, 2014, pp. 27–36 (2014)
26. Hinton, G.E.: training products of experts by minimizing contrastive divergence. Neural Comput. **14**(8), 1771–1800 (2002)
27. Jahani, A.: Forest landscape aesthetic quality model (FLAQM): a comparative study on landscape modelling using regression analysis and artificial neural networks. J. Forest Sci. **65**, 61–69 (2019)
28. Thiemermann, S., Braun, G., Klüver, C.: Homogeneity testing of LED light guides by neural networks. In Klüver, C, Klüver, J. (eds.): New algorithms for practical problems: variations on artificial intelligence and artificial life, pp. 325–339. Wiesbaden: Springer Fachmedien Wiesbaden (2021). (in German)

An Empirical Study of Adversarial Domain Adaptation on Time Series Data

Sarah Hundschell[(✉)], Manuel Weber[ID], and Peter Mandl[ID]

Munich University of Applied Sciences, Munich, Germany
`sarah.hu@web.de`

Abstract. Domain-adversarial learning allows a machine learning model to be trained with supplementary data from a different domain. This enables applications in various time series domains. Although several domain-adversarial models have been proposed in the past, there is a lack of empirical results with different types of time series. This paper provides an empirical analysis with multiple models, datasets and evaluation objectives. Two models known from literature are evaluated in combination with four public datasets: An RNN-based model (VRADA) is contrasted with a newer CNN-based one (CoDATS). The datasets include indoor climate, gas sensors, human activity and physiological data. Our experiments explicitly consider different dataset sizes, similarities between domains and the scenario of multisource training. It is found that CoDATS is very suitable for univariate datasets and performs well even with small datasets. Multivariate datasets can only benefit from the adversarial domain adaptation if the number of data points is large enough. VRADA was found to outperform CoDATS in modeling multivariate datasets. The multisource training available in CoDATS appears promising. A correlation is shown between the similarity of domains and prediction performance.

Keywords: Domain adaptation · Adversarial learning · Time series

1 Introduction

For the longest time, time series data was analyzed using traditional methods. In recent years the focus has shifted to the use of deep learning models. This is partly due to the fact that for the meaningful use of neural networks an accurate training is necessary. An example in this regard is indoor climate data. In general, if data such as humidity or CO_2 values are recorded in a certain environment, machine learning models for a different environment cannot be trained with that data. Transfer Learning (TL) can provide a remedy. TL refers to a subfield of machine learning, in which training and test data can come from different domains. Reference is made to the statement that, in general, existing knowledge can be used to tackle new problems faster or with better approaches to solve them [13, p. 1346].

L. Rutkowski et al. (Eds.): ICAISC 2022, LNAI 13588, pp. 39–50, 2023.
https://doi.org/10.1007/978-3-031-23492-7_4

Using TL, it is possible, given a particular source domain D_S and the associated problem T_S, to find a target function $f_T(\cdot)$ for the target domain D_T and the associated task T_T [13, p. 1347]. Transfer learning can be defined according to Tan et al. [17] as follows: Given a target problem T_T, based on the target domain $\mathcal{D}_T$, Transfer Learning, with the help of the source domain $\mathcal{D}_S$ for the source problem T_S, aims to improve the performance of the prediction function $f_T(\cdot)$ for the target problem T_T. This is done by determining and transferring latent knowledge of $\mathcal{D}_S$ and T_S, where $\mathcal{D}_S \neq \mathcal{D}_T$ and/or $T_S \neq T_T$ holds. Additionally, in most cases, the size N of the domain $\mathcal{D}_S$ is much higher than that of $\mathcal{D}_T$, so $N_S \gg N_T$ holds. This results in conserving resources by eliminating the time and financial burden of having to perform a recollection of data for specific scenarios.

Within the last five years, there has been a vast increase in research on TL with time series [19]. Frequently addressed approaches are pre-training a source prediction model and fine-tuning it in the target domain, or training an autoencoder to transform between feature spaces in advance to the actual model training. Beside these, another trending approach is domain-adversarial learning. It is a model-based TL technique that allows a model to be trained with data from different domains in the same time [6]. Although this approach is still young, according to a recent literature review on time series TL [19], it has been used more frequently than other model-based approaches supporting joint-training. Alternative approaches are ensemble-based transfer or the construction of a transfer-dedicated model objective function. In the last years, new models depending on adversarial domain adaptation have been introduced, see Variational Recurrent Adversarial Deep Domain Adaptation (VRADA) [14], Joint Adversarial Domain Adaptation (JADA) [22] or Domain-Adversarial Neural Network (DANN) [7]. However, to the best of our knowledge, there is no empirical review of existing models. Therefore, this paper presents an evaluation of existing models in different scenarios.

2   Background

Domain Adaptation (DA). There are several approaches to TL. In this paper, we discuss DA. In this case, source and target problems are identical, but differ in source and target domains. This is possible both when the target domain data is unlabeled and when there is little labeled data in the target domain. Thus, Neural Networks (NNs) can be trained with data different from the target data.

Generative Adversarial Learning. Adversarial learning is inspired by Generative Adversarial Networks (GANs) [1]. Two NNs are trained simultaneously. One of the networks is a generative network, the generator G. The discriminator D, on the other hand, is a discriminative network. Here, G is supposed to represent the probability distribution of a given dataset, while D computes the probability that a sample was not generated by G. Thus, the two networks are constantly optimizing each other. This is defined by the following function [8]:

$$min_G max_D V(D; G)$$
$$= \mathbb{E}_{x \sim p_{data}(x)}[\log D(x)]$$
$$+ \mathbb{E}_{z \sim p_z(z)}[\log(1 - D(G(z)))] \qquad (1)$$

x describes the data whose probability should be learned, z the input noise variables and p the probability over those variables.

Domain Adversarial Leaning. The first use of adversarial NNs in conjunction with DA was presented in 2014 by Ajakan et al. [1]. This is the DANN, which was developed for image processing and has better performance than a standard NN or even Support Vector Machines. The approach is based on the assumption that a representation of the data has to be found, which cannot distinguish between training and test data [1]. It must be ensured that the internal representation no longer contains any discriminative information that can be used to infer from which domain the data was obtained.

The problem in [1] is described by

$$\min_{W,V,b,c} \left[\frac{1}{m} \sum_{i=1}^{m} \mathcal{L}(f(x_i^s), y_i^s) \right.$$
$$\left. + \lambda \cdot \max_{u,d} \left(-\frac{1}{m} \sum_{i=1}^{m} \mathcal{L}^d(o(x_i^s), 1) - \frac{1}{m'} \sum_{i=1}^{m'} \mathcal{L}^d(o(x_i^t), 0) \right) \right] \qquad (2)$$

with

$$\mathcal{L}^d(o(x), z) = -z \cdot \log(o(x)) - (1 - z) \cdot \log(1 - o(x)) \qquad (3)$$

The hyperparameter λ must be greater than zero and describes the weighting of the DA regularization term. The adversarial part of the network is described by the NN, parameterized by $\{W, V, b, c\}$, and the domain regressor, parameterized by $\{u, d\}$, which continuously optimize each other during the learning process [1].

3 Related Work

VRADA was one of the first models introduced using time series. Since then more have been added, differing in the architecture used, task to be solved and data to be addressed. Table 1 compares prominent models concerning adversarial DA in relation to time series data. Learning tasks include regression and classification and models are available for both univariate and multivariate time series data. As shown in Table 1, adversarial DA is only introduced since 2017 in relation to time series data. Since then mainly new models have been introduced and empirical studies are hard to find.

Table 1. Published models

Model	Year	Architecture	Data	Task
CoDATS [20]	2020	CNN	Multivariate	Classification
DATSING [9]	2020	CNN	Univariate	Regression
CDAN [18]	2020	CNN	Multivariate	Classification
MTS-ADNN [21]	2020	CNN	Multivariate	Classification
LSTM-DANN [3]	2019	LSTM	multivariate	Regression
ADA [11]	2019	LSTM	Univariate	Regression
DANN [7]	2018	CNN	Univariate	Regression
ADAN [4]	2018	LSTM	Multivariate	Regression
JADA [22]	2018	CNN	Univariate	Classification
VRADA [14]	2017	RNN	Multivariate	Classification

4   Methodology

For the evaluation of the models, elements of implementation are presented below.

4.1   Models

CoDATS. The Convolutional Deep Domain Adaptation Model for Time Series Data (CoDATS) was developed to fill the gap of unsupervised learning models that work with time series data [20]. It was developed with the intention of using Convolutional Neural Networks (CNNs) to train and evaluate models more quickly. The three most important aspects of CoDATS in doing so are the use of existing DA methods, better performance than existing time series fitting methods with respect to individual sources, and ease of extension to varying situations. Additionally, the network can be used when multiple source domains are available, making it usable for complex scenarios. This is referred to as multisource training in the following.

The optimization steps for the CoDATS can be described via Eq. 4 [20].

$$\underset{\theta_f, \theta_c, \theta_d}{argmin} \sum_{i=1}^{n} \mathbb{E}_{(x,y) \sim \mathcal{D}_{S_i}} [\mathcal{L}_y(C(F(\boldsymbol{x})), y) + \mathcal{L}_d(D(\mathcal{R}(F(\boldsymbol{x})), d_{S_i})]$$
$$+ \mathbb{E}_{X \sim \mathcal{D}_T^X} [\mathcal{L}_d(D(\mathcal{R}(F(\boldsymbol{x})), d_T)] \tag{4}$$

Underlying this is a discriminator $D(;\theta_d)$ with parameters θ_d, a feature extractor $F(;\theta_f)$ with parameters θ_f, and a problem classifier $C(\cdot; \theta_c)$ with parameters θ_c. $\mathcal{D}_{S_i}$ corresponds to the labeled data from the source domain and $\mathcal{D}_T^X$ corresponds to the unlabeled data from the target domain. $\mathcal{L}_y$ and $\mathcal{L}_d$ correspond to the multiclass cross-entropy losses of the labels and domains. Using the Gradient Reversal Layer (GRL) represented by $\mathcal{R}(x)$, the forward- and backpropagation is represented as $\mathcal{R} = x; \frac{d\mathcal{R}}{dx} = \lambda \boldsymbol{I}$ where $\boldsymbol{I}$ corresponds to the identity matrix and λ to a constant, defining if forward or backpropagation is used.

Considering the lack of labeled data in time series Wilson, Doppa and Cook developed Domain Adaptation with Weak Supervision (DA-WS). Here, the few parameters that exist for the given dataset are used as a constraint to optimally search for model parameters, described by Equation [20].

$$\begin{aligned}\mathcal{L}_{WS} &= D_{KL}(Y_{true}(y)\|\tilde{Y}_{pred}(y)) \\ &= D_{KL}(Y_{true}\|\mathbb{E}_{X \sim \mathcal{D}_T^X}[C(F(\boldsymbol{x}))])\end{aligned} \tag{5}$$

Combining the DA-WS, shown in Eq. 5 with the presented optimization function of the CoDATS, shown in Eq. 4, yields to the Convolutional Deep Domain Adaptation Model for Time Series Data with Weak Supervision (CoDATS-WS).

VRADA. The VRADA was developed in 2017 by Purushotham et al. to be one of the first models to apply adversarial DA to time series data [14]. The NN in this case is a Recurrent Neural Network (RNN). Using the VRADA model, a Variational Autoencoder (VAE) is conditioned on itself at each time point via recurrent lines. VRADA can be described by Equation [14].

$$\begin{aligned}\mathbb{E}(\theta_e, \theta_g, \theta_y, \theta_d) = {} &\frac{1}{N}\sum_{i=1}^{N}\frac{1}{T^i}\mathcal{L}_r(\boldsymbol{x}^i; \theta_e, \theta_g) + \frac{1}{n}\sum_{i=1}^{n}\mathcal{L}_y(\boldsymbol{x}^i; \theta_y) \\ &- \lambda\left(\frac{1}{n}\sum_{i=1}^{n}\mathcal{L}_d(\boldsymbol{x}^i; \theta_d) + \frac{1}{n'}\sum_{i=n+1}^{N}\mathcal{L}_d(\boldsymbol{x}^i; \theta_d)\right)\end{aligned} \tag{6}$$

All points in the time series dataset are described using $\boldsymbol{x}^i = (x_t^i)_{t=1}^{T^i}$. For each x^i, the term $\tilde{z}^i \sim q_{\theta_e}(z_{T^i}^i | x_{\leq T^i}^i, z_{<T^i}^i)$ is used as the feature representation for the source domain classification. Equation 6 combines the optimization of Variational Recurrent Neural Network (VRNN) and a regulator for the parameters of the VRNN encoder. λ corresponds to balancing the optimization of the domain invariant representation and the optimization of the accuracy of the source classification.

4.2 Datasets

For the evaluation four different datasets are used. A brief summary of the datasets can be found in Table 2.

The first univariate dataset of Arendt et al. from the University of Southern Denmark [2] is used to predict room occupancy based on carbon dioxide data. Data was acquired in four rooms of the premises of the university in 2017, whereby the usage of the different rooms (study and educational purpose) represent the different domains. Occupancy is categorized into "empty", "occupied" and "full", which depict the classes. The occupancy was measured by the PC2 3D Stereo Vision camera by Xovis, with which the rooms were watched for 5713 h

Table 2. Datasets

Dataset	Number of				Fraction of data used
	Domains	Classes	Features	Datapoints per domain	
Room occupation [2]	4	3	1	259200	1
Gas sensors [10]	8	3	1	3715964	$\frac{1}{6}$
WESAD [16]	15	5	8	3749759–5055419	$\frac{1}{180}$
sEMG [12]	36	6	8	343540–507536	$\frac{1}{6}$

hours. The feature used for training was solely CO_2 concentration since the values were used as a representative of a univariate dataset.

The second dataset was collected in 2016 for online decorrelation of humidity and temperature in chemical sensors for continuous monitoring [10]. Classes are representations of stimuli used on the gas sensors, which are "wine", "banana" and "background". The domains are the eight different sensors whereby data was acquired by stimulating the sensors for different time periods (from seven minutes to two hours) over a course of 79 days. As a feature, the sensor values were used, which depict the sensor resistance in Ohm. Only $\frac{1}{6}$ of the available data is used to create a better comparison between datasets.

As a multivariate dataset, Wearable Stress and Affect Detection (WESAD) is presented. It contains various physiological data over which specific emotional states are to be determined, which were acquired in 2018 [16]. The emotional states depict different classes: "transient", "baseline", "stress", "amusement" and "meditation". The different features used of the multivariate dataset were represented by the sensors of the RespiBAN Professional, a wrist-worn device. It measures acceleration in x, y and z direction, ECG-, EDA-values, respiration and temperature. Those values were acquired by 17 people, hereby used as domains, were only 15 could be used because of problems with the used sensors. Since so many data points are available, only $\frac{1}{180}$ is used.

As a second multivariate dateset, Surface Electromyography (sEMG) is used, which is composed of eight channels of a MYO Thalmic wristband, which recorded hand movements of 37 people in the form of myographic signals in 2018 [12]. The individuals are used as domains, whereby the features are represented by the eight different sensors of the wristband. Different movements are used as classes: "hand at rest", "hand clenched in a fist", "wrist flexion", "wrist extension", "radial deviations" and "ulnar deviations". For better comparison, only $\frac{1}{6}$ of the available data is used.

4.3 Experimental Setup

We conduct three different experiments considering dataset size, similarity and a multisource training scenario. The experiments on dataset size and domain similarity are applied to CoDATS-WS and VRADA. These networks represent on the one hand the most current state of the art of adversarial DA, since the CoDATS is one of the most recent developments of the adversarial DA in terms of time series data. On the other hand, by additional investigation of the VRADA, both RNNs and CNNs are evaluated. The CoDATS is representative of the use of CNNs in adversarial domain adaptation and the for the use of Long Short-Term Memorys (LSTMs) and therefore of an RNN.

To obtain an upper and lower bound, the experimental scenarios are performed on additional models. The lower bound is the base model of CoDATS without DA. This is based on the assumption that models with DA always perform better than models without DA. The same model is used as the upper bound, whereby the target domain serves as the source domain. Therefore the same data is used during training as well as in testing. The assumption in this case is, that since data during training and testing is the same, no other model will outperform this upper bound.

Dataset Sizes. For the consideration of the effect of the dataset size on the models to be examined, three different datasets are created from the given data. Small datasets are represented by the usage of 30 datapoints in the training phase. 2500 datapoints are used to represent medium sized datasets and 16000 datapoints represent large datasets. To obtain comparable results, an identical number of 3000 test data points is used in each case. In the room occupancy dataset, room 1 serves as the source domain and room 4 serves as the target domain. In the gas sensor dataset, sensor 5 serves as the source domain and sensor 6 serves as the target domain. Subject 6 represents the source domain and subject 8 represents the target domain in the WESAD dataset. The sEMG dataset uses the dataset of proband 1 as source domain and proband 2 as target domain.

Similarity of Domains. To determine the similarity between domains, we use Dynamic Time Warping (DTW), as implemented by S. Salvador and P. Chan [15]. This allows a comparison between the individual domains in a dataset. For a comparison of datasets DTW distances are determined after data normalization. This leads to the fact that all examined values lie between 0 and 1 and thus a better comparison between the datasets is possible. However, important information is lost. Conducting the experiment, different combinations of source and target domains are examined in each case and then related to the corresponding DTW distance. In each scenario, 80% of the available data is used to train the respective model. 20% of the available data is used for testing.

Multisource Training. The purpose of investigating the multisource approach is to determine the extent to which training with multiple source domains affects prediction accuracy. It is only applied to the CoDATS-WS since, according to the authors, it is the first NN in the field of adversarial DA for time series data that supports training with multiple source domains [20]. For testing each dataset, the model is trained in stages with the possible source domains. This results in $N - 1$ possible source domains for each dataset with N domains and thus

$N-1$ experimental scenarios. The last domain is used as the target domain. For training 80% and for testing 20% of the data in the selected domain are used.

5   Results

Dataset Sizes. In Fig. 1 the Area under a Receiver Operating Characteristic Curve (AUROC) value of the performed experiments is shown. It can be seen that adversarial DA does not always provide better results than models without DA. This seems true for the small WESAD dataset, i.e., a multivariate dataset, with only few data in the training phase. The undercutting of the lower bound is due to the fact that learning invariant features is significantly more difficult in multivariate datasets than in univariate datasets.

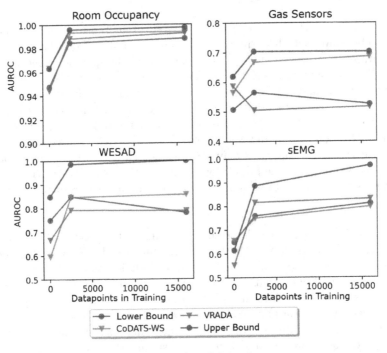

Fig. 1. Results of data sizes

Training of NNs benefits from a larger amount of test data in almost every case. It should be noted that for multivariate datasets, where little data is available in training, machine learning does not always benefit from adversarial DA methods. For univariate datasets, DA methods significantly improve training in all cases, since they are easier to model. CoDATS-WS provides particularly good values for univariate datasets.

Similarity of Data Domains. In Fig. 2, AUROC values are shown in dependence on the DTW distance in percent. This can be calculated by

$$W_{\text{DTW}} = \frac{DTW_x - DTW_{\min}}{DTW_{\max} - DTW_{\min}} \tag{7}$$

where $DTW_{\min}$ and $DTW_{\max}$ describe the minimal and maximal distance acquired in a dataset. DTW_x describes the distance of the corresponding domain combination. Considering Fig. 2, it does not make much of a difference which adversarial methods are used to model univariate datasets. In addition, the similarity of the source and target domains has much less of an effect than assumed. In most cases, the adversarial DA methods perform better than the model without adversarial DA. The results of WESAD in Fig. 2 show very large variations between the individual DTW distances. No relationship between AUROC and domain similarity is apparent. It can be seen that VRADA produces better results than CoDATS in many cases, especially at a low DTW distance, i.e., a high similarity. The adversarial DA is again superior to the model without DA in most cases. Nevertheless, no general statement can be made about the influence of the similarity of the source and target domains. Regarding the AUROC of the sEMG dataset, CoDATS-WS does not perform better than the model without DA or the VRADA in any case. Due to the constant results when training with the test data, again an investigation of the effect of domain similarity is basically very well done. In Fig. 2, a downward trend in all models is additionally evident as the DTW distance increases. Although there are some deviations of the negative slope in the results these combinations also perform better than average when trained with the test data. The DTW distance, which is determined over the normalized samples of a dataset, provides a good estimate of the quality of the dataset in a machine modeling task. However, it does not provide a meaningful estimate of the similarity of domains within a dataset.

As opposed to Fawaz et al.'s conclusion that DTW distance may be used as an identifier to detect negative transfer [5] it can be stated that this does not necessarily seem to affect classification performance in domains of the same application field since they may be already similar.

Multisource Training. In Fig. 3 the precision of the results is shown. It can be seen that multisource training can improve classification performance, compare data of room occupation and sEMG. But it can also be seen that adding domains does not always improve results. Therefore we compare DTW distances, shown in Table 3, with the acquired results.

It is noticeable that a lower DTW distance leads to an improvement in multisource training over time. Since new domains are only a small fraction in the training data, the model does not immediately improve when applying a source domain more similar to the target domain. However, they generally lead to a better mapping. In summary, this means that the adversarial DA benefits from the possibility of multisource training when the domains have a certain similarity.

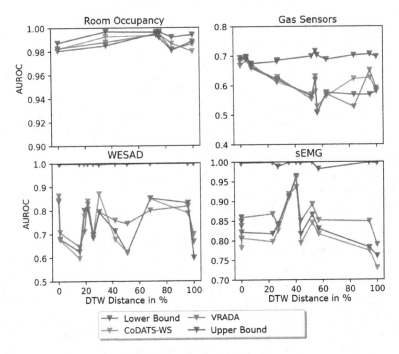

Fig. 2. Similarity of domains

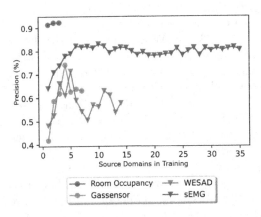

Fig. 3. Multisource training

Table 3. DTW distances

Source domain	DTW distance to domain 17
2	15,51
3	**9,57**
4	**12,07**
5	8,72
6	**10,02**
7	18,87
8	10,01
9	12,01
10	**14,44**
11	8,30
13	**11,30**
14	12,15
15	11,79
16	**12,88**

6   Conclusion

As a part of this work, both VRADA and CoDATS were applied to a number of multi- and univariate datasets. When examining dataset sizes, it was found that adversarial DA does not always produce better results than methods without DA. For univariate datasets, where only few data are available for the training phase, the CoDATS-WS provided good results, but for multivariate datasets the adversarial DA did not generate satisfactory results. Alternative methods should be considered here. It is assumed that the DTW distance mainly influences the modeling of domains of datasets that generally have higher similarity to each other. The lower the DTW distance, the more likely a statement can be made about how well a model can represent these domains. However, this statement should be investigated further. It can be stated that negative transfer in domains from the same application field can not be detected by DTW distance as proposed by Fawaz et al. [5] since it does not affect modeling in this case. The VRADA model was found to be significantly superior to the CoDATS-WS model in modeling multivariate datasets. However, in terms of similarity, determined by DTW distance, it may not make a difference which model is applied. Prediction performance benefits from multisource training provided by the CoDATS-WS model. To guarantee this improvement during training, it should be considered which domains are similar. As a indicator of prediction performance regarding different datasets the DTW distance can be applied to normalized time series.

References

1. Ajakan, H., Germain, P., Larochelle, H., Laviolette, F., Marchand, M.: Domain-adversarial neural networks. arXiv: 1412.4446v2 [stat.ML] (2014). http://arxiv.org/pdf/1412.4446v2
2. Arendt, K., Johansen, A., Jørgensen, B. N., et al.: Room-level occupant counts, airflow and CO_2 data from an office building. In: Gao, J., Zhang, P., Pan, S., Ni, C.-C. (eds.) First Workshop on Data Acquisition to Analysis, New York, USA, pp. 13–14 (2018). https://doi.org/10.1145/3277868.3277875
3. Da Costa, P.R.d.O., Akcay, A., Zhang, Y., Kaymak, U.: Remaining useful lifetime prediction via deep domain adaptation. arXiv:1907.07480v1 [cs.LG] (2019). http://arxiv.org/pdf/1907.07480v1
4. Farshchian, A., Gallego, J.A., Cohen, J.P., Bengio, Y., Miller, L.E., Solla, S.A.: Adversarial domain adaptation for stable brain-machine interfaces. arXiv: 1810.00045v2 [cs.LG] (2019). http://arxiv.org/pdf/1810.00045v2
5. Fawaz, H., Forestier, G., Weber, J., Idoumghar, L., Muller, P.-A.: Transfer learning for time series classification. In: 2018 IEEE International Conference on Big Data (Big Data), pp. 1367–1376 (2018). https://doi.org/10.1109/BigData.2018.8621990
6. Ganin, Y., Ustinova, E., Ajakan, H. et al.: Domain-adversarial training of neural networks. arXiv: 1505.07818 [stat.ML] (2016). https://arxiv.org/pdf/1505.07818.pdf
7. Getmantsev, E., Zhurov, B., Pyrkov T. V., Fedichev P.O.: A novel health risk model based on intraday physical activity time series collected by smartphones. arXiv: 1812.02522v1 [cs.HC] (2018). http://arxiv.org/pdf/1812.02522v1

8. Goodfellow I.J., Pouget-Abadie, J., Mirza M., et al.: Generative adversarial nets. In: NIPS'14: Proceedings of the 27th International Conference on Neural Information Processing Systems, vol. 2, pp. 2672–2680 (2014)

9. Hu, H., Tang, M., Bai, C.: DATSING: data augmented time series forecasting with adversarial domain adaptation. In: d'Aquin, M., Dietze, S., Hauff, C., Curry, E., Cudre, P., Mauroux (eds.) Proceedings of the 29th ACM International Conference on Information and Knowledge Management, New York, USA, pp. 2061–2064 (2020). https://doi.org/10.1145/3340531.3412155

10. Huerta, R., Mosqueiro, T.S., Fonollosa, J., Rulkov, N.F., Rodriguez-Lujan, I.: Online decorrelation of humidity and temperature in chemical sensors for continuous monitoring. Chemometr. Intell. Lab. Syst. **157**(3), 169–176 (2016). https://doi.org/10.1016/j.chemolab.2016.07.004

11. Liu, Y., Hu, X., Jin, J: Remaining useful life prediction of cutting tools based on deep adversarial transfer learning. In: ICCPR '19: Proceedings of the 2019 8th International Conference on Computing and Pattern Recognition, pp. 434–439. Association for Computing Machinery, New York (2019). https://doi.org/10.1145/3373509.3373543

12. Lobov, S., Krilova, N., Kastalskiy, I., Kazantsev, V., Makarov, V.A.: Latent factors limiting the performance of sEMG-interfaces. Sensors **18**(4) (2018). https://doi.org/10.3390/s18041122

13. Pan, S.J., Yang, Q.: A survey on transfer learning. IEEE Trans. Knowl. Data Eng. **22**, 1345–1359 (2010). https://doi.org/10.1109/TKDE.2009.191

14. Purushotham, S., Carvalho, W., Nilanon, T., Liu Y.: Variational recurrent adversarial deep domain adaptation. In: International Conference on Learning Representations (ICLR) (2017). https://openreview.net/pdf?id=rk9eAFcxg

15. Salvador, S., Chan P.: FastDTW: toward accurate dynamic time warping in linear time and space. Intell. Data Anal. **11**(5), 561–580 (2007)

16. Schmidt, P., Reiss, A., Duerichen, R., Marberger, C., van Laerhoven, K.: Introducing WESAD, a multimodal dataset for wearable stress and affect detection. In: D'Mello, S. K., Georgiou, P., Scherer, S., Provost, E.M., Soleymani, M., Worsley, M. (eds.) 20th ACM International Conference on Multimodal Interaction, New York, USA, pp. 400–408 (2018). https://doi.org/10.1145/3242969.3242985

17. Tan, C., Sun, F., Kong, T., Zhang, W., Yang, C., Liu, C.: A survey on deep transfer learning. arXiv: 1808.01974v1 [cs.LG] (2018). http://arxiv.org/pdf/1808.01974v1

18. Tang, X., Zhang, X.: Conditional adversarial domain adaptation neural network for motor imagery EEG decoding. In: Entropy 2020, vol. 22 (2020). https://doi.org/10.3390/e22010096

19. Weber, M., Auch, M., Doblander, C., Mandl, P., Jacobsen, H.-A.: Transfer learning with time series data: a systematic mapping study. IEEE Access **9**, 165409–165432 (2021). https://doi.org/10.1109/ACCESS.2021.3134628

20. Wilson G., Doppa, J.R., Cook, D.J.: Multi-source deep domain adaptation with weak supervision for time-series sensor data. arXiv: 2005.10996v1 [cs.LG] (2020). http://arxiv.org/pdf/2005.10996v1

21. Xie, Y., Murphey, Y.L.: Unsupervised driver workload learning through domain adaptation from temporal signals. In: IEEE Conference on Evolving 2020, pp. 1–8 (2020). https://doi.org/10.1109/EAIS48028.2020.91227

22. Zou, H., Yang, J., Zhou, Y., Spanos, C.J.: Joint adversarial domain adaptation for resilient WiFi-enabled device-free gesture recognition. In: 17th IEEE International Conference 2018, pp. 202–207 (2018). https://doi.org/10.1109/ICMLA.2018.00037

Human-AI Collaboration to Increase the Perception of VR

Antoni Jaszcz[1], Katarzyna Prokop[1], Dawid Połap[1(✉)], Gautam Srivastava[2], and Jerry Chun-Wei Lin[3]

[1] Faculty of Applied Mathematics, Silesian University of Technology,
Kaszubska 23, 44-101 Gliwice, Poland
{aj303181,katapro653}@student.polsl.pl, Dawid.Polap@polsl.pl
[2] Department of Mathematics and Computer Science, Brandon University,
Brandon, MB R7A 6A9, Canada
SrivastavaG@brandonu.ca
[3] Western Norway University of Applied Sciences, Bergen, Norway

Abstract. Virtual reality (VR) is gaining popularity very fast due to newer solutions that increase user perception. Glasses, sensors, and treadmills are the basic functionality for immersing yourself in a virtual environment. In this paper, we propose a human-AI collaboration for analyzing the newly generated images that can be used for creating worlds. The presented method is based on analyzing different scenes (from simulation and real environment) using generative adversarial networks (GAN) and the communication with the user for assessments of the created new environment. User's information contributes to the analysis of sample quality and possible rebuilding or retraining of the GAN model. The proposal increases the perception of VR by taking the user's feelings in creating new environments. For this purpose, we combine GAN with fuzzy soft sets inference to gain the possibility of retraining/remodeling the used neural network. It was examined in theoretical simulation and real-environment case study.

Keywords: Neural networks · Convolutional · Soft sets · Virtual reality · Human-AI · Collaboration · Machine learning

1 Introduction

The era of virtual reality (VR) began with simple blue and red filter glasses. The next decades allowed the use of smartphones to emit an image with the help of special goggles. Initial immersion consisted of moving the head with goggles. This type of activity focused mainly on simple games and video projections. Then, additional functionality was introduced, such as allowing additional movements. To enable this, the operation of the joysticks/controllers was added. Their operation was based on pairing them with a smartphone or goggles, which took the position of controllers and make a projection of it in VR's application. The result was gaining huge possibilities through additional manipulation of objects in VR and creating smart classes [2,11].

Unfortunately, the perception of VR did not meet the expectations of users. There were problems with the operation, quality, and possible safety of people immersed in the created world. The first two aspects, i.e. the operation of the application and their quality, are elements that can be improved through additional computational effort. It is especially possible with much better technical parameters of the equipment. However, the inability to interact with other users or with artificial intelligence was also noticed. These problems caused that to some extent, VR has been rejected by many users. An additional element was the lack of safety during the immersion. To get the best experience in analyzed technology, goggles, controllers, and headphones are used [4]. As a consequence, the user lost the ability to analyze sounds, images, and interactions through the hands with the surrounding environment. In the case when some unexpected object would have appeared in the area, the user may be putting himself at risk. Despite cooling enthusiasm for the VR technology itself, recent years have brought some interesting solutions to eliminate the existing problems.

It should be noted that the times of the global pandemic caused by covid-19 have contributed to a renewed wave of popularity of VR. This is especially visible in virtual walks around centers or museums [14]. The authors focused on the problem of such immersion and the possibility of movement between shows. It was noticed that despite the lack of realistic objects or the atmosphere, peace and the possibility of analyzing works of art were obtained. The reason for this is the lack of talks by other visitors. Moreover, this technology is used in schools and colleges [18]. The inability to conduct chemical or physical experiments can be compensated for by a virtual simulation of this phenomenon. In [5], the idea of molecular visualization was presented and this is a perfect application of VR. Very small objects can be enlarged and modeled in 3D for better understanding and manipulation of them. In addition to the practical applications of the current state of VR, scientists are developing technology and pointing out future directions. One of such directions is the adaptation and creation of games to strengthen/increase movement by training [7,17]. Moreover, these developments are used for analyzing neural activity and behavior of animals like adult zebrafish [10]. The scientist proposes to create a VR system that helps to analyze behavior by an automatic mechanism. Such solutions can be helpful in many areas of our life because it helps to understand some phenomena in virtual simulation (when it is impossible in a real one). The next step of this is the merge of VR into the Internet of Things. It can be very helpful for automatization, analyzing actions occurring in the communication and operation of intelligent things and other aspects of smart technologies and machine learning approaches [1,13,15,16].

Based on these technological changes and various additional aspects to increase perception and safety, we propose a method for increasing the immersing quality. Our solution is based on a model of collaboration between the user and artificial intelligence, which uses a player's feelings about the created scenario for possible retraining or rebuilding the model. The main contribution of this paper are: the human-AI-collaboration method for developing the machine learning model, a mechanism for evaluating the information received from the user, the use of generative models during immersion to increase user perception, combine GAN and soft sets for VR applications.

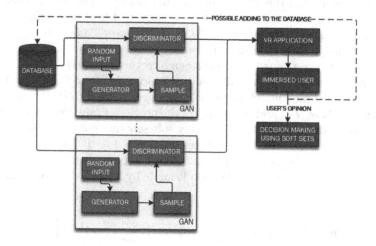

Fig. 1. Visualization of the proposed framework for building VR applications using AI-human collaboration.

2   AI Model for Virtual Reality Scenario Creation

The proposal is based on the use of generative adversarial networks (GAN) [12,21] for generating a new simulation environment depending on the user's preferences whether it should be as close to reality or simulation as possible. For this purpose, we propose using two neural networks that compete with each other in a zero-sum game. Such a model is named GAN. In this approach, the first neural network is called the discriminator network and is marked as a function D. The purpose of this network is to evaluate incoming images x to distinguish real from generated. The second network is called a generator defined as a function G and trying to train the model to deceive the other network. The idea of using such a model is to create realistic new images. In such a model, the training is a *min-max* optimization task, where G is minimized and D is maximized. It can be described as:

$$\min_G \max_D V(D, G) = \mathbf{E}_{x \sim p_{data}(x)}[\log D(x)]$$
$$+ \mathbf{E}_{z \sim p_z(z)}[log(1 - D(G(z)))], \tag{1}$$

where p_{data}, p_z are respectively the probability distribution of database, and of latent space (a random Gaussian distribution).

2.1 Human-AI-Collaboration

Having a database with k scenarios, a k GAN is also created. At first, each instance is trained by a training algorithm until some constant number of iterations T_{max} would not be reached. When all GAN are trained, the newly generated samples can be used in the VR simulation environment. The user is testing the newly generated sample and assessing the quality γ of his immersion/quality of immersion, etc. This value is in a set $\{0,1\}$. In the user's assessment, the value of the loss function described in equation (1) is taken into account whether the classifier should be retrained or even the architecture changed.

The analysis of the human verdict and all information from GAN are analyzed by the probability module which takes mentioned γ_1, $V_1(D,G)$, the number of samples in database $|P_1|$, and the results of the previous evaluation (γ_0, $V_0(D,G)$, $|P_0|$). The decision is made based on fuzzy soft sets inference idea [6].

A pair (F,E) is called a fuzzy soft set when $F : A \mapsto I^U$ and $E \subset A$, where U is universal set, A is a subset of U and I^U is the power set of all U. Inference using fuzzy soft sets to choose a decision is made by finding the maximum value according to:

$$\max\left\{ \sum_{i=0}^{o} w_i \cdot r_i^{(0)}, \sum_{i=0}^{o} w_i \cdot r_i^{(1)}, \ldots, \sum_{i=0}^{o} w_i \cdot r_i^{(j-1)} \right\}, \qquad (2)$$

where $r_i^{(j)}$ is i-th value from j-th decision, and w_i means the weights (importance) of a a given value identified with calculated value.

In the proposed approach, we have a set of three values for each GAN instance: $\{\gamma_1 - \gamma_0, V_0(D,G) - V_1(D,G), |P_1| - |P_0|\}$ and three decisions $\{retrain, remodel, noAction\}$. Retrain means retraining a GAN by additional 10 iterations, remodeling means changing the architecture of used networks and the last decision does not make any changes in the current state of the model. This two sets can be formulated as a table (see Table 1) with r_i values. In the case of *retrain*, the most important value is the user's assessment and in the case of *remodel*, the loss difference is more important. It must be noticed, that the last value which is the increased number of samples is larger than 0 only in one case. If the user's assessment is higher or even to 7, the generated sample is added to the database.

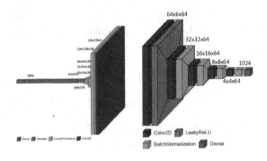

Fig. 2. Generator and discriminator

Table 1. Proposed soft set table for decision making.

	$\gamma_1 - \gamma_0$	$V_0 - V_1$	$|P_1| - |P_0|$
Retrain	0.45	0.4	0.15
Remodel	0.4	0.5	0.1
No action	0.3	0.3	0.4

Algorithm 1: Proposition of human-AI collaboration.

Input: GAN instances, the number of images evaluation M
1 Train all instances of GAN;
2 **for** *each GAN instance* **do**
3     $m := 0$;
4   **for** $m < M$ **do**
5   Create a sample image;
6   Load image to VR application;
7   Get user's assessments;
8        $m{+}=1$;
9   Average user's assessments;
10   **if** *there are previous measurements* **then**
11   Create a soft set table;
12   Calculate weights values;
13   Make a decision using Eq. (2);
14   **if** *decision == retrain* **then**
15   Retrain model;
16   **else if** *decision==remodel* **then**
17   Send information that GAN architecture should be remodeled;
18

This inference system is running after a constant value of the user's evaluation M. The whole idea is shown in the form of pseudo-code in Algorithm 1 and Fig. 1, respectively.

3 Experiments

Our proposal was examined in simulation for analyzing the best architectures, to find an optimal model which can be applied in VR applications. For this reason, we used a database called *Scene Classification: Simulation to Reality* [3] that contains 6 classes: field (1085 images), forest (1007 images), bathroom (1034 images), computer lab (1029 images), living room (1065 images), and stairs (708 images). At first, we analyzed the use of different transfer learning like VGG19 [19], Inception [20], Xception [8], and ResNet50 [9], and proposed architecture shown in Fig. 2. The main metric that was taken into consideration was accuracy calculated as:

$$accuracy = \frac{TP + TN}{TP + TN + FP + FN}, \tag{3}$$

where TP, TN, FP, FN successively means true positive/negative and false positive/negative.

In the next step of conducting tests, we used the selected model of GAN and asked 20 people to test the proposed method. For this purpose, we used Goggle VR Dell Visor VR with controls and a simple application that shows true images and created ones. The main task of the user was to decide if this is a real or fake image of some scenery The user's assessment is used in our approach to give some information and data for the AI mechanism.

Table 2. Comparison of different models for classification database.

Method	Accuracy [%]
Proposed architecture	0,7802
VGG19	0,7801
VGG19 with frozen first 2 blocks	0,6791
VGG19 with frozen first 4 blocks	0,6901
Xception	0,7401
Xception with frozen first 2 blocks	0,7134
Xception with frozen first 4 blocks	0,7323
Inception	0,7802
Inception with frozen first 2 blocks	0,7831
Inception with frozen first 4 blocks	0,7763
ResNet50	0,7742
ResNet50 with frozen first 2 blocks	0,7313
ResNet50 with frozen first 4 blocks	0,5441

3.1 Simulations

Used datasets were split into 70% to 30% (training set to validation) - equally for each class. Then all mentioned before models were trained without the freezing layers technique, and learning transfer architectures have been subjected to freezing operation the first two/four blocks for more accurate comparison. This experiment was made for choosing architecture for the discriminator in our proposal. The obtained results with calculated accuracy are shown in Table 2. Based on these values, all cases where layer freezing was applied have the worst results. This is because the frozen blocks have predetermined filters to extract features. Transferring ready-made filters is not always beneficial for the new dataset. Here was the case where transfer learning models were pre-trained for other databases and here used for analyzing six classes of scenery. However, the best models were the proposed architecture, VGG19, and Inception, which reached over 78% of accuracy. This result shows that there is no one and the best model for the selected database. Having such selected models, we choose a new architecture trained from the beginning for the discriminator. Moreover, we evaluate this trained model (after 800 iterations) to check the whole database and analyze the classification results. It is shown in the form of a confusion matrix in Fig. 3a. It can be seen that the main problems occurred in the case of forest/field and bathroom/computer lab. Some problems were also in the classification living room and stairs. These problems can be caused because of similar features, but for the most part, the results are correct.

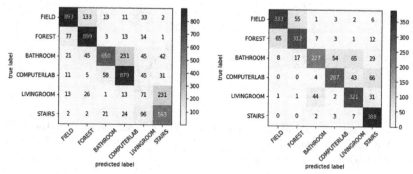

(a) Classification of all classes in analyzed database

(b) Classification of 400 created samples by generator for each class.

Fig. 3. Confusion matrices

Having chosen a discriminator, we added a generator shown in Fig. 2. Before we trained a GAN architecture, we split the dataset into pairs of field/forest, bathroom/computer lab, and living room/stairs. Then we created three instances of GAN - one for each set of pair classes. And trained by 800 iterations. The loss values of generator and discriminator were very similar for all instances. The evaluation of such structure of many GAN was made as follows: generated samples were analyzed by all discriminators and the highest result was taken. In such a way, we created 400 images for each class and evaluated them by discriminators. The results are presented as a confusion matrix in Fig. 3b. It can be seen that the generated samples have high accuracy, exactly 77,8%. This result is very high because the classifier trained for this database has a similar value of accuracy. So, the generator can reach high efficiency in deceiving the discriminator, so the generated samples should be quite realistic. Such trained models indicate a good adaptation to generate new samples that will be used in a practical test using the already proposed collaboration technique between the user and AI.

3.2 Real-Time Experiments

The second step of the research was aimed at a practical analysis of the proposed collaboration technique. We asked 20 people (the age of the participants ranged from 18–30 years) to wear the goggles and controls and use a simple VR app. The main idea was to give information about is it real scenery or created by AI. Each user has to analyze 10 real images and 10 generated by GAN. The answer was given by moving the hand with the controller to the left (real), and the right (artificially generated).

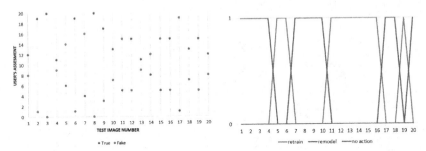

(a) Users' decisions about images (the first 10 images were real and the rest were GAN-created).

(b) The soft set decision makes after all user's assessments on a given image.

Fig. 4. Decision results using user assessment

In Fig. 4a, we can see a sum of all votes on each image. An interesting observation is that in the case of the original images, there were times when users were unsure if it was an image created by a neural network. However, most of the users were deciding on the image correctly. In the case of image no 7 (that was real), 16 people thought it was generated. The image presented only a field and sky. The simplicity of the image itself contributed to the conclusion that the image was generated. For the fake images, opinions were quite mixed. Generated images most often presented fragments or slightly blurred shapes. In particular, the most problematic were images number 16 and 19, which depicted a fragment of the forest (see Fig. 3b). On the other hand, the network also generated samples that were difficult to assign to any class, which can be seen in the evaluation of image number 17. The general opinion was that the quality of images was not the best so it was difficult to analyze if it is real or fake. This issue is based on the fact, that network returned an image of size 128 × 128. In this case, if there would be a need for higher quality, an autoencoder can be applied.

After the analysis of one image by all users, the proposed model analyzed the current situation using the soft set idea (see Table 1). The soft set table decides about three actions - retrain, remodel, and no action. If the decision is retrained, then all GAN were trained by additional 10 iterations which should improve the next generated images. The second situation that is remodeling stops the GAN from working and waits for remodel by the programmer. Obtained results during conducting the experiments are shown in Fig. 4b. The case of remodeling does not appear because, the model was trained to a good accuracy level and if the sum of the user's assessment was higher than 10 on the fake image, then this image was added to the database. Consequently, the database was larger with one more sample after deceiving the user. The other two actions were more frequent, as can be seen in Fig. 4b. In most cases, the method determined the lack of additional operations. This means that the classifier is well suited for practical use. In other situations, the retrain option allowed for increasing the accuracy, which affects the quality of subsequent images. The proposed method shows that in the case when the users are not satisfied with the current effect, the AI decides about additional retraining or no action. It is a simple mechanism of collaboration that results in a better quality of AI method and better perception of VR by the user.

4   Conclusions and Future Works

In this paper, we propose a method for collaboration of an AI method with users in VR. The proposal is based on GAN and a soft set inference mechanism that takes the users' assessment. The main idea was to adapt the AI-human collaboration for increasing the quality of generated images and the reception of the user, or even the immersion itself. The evaluation of the method shows that it can be used in VR apps and provide an automatic analysis of the quality of the used AI technique. The low assessment provides a retrain of classifiers, so there is no need to analyze the number of training iterations. This value will be automatically increased when the result will be badly received by the user.

In future works, we will focus on the analysis of the soft set table that is used in decision-making. In this research, this table was filled with values based on some simulation tests, but in the case of other applications or used databases, these values might be not so efficient. For this purpose, we plan to extend this idea to automatically adjust these values during use.

Acknowledgements. This work is supported by the Rector proquality grant at the Silesian University of Technology, Poland No. 09/010/RGJ22/0067.

References

1. Artiemjew, P., Rudikova, L., Myslivets, O.: About rule-based systems: single database queries for decision making. Future Internet **12**(12), 212 (2020)
2. Bhargava, A., et al.: Revisiting affordance perception in contemporary virtual reality. Virtual Reality, 1–12 (2020)
3. Bird, J.J., Faria, D.R., Ekárt, A., Ayrosa, P.P.: From simulation to reality: CNN transfer learning for scene classification. In: 2020 IEEE 10th International Conference on Intelligent Systems (IS), pp. 619–625. IEEE (2020)
4. Cassani, R., Moinnereau, M.A., Ivanescu, L., Rosanne, O., Falk, T.H.: Neural interface instrumented virtual reality headsets: toward next-generation immersive applications. IEEE Syst. Man Cybern. Mag. **6**(3), 20–28 (2020)
5. Cassidy, K.C., Šefčík, J., Raghav, Y., Chang, A., Durrant, J.D.: ProteinVR: web-based molecular visualization in virtual reality. PLoS Comput. Biol. **16**(3), e1007747 (2020)

6. Chandrasekhar, U., Mathur, S.: Decision making using fuzzy soft set inference system. In: Vijayakumar, V., Neelanarayanan, V. (eds.) Proceedings of the 3rd International Symposium on Big Data and Cloud Computing Challenges (ISBCC – 16'). SIST, vol. 49, pp. 445–457. Springer, Cham (2016). https://doi.org/10.1007/978-3-319-30348-2_37

7. Checa, D., Bustillo, A.: A review of immersive virtual reality serious games to enhance learning and training. Multimedia Tools Appl. **79**(9), 5501–5527 (2020)

8. Chollet, F.: Xception: Deep learning with depthwise separable convolutions. In: Proceedings of the IEEE Conference on Computer Vision and Pattern Recognition, pp. 1251–1258 (2017)

9. He, K., Zhang, X., Ren, S., Sun, J.: Deep residual learning for image recognition. In: Proceedings of the IEEE Conference on Computer Vision and Pattern Recognition, pp. 770–778 (2016)

10. Huang, K.H., Rupprecht, P., Frank, T., Kawakami, K., Bouwmeester, T., Friedrich, R.W.: A virtual reality system to analyze neural activity and behavior in adult zebrafish. Nat. Methods **17**(3), 343–351 (2020)

11. Ikedinachi, A., Misra, S., Assibong, P.A., Olu-Owolabi, E.F., Maskeliūnas, R., Damasevicius, R.: Artificial intelligence, smart classrooms and online education in the 21st century: implications for human development. J. Cases Inf. Technol. (JCIT) **21**(3), 66–79 (2019)

12. Jabbar, A., Li, X., Omar, B.: A survey on generative adversarial networks: variants, applications, and training. ACM Comput. Surv. (CSUR) **54**(8), 1–49 (2021)

13. Kazimierski, W., Wawrzyniak, N., Wlodarczyk-Sielicka, M., Hyla, T., Bodus-Olkowska, I., Zaniewicz, G.: Mobile river navigation for smart cities. In: Damaševičius, R., Vasiljevienė, G. (eds.) ICIST 2019. CCIS, vol. 1078, pp. 591–604. Springer, Cham (2019). https://doi.org/10.1007/978-3-030-30275-7_45

14. Lee, H., Jung, T.H., tom Dieck, M.C., Chung, N.: Experiencing immersive virtual reality in museums. Inf. Manag. **57**(5), 103229 (2020)

15. Lv, Z.: Virtual reality in the context of internet of things. Neural Comput. Appl. **32**(13), 9593–9602 (2020)

16. Nowicki, R.K., Seliga, R., Żelasko, D., Hayashi, Y.: Performance analysis of rough set-based hybrid classification systems in the case of missing values. J. Artif. Intell. Soft Comput. Res. **11** (2021)

17. Polap, D., Kesik, K., Winnicka, A., Wozniak, M.: Strengthening the perception of the virtual worlds in a virtual reality environment. ISA Trans. **102**, 397–406 (2020)

18. Radianti, J., Majchrzak, T.A., Fromm, J., Wohlgenannt, I.: A systematic review of immersive virtual reality applications for higher education: design elements, lessons learned, and research agenda. Comput. Educ. **147**, 103778 (2020)

19. Simonyan, K., Zisserman, A.: Very deep convolutional networks for large-scale image recognition. arXiv preprint arXiv:1409.1556 (2014)

20. Szegedy, C., Vanhoucke, V., Ioffe, S., Shlens, J., Wojna, Z.: Rethinking the inception architecture for computer vision. In: Proceedings of the IEEE Conference on Computer Vision and Pattern Recognition, pp. 2818–2826 (2016)

21. Yang, X., Huo, H., Li, J., Li, C., Liu, Z., Chen, X.: DSG-fusion: infrared and visible image fusion via generative adversarial networks and guided filter. Expert Syst. Appl. 116905 (2022)

Portfolio Transformer
for Attention-Based Asset Allocation

Damian Kisiel[✉] and Denise Gorse

Department of Computer Science, University College London, London, UK
{d.kisiel,d.gorse}@cs.ucl.ac.uk

Abstract. Traditional approaches to financial asset allocation start with returns forecasting followed by an optimization stage that decides the optimal asset weights. Any errors made during the forecasting step reduce the accuracy of the asset weightings, and hence the profitability of the overall portfolio. The *Portfolio Transformer* (PT) network, introduced here, circumvents the need to predict asset returns and instead directly optimizes the Sharpe ratio, a risk-adjusted performance metric widely used in practice. The PT is a novel end-to-end portfolio optimization framework, inspired by the numerous successes of attention mechanisms in natural language processing. With its full encoder-decoder architecture, specialized time encoding layers, and gating components, the PT has a high capacity to learn long-term dependencies among portfolio assets and hence can adapt more quickly to changing market conditions such as the COVID-19 pandemic. To demonstrate its robustness, the PT is compared against other algorithms, including the current LSTM-based state of the art, on three different datasets, with results showing that it offers the best risk-adjusted performance.

Keywords: Transformers · Deep learning · Portfolio optimization

1 Introduction

Portfolio optimization algorithms aim to select the optimal weighting of financial assets in a given portfolio as a means to maximize or minimize some specific metric of interest. It is arguably the most important phase in the entire investment lifecycle, without which investors would be exposed to unacceptable levels of risk. Markowitz formally formulated this problem in what is now known as *Modern Portfolio Theory* (MPT) [14]. The risk-return trade-off pioneered by Markowitz was very influential at the time and became in effect a go-to tool for the vast majority of industry practitioners. However, despite its rigorous theoretical foundations and wide popularity, the MPT has significant shortcomings when applied in practice. One of these limitations is the assumption that future investment returns of individual assets can be predicted with a reasonable level of precision. This task has, however, been shown to be extremely difficult due to the highly stochastic nature of financial markets [15].

L. Rutkowski et al. (Eds.): ICAISC 2022, LNAI 13588, pp. 61–71, 2023.
https://doi.org/10.1007/978-3-031-23492-7_6

Recent advances in computational power and the wider availability of market data have allowed machine learning architectures to be used for portfolio optimization. For example, by using ensembles of gradient-boosted trees, one can reduce estimation errors in the returns predictions of standard Markowitz-style optimization [3], and the XGBoost model has been successfully used within a meta-allocation framework to switch between different risk-based strategies in order to achieve better risk-adjusted performance [12]. Most recently, deep learning architectures have also started to play a major role [7]. However, a substantial drawback of the above portfolio selection methodologies is that they follow the classical two-step procedure in which errors in the parameter estimations of the first step are translated into inaccurate asset weightings in the second step.

Moody et al. [16] pioneered the contrasting idea of combining prediction and performance optimization in a single step. This work was later extended by that of Zhang et al. [24], who introduced an LSTM-based architecture that showed significant performance improvements over classical asset allocation techniques. In this paper, we introduce the *Portfolio Transformer* (PT), which combines prediction and optimization in a novel end-to-end deep learning architecture based on an attention mechanism, directly outputting portfolio weights that optimize the Sharpe ratio, a measure of risk-adjusted return widely used in practice, under the specified transaction cost penalties. Additionally, the PT makes use of specialized gating mechanisms to determine the ideal level of non-linearity when optimizing each portfolio. Results demonstrate that the Portfolio Transformer is able to outperform a number of other methodologies, ranging from a classical optimization method to the current LSTM-based state of the art [24], on three different datasets encompassing ETFs, commodities, and stocks.

2   Background and Related Work

2.1   Long Short-Term Memory (LSTM)

There exists a large volume of literature applying recurrent neural networks, such as simple RNNs [18] or Gated Recurrent Unit (GRU) networks [17], to financial time series prediction problems. However, the main drawback of standard recurrent neural networks, observed in multiple domains, including finance, is the so-called 'vanishing gradient problem' [8], whereby gradients corresponding to long-term dependencies become very small, effectively preventing the model from further training. LSTM networks [9] tackle this problem by introducing gate mechanisms that allow gradients to flow unchanged; through a process of filtering and summarizing they can ignore irrelevant past information. The vanilla LSTM architecture proposed by Zhang et al. [24] for portfolio optimization is used in this work as one of the benchmark algorithms.

2.2   The Transformer Model

Despite their efficacy at learning time-localized patterns, LSTMs struggle to capture meaningful dependencies when the length of a sequence is relatively large. The Transformer architecture [19] was developed to address this issue and has now been established as state of the art in most work in natural language processing [20]. Several studies also show the successful application of this architecture in the financial domain. For example, Wood et al. [21] show that their Transformer model outperforms an LSTM network in time-series momentum strategies, and Xu et al. [22] apply the model to portfolio policy learning in a reinforcement learning setting (though to maximize cumulative return rather than the industry-preferred Sharpe ratio).

At the heart of every Transformer architecture lies a mechanism called 'self-attention', which replaces recurrence and allows for simultaneous processing of all sequence elements. The Portfolio Transformer implements 'scaled dot-product attention' [19], as given by

$$Attention\,(Q, K, V) = softmax\left(\frac{QK^T}{\sqrt{d_{model}}} + M\right)V. \tag{1}$$

The *value matrix* $V \in \mathbb{R}^{\tau \times d_v}$ of equation (1) is weighted by a set of 'scores' obtained from the softmax operation, which determines how much emphasis each time step from the *key matrix* $K \in \mathbb{R}^{\tau \times d_k}$ should receive when encoding sequence positions from the *query matrix* $Q \in \mathbb{R}^{\tau \times d_k}$. The dot product of Q and K is divided by the square root of the encoding dimension (d_{model}) to counteract problems associated with small gradients. Additionally, the Portfolio Transformer implements masking via matrix $M \in \mathbb{R}^{\tau \times \tau}$ in the first attention block of each decoder layer to ensure it can only attend to preceding time steps and hence maintain its autoregressive property. The operation defined by equation (1) is repeated h times in what is known as *multi-headed attention* (MHA),

$$MHA\,(Q, K, V) = Concatenate\,(head_1, \cdots, head_h)\,W^O, \tag{2}$$

$$head_i = Attention\left(QW_i^Q, KW_i^K, VW_i^V\right), \tag{3}$$

which allows the model to extract information from multiple representation subspaces, where each $head_i$ is implemented using its own set of learned linear projection matrices $W_i^Q \in \mathbb{R}^{d_{model} \times d_k}$, $W_i^K \in \mathbb{R}^{d_{model} \times d_k}$ and $W_i^V \in \mathbb{R}^{d_{model} \times d_v}$. Outputs from all attention heads are then concatenated and again linearly projected using a learned parameter matrix $W^O \in \mathbb{R}^{hd_v \times d_{model}}$ to obtain final values.

2.3   Gated Residual Network (GRN)

In the original Transformer implementation of [19] each attention layer is followed by a simple feed-forward network. The Portfolio Transformer adopts a more flexible approach and instead makes use of a *Gated Residual Network* (GRN) [13], acting as a gating mechanism that determines the extent of non-linear processing required for a particular portfolio, defined by

$$GRN\,(z) = LayerNorm\,(z + GLU\,(g_1)), \tag{4}$$

$$g_1 = W_1 g_2 + b_1, \tag{5}$$

$$g_2 = ELU\left(W_2 z + b_2\right), \tag{6}$$

in which the GRN's input is given by vector z, g_1 and g_2 are intermediate layers, and ELU is the Exponential Linear Unit [4] activation function, and in which the process of filtering non-linear contributions is carried out via a Gated Linear Unit (GLU) [5], which provides the Portfolio Transformer with the ability to scale down the amount of non-linear processing and default to a simpler model when, for example, the dataset is small or highly noisy.

2.4 Time2Vec Embedding

Since the attention layers of a Transformer do not make use of recurrence, they cannot inherently capture any information about the relative position of each element in a sequence. In the original model [19] this information is injected via positional encoding. The Portfolio Transformer, however, implements the time encoding proposed by Kazemi et al. [10], that takes the following form:

$$Time2Vec\left(t\right)[i] = \begin{cases} \omega_i t + \varphi_i & \text{if } i = 0 \\ sin\left(\omega_i t + \varphi_i\right) & \text{if } 1 \leq i \leq k. \end{cases} \tag{7}$$

The temporal signal represented by t in Eq. (7) is decomposed into a set of frequencies ω and phase shifts φ. This time decomposition technique is closely related to Fourier transforms, but instead of using a fixed set of values, all frequencies and phase shifts are learnable parameters. It should be noted that the use of *sine* as the activation function enables the Portfolio Transformer to capture periodic behaviors in data.

3   Methodology

3.1 Portfolio Transformer Architecture

The network architecture of the Portfolio Transformer is shown in Fig. 1. It consists of four main building blocks: input layer (Time2Vec embedding, Sect. 2.4), encoder and decoder (Sects. 2.2 and 2.3), and output layer. Each of these blocks will now be discussed in turn.

Input Layer. Each sequence position, denoted by vector x_t in Fig. 1, contains concatenated returns of all N assets in a given portfolio on day t. There are τ such vectors (per encoder and decoder block) stacked together to form the input matrix X of dimension (τ, N). Time2Vec embedding is then used to extract time-encoded cross-sectional features.

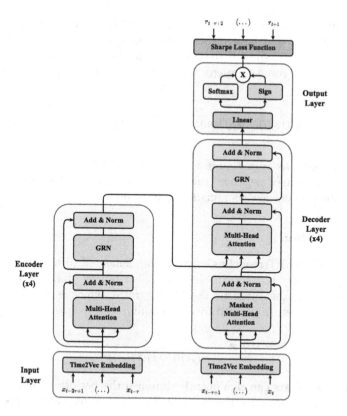

Fig. 1. The Portfolio Transformer: model architecture.

Encoder. The Portfolio Transformer uses a stack of four identical encoder layers. Inside each of these layers, the time-encoded input is first processed by a multi-headed attention mechanism, where the number of attention heads h is selected during hyperparameter optimization. A series of gating mechanisms are then applied, using a GRN module that determines the ideal amount of non-linearity. A residual connection [6], followed by layer normalization [1], are additionally applied to these two sub-layers, as shown in Fig. 1.

Decoder. The decoder block is also composed of four identical layers, each of which contains two multi-headed attention modules. The first one uses masking to ensure predictions made by the Portfolio Transformer depend only on data from preceding time steps. The second one allows the decoder to attend to the output of the encoder stack, which provides a much more nuanced representation of the data and the ability to learn longer-term dependencies among portfolio assets. As in the case of the encoder, a GRN is used to remove any unnecessary complexity and there is a residual connection around each sub-layer followed by layer normalization.

Output Layer. The Portfolio Transformer allows for short-selling via a specialized output layer that implements a compound function proposed by Zhang et al. [23]. First, the output of the final decoder layer is processed by a fully-connected layer. The resulting vector $(s_{i,t})$ is then used to compute final portfolio weights, given by

$$w_{i,t} = sign\,(s_{i,t}) \times softmax\,(s_{i,t}) \triangleq sign\,(s_{i,t}) \times \frac{e^{s_{i,t}}}{\sum_{j=1}^{N} e^{s_{j,t}}}. \tag{8}$$

The use of the softmax operation in Eq. (8) ensures that, while portfolio weights can be either positive or negative (the latter allowing for short-selling), the sum of their absolute values always remains equal to one.

3.2 Loss Function

The objective of the PT model as currently implemented (other objective functions being possible) is to learn the asset distribution that maximizes risk-adjusted returns as measured by the Sharpe ratio, which is defined below as expected portfolio return divided by its volatility:

$$SR = \frac{E(R_P)}{\sqrt{E(R_P^2) - (E(R_P))^2}}. \tag{9}$$

Since transaction costs can significantly diminish the performance of allocation strategies with high turnover, the Portfolio Transformer uses cost-adjusted portfolio returns

$$R_{P,t} = \sum_{i}^{N} w_{i,t-1} \times r_{i,t} - C \times \sum_{i}^{N} |w_{i,t-1} - w_{i,t-2}| \tag{10}$$

in order to find solutions that account for trading costs, where C is a constant cost rate, set to a realistic value of two basis points (2 bps), $w_{i,t-1}$ represents the weight of asset i on day $t-1$, and $r_{i,t}$ denotes the realized arithmetic return of asset i from day $t-1$ to day t, computed using asset prices $P_{i,t}$ and $P_{i,t-1}$ as follows:

$$r_{i,t} = \frac{P_{i,t}}{P_{i,t-1}} - 1. \tag{11}$$

The expected portfolio return, denoted by $E\,(R_P)$ in Eq. (9), is obtained by taking an average of all portfolio returns over a trading period of length τ:

$$E\,(R_P) = \frac{1}{\tau}\sum_{t=1}^{\tau} R_{P,t}. \tag{12}$$

Finally, since the PT allows for short-selling but no leverage, the portfolio positions are constrained by $w_{i,t} \in [-1,1]$ and $\sum_{i}^{N} |w_{i,t}| = 1$, which is achieved through the use of the compound function in Eq. (8).

3.3 Training and Model Calibration

Data is split into training and test segments using an expanding window approach, where initially all data points before the end of 2015 are used for training
and the out-of-sample test is carried out on observations recorded in 2016. The
training window is then extended to include the year 2016 and the model is tested
on the subsequent year (2017), and so on. This way, the model is retrained every
year, with all available historical data being used to update the network parameters. Portfolio positions are adjusted on a daily basis, and a transaction cost
rate of 2 bps is used during performance evaluation.

The PT network is trained via mini-batch stochastic gradient descent (with
batch size being among the hyperparameters) using the Adam optimizer [11].
For model calibration purposes, and to control for overfitting, 10% of any training segment is set aside as a separate validation set. Hyperparameter optimization, conducted using 100 iterations of random grid search, is performed only
on the validation set, ensuring that the model has access to test data only during the performance evaluation stage. To further improve the model's ability to
generalize, early stopping is implemented. The Portfolio Transformer was developed using the TensorFlow framework and all experiments were conducted on
NVIDIA's Tesla P100 16 GB GPU with 55 GB of RAM memory.

3.4 Datasets Used

The efficacy of the Portfolio Transformer is demonstrated on three datasets
containing daily price observations. The first of these datasets starts in 2006
and is composed of the same four Exchange Traded Funds (ETFs) used in the
LSTM-based experiments carried out by Zhang et al. [24]: AGG (aggregate bond
index), DBC (commodity index), VIX (volatility index), and VTI (US stocks
index). The second dataset, which starts in 2002, is composed of 24 continuous
commodity futures contracts, including metals, agricultural products, and energy
commodities such as oil and natural gas. Finally, the PT model is tested on daily
observations of 500 stocks based in the US and listed on NASDAQ. This last
dataset starts in 1996 and aims to demonstrate the model's performance on a
large portfolio of hundreds of instruments.

3.5 Benchmark Models

Four algorithms are implemented as benchmarks: (1) mean-variance optimization (MV) [14], a classical two-step portfolio selection procedure with a moving
window of 50 days used to estimate expected asset returns and covariances; (2)
XGBoost [2], included because gradient-boosted decision trees perform very well
in applied machine learning competitions; (3) a multilayer perceptron (MLP),
as a universal function approximator that can capture highly non-linear dependencies; and (4) the LSTM architecture of [24], a high-performance recurrent
architecture that represents the previous state of the art.

4 Results

4.1 Performance Comparison: Full Investment Horizon

The performance of the Portfolio Transformer is compared to that of the benchmark algorithms of Sect. 3.5 using a number of metrics that aim to capture portfolio risk level through annualized volatility (Vol.) and maximum drawdown (MDD), profitability via annualized returns (Returns) and percentage of positive returns (% of + Ret), and risk-adjusted performance using annualized Sharpe, Sortino and Calmar ratios.

Table 1. Experimental results for different algorithms and datasets.

	Returns	Vol.	Sharpe	Sortino	MDD	Calmar	% of +Ret
Panel A: ETFs							
MV	0.004	0.122	0.012	0.270	0.120	0.836	0.497
XGBoost	0.100	0.140	0.657	1.240	0.116	2.591	0.496
MLP	0.135	0.128	0.923	1.789	0.087	2.225	0.498
LSTM	**0.215**	0.133	1.539	2.830	0.096	3.621	0.535
Portfolio Transformer	0.138	**0.067**	**2.252**	**4.093**	**0.036**	**4.773**	**0.548**
Panel B: Commodities							
MV	0.008	0.053	0.174	0.342	0.059	0.682	0.519
XGBoost	0.015	0.059	0.277	0.420	0.063	0.267	0.505
MLP	0.026	0.056	0.479	0.727	0.057	0.797	0.515
LSTM	0.038	**0.031**	1.182	1.852	**0.023**	2.108	0.528
Portfolio Transformer	**0.174**	0.108	**1.506**	**2.304**	0.077	**2.272**	**0.543**
Panel C: Stocks							
MV	0.079	0.126	0.694	1.126	0.106	1.386	0.523
XGBoost	0.101	0.118	0.923	1.352	0.080	1.491	0.533
MLP	0.089	0.121	0.767	1.102	0.087	1.236	0.534
LSTM	0.111	**0.077**	1.456	2.155	**0.056**	2.561	0.565
Portfolio Transformer	**0.334**	0.147	**2.001**	**3.440**	0.091	**4.824**	**0.566**

Panel A in Table 1 shows the results for the portfolio of ETFs, where the Portfolio Transformer outperforms the benchmarks on all but one metric, annualized returns. These are higher for the LSTM model, but this comes at a cost of increased volatility; when this is accounted for, by using the Sharpe ratio, the PT offers much higher *risk-adjusted* returns. Results presented in Panels B (commodities) and C (stocks) show that in both cases the PT achieves the best risk-adjusted performance, while also delivering the highest annualized returns. The PT is beaten by the LSTM in these cases on only two out of seven metrics, volatility and maximum drawdown. In the former case, this is compensated by the PT's higher Sharpe ratio and in the latter by the higher Calmar ratio, relative to the LSTM; the Sharpe ratio is a measure of portfolio return adjusted by

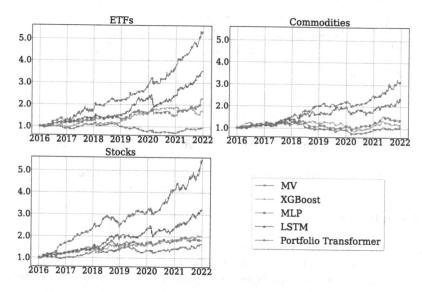

Fig. 2. Comparison of cumulative returns.

volatility, and Calmar ratio of portfolio return adjusted by maximum drawdown, arguably of more relevance than volatility and maximum drawdown per se.

The cumulative return plots of Fig. 2 demonstrate the superior performance of the Portfolio Transformer over the whole investment horizon. The PT generates the highest cumulative returns and offers a reasonable risk profile for all three datasets. The second-best performing model is the LSTM, suggesting that time dependencies learned through recurrence, in case of the LSTM model, or through an attention mechanism, in case of the PT, are very useful in a portfolio optimization setting. However, it can be seen that the LSTM model struggles during the COVID-19 crisis (first quarter of 2020), while the attention-driven Portfolio Transformer shows a much quicker response to this sudden market regime change. The performances of the XGBoost and MLP models are comparable, but lag considerably behind those of the LSTM and the PT, while mean-variance optimization (MV) is by far the worst-performing algorithm, suggesting that highly inaccurate asset weightings are generated by this classical two-step procedure.

4.2 Performance Comparison: COVID-19 Crisis

The above-mentioned difference between the performances of the PT and LSTM models during this period of extreme market volatility is further illustrated in Fig. 3, which shows the 12-month rolling Sharpe ratio of the two models on the ETF dataset. The LSTM suffers a large fall in its risk-adjusted returns during this period and there are times when it even drops down below zero. The Portfolio Transformer, on the other hand, shows a more stable behavior, with its rolling Sharpe ratio staying mostly well above one and delivering outstanding risk-adjusted performance during the Bull market that followed the crisis.

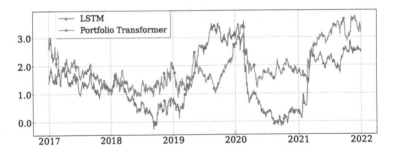

Fig. 3. Rolling Sharpe ratio (12-month) on the ETF dataset.

5 Conclusions

This work has introduced the Portfolio Transformer (PT), which directly optimizes risk-adjusted returns using a novel end-to-end attention-based architecture with specialized time-encoding layers and gating mechanisms. By incorporating transaction costs directly into its loss function the PT model can account for trading cost constraints faced by investors. The results demonstrate that the PT model delivers exceptional risk-adjusted performance, in this respect outperforming all benchmark algorithms on three different datasets with varying portfolio sizes. Due to its full encoder-decoder configuration and its attention mechanism the Portfolio Transformer is able to learn long-term dependencies and as a result can react more quickly to changing market regimes, as demonstrated by its response to the COVID-19 crisis. Turning to future work, one extension of the current model could study the Portfolio Transformer's performance under an attention mechanism different from the scaled dot-product attention currently used. In addition, while it is popular with industry practitioners, the Sharpe ratio is only one of many possible objective functions that could be optimized, and subsequent work will consider alternative metrics and their effect on the overall portfolio performance.

References

1. Ba, J.L., Kiros, J.R., Hinton, G.E.: Layer normalization. arXiv preprint arXiv:1607.06450 (2016)
2. Chen, T., Guestrin, C.: XGBoost: a scalable tree boosting system. In: Proceedings of the 22nd ACM SIGKDD International Conference on Knowledge Discovery and Data Mining, pp. 785–794 (2016)
3. Chen, W., Zhang, H., Mehlawat, M.K., Jia, L.: Mean-variance portfolio optimization using machine learning-based stock price prediction. Appl. Soft Comput. **100**, 106943 (2021)
4. Clevert, D.A., Unterthiner, T., Hochreiter, S.: Fast and accurate deep network learning by exponential linear units (elus). arXiv preprint arXiv:1511.07289 (2015)
5. Dauphin, Y.N., Fan, A., Auli, M., Grangier, D.: Language modeling with gated convolutional networks. In: International Conference on Machine Learning, pp. 933–941 (2017)

6. He, K., Zhang, X., Ren, S., Sun, J.: Deep residual learning for image recognition. In: Proceedings of the IEEE Conference on Computer Vision and Pattern Recognition, pp. 770–778 (2016)

7. Heaton, J.B., Polson, N.G., Witte, J.H.: Deep learning for finance: deep portfolios. Appl. Stoch. Model. Bus. Ind. **33**(1), 3–12 (2017)

8. Hochreiter, S., Bengio, Y., Frasconi, P., Schmidhuber, J.: Gradient flow in recurrent nets: the difficulty of learning long-term dependencies. In: A Field Guide to Dynamical Recurrent Neural Networks, chap. 14, pp. 237–374. IEEE Press (2001)

9. Hochreiter, S., Schmidhuber, J.: Long short-term memory. Neural Comput. **9**(8), 1735–1780 (1997)

10. Kazemi, S.M., Goel, R., Eghbali, S., Ramanan, J., Sahota, J., Thakur, S.: Time2Vec: learning a vector representation of time. arXiv preprint arXiv:1907.05321 (2019)

11. Kingma, D.P., Ba, J.: Adam: a method for stochastic optimization. arXiv preprint arXiv:1412.6980 (2014)

12. Kisiel, D., Gorse, D.: A meta-method for portfolio management using machine learning for adaptive strategy selection. In: International Conference on Computational Intelligence and Intelligent Systems 2021, pp. 67–71. ACM (2021)

13. Lim, B., Arık, S.Ö., Loeff, N., Pfister, T.: Temporal fusion transformers for interpretable multi-horizon time series forecasting. Int. J. Forecast. **37**(4), 1748–1764 (2021)

14. Markowitz, H.: Portfolio selection. J. Finance **7**(1), 77–91 (1952)

15. Michaud, R.O., Michaud, R.O.: Efficient Asset Management: A Practical Guide to Stock Portfolio Optimization and Asset Allocation. Oxford University Press, Oxford (2008)

16. Moody, J., Wu, L., Liao, Y., Saffell, M.: Performance functions and reinforcement learning for trading systems and portfolios. J. Forecast. **17**(5–6), 441–470 (1998)

17. Shen, G., Tan, Q., Zhang, H., Zeng, P., Xu, J.: Deep learning with gated recurrent unit networks for financial sequence predictions. Procedia Comput. Sci. **131**, 895–903 (2018)

18. Tino, P., Schittenkopf, C., Dorffner, G.: Financial volatility trading using recurrent neural networks. IEEE Trans. Neural Netw. **12**(4), 865–874 (2001)

19. Vaswani, A., Shazeer, N., Parmar, N., Uszkoreit, J., Jones, L., Gomez, A.N.: Attention is all you need. In: Advances in Neural Information Processing Systems, vol. 30 (2017)

20. Wolf, T., Debut, L., Sanh, V., Chaumond, J., Delangue, C., Moi, A.: Huggingface's transformers: State-of-the-art natural language processing. arXiv preprint arXiv:1910.03771 (2019)

21. Wood, K., Giegerich, S., Roberts, S., Zohren, S.: Trading with the momentum transformer: an intelligent and interpretable architecture. arXiv preprint arXiv:2112.08534 (2021)

22. Xu, K., Zhang, Y., Ye, D., Zhao, P., Tan, M.: Relation-aware transformer for portfolio policy learning. In: Proceedings of the Twenty-Ninth International Conference on International Joint Conferences on Artificial Intelligence, pp. 4647–4653 (2021)

23. Zhang, C., Zhang, Z., Cucuringu, M., Zohren, S.: A universal end-to-end approach to portfolio optimization via deep learning. arXiv preprint arXiv:2111.09170 (2021)

24. Zhang, Z., Zohren, S., Roberts, S.: Deep learning for portfolio optimization. J. Financ. Data Sci. **2**(4), 8–20 (2020)

Transfer Learning with Deep Neural Embeddings for Music Classification Tasks

Mateusz Modrzejewski[✉], Piotr Szachewicz, and Przemysław Rokita

Division of Computer Graphics, Institute of Computer Science,
The Faculty of Electronics and Information Technology,
Warsaw University of Technology, Nowowiejska 15/19, 00-665 Warsaw, Poland
{mateusz.modrzejewski,piotr.szachewicz.stud,przemyslaw.rokita}@pw.edu.pl

Abstract. In this paper we present an approach for transfer learning with deep neural embeddings applied to a selection of music information retrieval (MIR) classification tasks with several datasets. The tasks include genre recognition, speech/music distinguishing, predominant instrument recognition and performer identification. We propose the usage of pre-trained L^3 neural networks for feature extraction and apply several supervised classification algorithms to the obtained feature representations in order to compare their performance. The deep neural embedding representations are compared with traditional, hand-crafted features and are shown to outperform the baselines.

Keywords: Artificial intelligence · Music information retrieval · Music classification · Deep embeddings · Neural networks · Transfer learning

1 Introduction

Music information retrieval (MIR) is a well established, interdisciplinary research area concerned with meaningful analysis and generation of musical content using computational methods. Over the years it has gained significant attention from researchers as well as industries, like streaming services and audio equipment manufacturers. However, music is a particularly tricky data type for artificial intelligence algorithms: the sequential structure of music and its deeply human, often abstract qualities, introduce several unique problems into the MIR domain. Aside from more traditional digital signal processing methods [1–3], various deep learning solutions inspired by computer vision and natural language processing have already been proposed in the field in tasks like music tagging [4–6], musical content analysis [7–10] and music generation [11–15]. These methods are often data and compute hungry, which poses a significant challenge for MIR due to dataset copyright reasons and overall data sparsity in comparison with the vast variety of datasets used in other fields. Also, obtaining large annotated datasets requires very specific human expertise and is in most cases not feasible.

© The Author(s), under exclusive license to Springer Nature Switzerland AG 2023
L. Rutkowski et al. (Eds.): ICAISC 2022, LNAI 13588, pp. 72–81, 2023.
https://doi.org/10.1007/978-3-031-23492-7_7

This has lead to the introduction of new methods, such as applications of transfer learning, self-supervised learning and semi-supervised learning for various tasks of music classification. Models are trained on large amounts of available data and later applied to subsequent, downstream tasks on smaller datasets. Development of such solutions has lead to the usage of deep audio embeddings computed with the usage of neural networks such as VGGish [16] and L^3 (*Look, Listen and Learn*) net [17,18], which have been shown to outperform other feature representations in tasks of environmental sound classification [19] and music emotion recognition [20]. Furthermore, the usage of embeddings allows for a decrease in compute and enables the training of lightweight classifiers on CPU, as opposed to deep learning models based on image representations of music such as spectrograms.

In this paper we propose and analyze the usage of deep audio embeddings for representation of selected musical qualities such as genre and instrumentation, as applied to downstream tasks of genre recognition, instrument recognition and performer recognition. We train several classification models for each of the tasks and present experimental results and evaluation of the results in comparison with baseline Mel-frequency cepstral coefficient (MFCC) features.

2   Related Work

Transfer learning, in general, is the idea of training a model on large amounts of available data and applying them to downstream tasks on smaller, previously unseen datasets. This can be done either via fine-tuning the pre-trained model on a specific dataset or using the model to extract new representations (embeddings) on the specific dataset.

This idea has already shown promising results in the MIR domain [21]. Kim et al. propose methods of analyzing and interpreting deep audio embeddings in terms of their consistency [22] and apply a transfer learning framework for artist-related information in order to predict musical genre [23]. Choi et al. [24] propose embeddings built of activations of feature maps of a convolutional neural network. This representation was then further used for several downstream tasks, like emotion prediction and ballroom dance classification. The usage of the convolutional embedding along with variants of SVMs was shown to outperform baseline MFCC models on all tasks. Pons et al. [25] show that very little model assumptions are needed for music tagging when operating with large amounts of data. Many experiments with musical data at scale are, however, performed with the usage of private, proprietary datasets, i.e. collected by streaming services. A recent work by Koh and Dubnov [20] shows deep neural embeddings outperform hand-crafted audio features on several music emotion recognition datasets. The authors present strong results with the usage of SVMs, random forests, bayesian classification and MLPs.

3   Experimental Design

Our experimental approach consists of two steps. In the first step, the input audio is passed to a deep embedding model to obtain feature representations. In the second step, a classification model is trained in a supervised way using the representation vectors from the first step. We make our code publicly available[1].

3.1   Audio Embeddings

VGGish. The VGGish embeddings were proposed by Hershey et al. in [16] and are based on the well-known VGG deep convolutional model for image classification presented in [26]. VGGish embeddings are 128-dimensional and are pre-trained on tens of millions of YouTube videos from a preliminary version of the YouTube-8M dataset [27].

L^3. The original L^3 network [18] is used for the audio-video correspondence task in order to detect correspondence between a single video frame and a 1s audio clip. The network consists of an audio and video convolutional subnetwork front-end for feature extraction and a fully-connected network backend for late fusion and learning whether the audio and video samples correspond to each other. This approach enables self-supervised training on large amounts of unlabelled data.

This idea was further improved for musical tasks in [17] with pre-training on AudioSet [28] musical performances. L^3 embeddings were already shown to outperform VGGish on emotion recognition tasks in music in [20]. We use the `openl3` implementation of the network for our experiments. We use the 512 dimensional embeddings extracted using mel-spectrogram input representation with 256 mel bands. A 1s window length and 0.5s hop size is used for full samples, while a 0.1s hop size is used for the ones which include cropping in the preprocessing stage, as described in further sections.

Baseline MFCC Feature. Mel-frequency cepstral coefficients (MFCCs) are a robust and widely used feature in speech recognition and a range of MIR tasks. MFCCs are derived from a cepstral representation of audio with the usage of the non-linear Mel scale. We use the `librosa` [29] implementation of MFCC extraction for our experiments. For each sample we craft a feature vector using the means and standard deviations of the first 20 MFCCs, along with means and standard deviations of the first and second derivatives of the MFCCs.

3.2   Tasks and Datasets

GTZAN - Genre Recognition. The GTZAN datasaet [30] has already been called "the MNIST of MIR" [31]. It is a musical genre recognition dataset with 1000 30s samples divided into 10 musical genres ("rock", "blues", "classical"

---

[1] https://github.com/pszachew/music_classification_deep_embeddings.

etc.). There are 100 samples per genre. It's the most widely used benchmark dataset in MIR, although it has its known flaws which already have been thoroughly described in literature [32]. Although the dataset is not recommended any more for drawing musically insightful conclusions in music tagging and genre recognition, it is still valuable as a reference for other tasks.

GTZAN Speech/Music - Speech vs. Music Recognition. The GTZAN speech/music [33] is a small dataset of 128 audio samples meant for a binary classification task of distinguishing between speech and music.

IRMAS - Predominant Instrument Recognition. The IRMAS dataset [34] consists of 9579 3 s samples of musical instruments divided into 11 classes ("cello", "saxophone", "human singing voice", "piano", "electric guitar" etc.) with a slight class imbalance. The authors of the dataset provide an official split with 6705 training and 2874 testing samples.

TinySOL - Instrument Recognition. The TinySOL dataset [35,36] consists of 2913 samples of 12 instruments in 3 dynamics (*pp, mf, ff*). It is a subset of the OrchideaSOL dataset of extended instrumental techniques. TinySOL contains only samples of the *ordinario* playing technique. The dataset has an interesting unbalance, with 691 accordion samples, around 300 samples of string instruments (violin, viola, cello, contrabass) and around 120 samples for each woodwind and brass instrument. The length varies between around 2 s and 10 s.

VocalSet - Performer Recognition. The VocalSet dataset [37] consists of over 10 h of monophonic vocal performances by professional singers, 11 male and 9 female. The recordings cover a variety of techniques (including arpeggios and long tones), dynamics and a wide range of pitches. In addition, the authors present classification results for the tasks technique classification and singer identification (performer recognition), the latter being the task described in this paper.

3.3 Data Processing

We use 10-fold cross validation with stratified splits in our experiments with GTZAN, GTZAN speech/music, IRMAS and TinySOL. We embed the full samples in all datasets except TinySOL and VocalSet. For TinySOL, we introduce additional difficulty for the classifiers by randomly selecting and embedding only a 1s chunk of each sample. For VocalSet, we attempt to closely recreate the original preprocessing and splitting method proposed by the authors in [37]. The preprocessing includes normalization, trimming of silence and partitioning into non-overlapping 3 s chunks. We use a 0.8 to 0.2 train test split with all singers present in both sets. We also disjoint the recordings between the training and test set.

3.4   Downstream Task Classifiers

We apply the following classification algorithms for our experiments: logistic regression (**LR**), k nearest neighbors [38] (**KNN**), support vector machines [39] [40] with a linear kernel function (**SVM**), decision trees [41] (**DT**), random forests [42] (**RF**), multi-layer perceptron [43] trained with ReLU activations and Adam [44] optimizer with a learning rate of 1e-3 and β_1 of 0.9 (**MLP**) and extreme gradient boosting (**XGB**). We use `scikit-learn` [45] implementations of classifires 1–6 and `xgboost` [46] for classifier 7.

4   Results

Upon training, we compare accuracy metrics for all of the experiments. Figures 1, 2, 3, 4 and 5 depict the results for L^3 embeddings as well as the baseline MFCC vectors. Support vector machines and logistic regression models consistently show the best performance in our experiments. The deep neural embedding has outperformed the baseline MFCC feature vectors by far in most tasks - the difference is especially noticeable in the IRMAS predominant instrument recognition task.

The figures do not include the regular decision tree, as its performance was consistently the poorest out of the considered classifiers, falling up to 50% accuracy when compared to SVMs. Interestingly, MFCC features have shown to work better with regular decision tree classifiers due to the lower dimensionality of the embedding vectors, however still falling short to the results obtained with other algorithms.

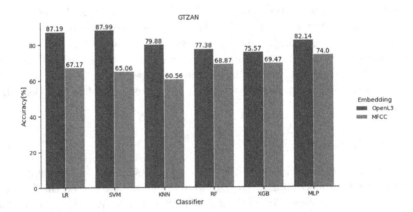

Fig. 1. Results for genre classification on GTZAN dataset.

The error rates on GTZAN speech/music are trivial due to the small size of the dataset. The deep neural embedding obtains generally better performance, however both representations achieve accuracy scores close to 100%, showcasing their ability to capture differences between musical and speech signals.

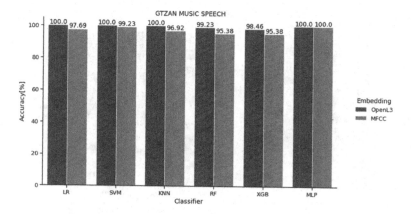

Fig. 2. Results for speech/music distinguishing on GTZAN speech/music dataset.

In the case of instrument recognition, in addition to the results presented in Figs. 3 and 4, we perform additional confusion matrix analysis for the SVM. In the case of IRMAS, human voice meets a 98% accuracy, further supporting the voice vs. music capabilities of L^3 observed on GTZAN speech/music. We notice that most errors occur between instances of the string instrument families, like cello and viola. With TinySOL, most misclassifications occur between alto sax and flute. Both of these cases may lie within the actual similarity of sound of these instruments within certain pitch ranges.

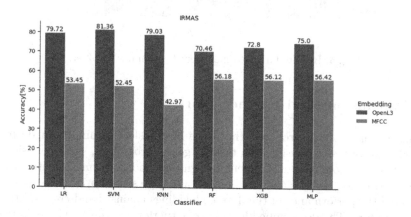

Fig. 3. Results for instrument classification on IRMAS dataset.

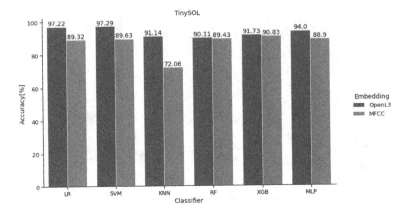

Fig. 4. Results for instrument classification on TinySOL dataset.

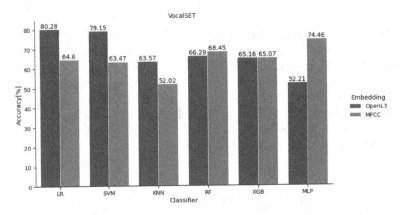

Fig. 5. Results for performer identification on VocalSet.

5   Conclusions and Further Work

In this paper we have proposed the usage of transfer learning with deep L^3 embeddings for music classification tasks of genre recognition, speech/music distinguishing, predominant instrument recognition and performer identification. We have trained several classifiers on 5 datasets, presented original results and compared the performance of the deep embeddings against hand-crafted MFCC feature vectors. The deep embeddings have outperformed the MFCC vectors on all considered tasks, giving further insight into the robustness of L^3 and its usefulness for various MIR tasks. Furthermore, the deep embedding approach has proved more than competent in fairly difficult tasks, like distinguishing between particular string instruments and female singing voice identification. The usage of the embeddings also allowed us to use CPU-trainable classification algorithms.

In further work, we would be very interested in performing a large-scale evaluation of other embeddings on a multitude of MIR tasks, as well as in an attempt of proposing an embedding of our own, perhaps using previously untested neural network architectures.

References

1. Casey, M.A., Veltkamp, R., Goto, M., Leman, M., Rhodes, C., Slaney, M.: Content-based music information retrieval: current directions and future challenges. Proc. IEEE **96**(4), 668–696 (2008)
2. Muller, M., Ellis, D.P.W., Klapuri, A., Richard, G.: Signal processing for music analysis. IEEE J. Sel. Top. Sig. Process. **5**(6), 1088–1110 (2011)
3. Müller, M.: Fundamentals of Music Processing: Audio, Analysis, Algorithms, Applications. Springer, Cham (2015). https://doi.org/10.1007/978-3-319-21945-5
4. Lee, J., Park, J., Kim, K.L., Nam, J.: SampleCNN: end-to-end deep convolutional neural networks using very small filters for music classification. Appl. Sci. **8**(1), 150 (2018)
5. Choi, K., Fazekas, G., Sandler, M., Cho, K.: Convolutional recurrent neural networks for music classification. In: 2017 IEEE International Conference on Acoustics, Speech and Signal Processing (ICASSP), pp. 2392–2396. IEEE (2017)
6. Won, M., Choi, K., Serra, X.: Semi-supervised music tagging transformer. arXiv preprint arXiv:2111.13457 (2021)
7. Böck, S., Davies, M.E.P.: Deconstruct, analyse, reconstruct: how to improve tempo, beat, and downbeat estimation. In: Proceedings of ISMIR (International Society for Music Information Retrieval). Montreal, Canada, pp. 574–582 (2020)
8. Korzeniowski, F., Widmer, G.: Feature learning for chord recognition: the deep chroma extractor. arXiv preprint arXiv:1612.05065 (2016)
9. Fuentes, M., McFee, B., Crayencour, H.C., Essid, S., Bello, J.P.: A music structure informed downbeat tracking system using skip-chain conditional random fields and deep learning. In: ICASSP 2019–2019 IEEE International Conference on Acoustics, Speech and Signal Processing (ICASSP), pp. 481–485. IEEE (2019)
10. Böck, S., Davies, M.E.P., Knees, P.: Multi-task learning of tempo and beat: learning one to improve the other. In: ISMIR, pp. 486–493 (2019)
11. Ji, S., Luo, J., Yang, X.: A comprehensive survey on deep music generation: multi-level representations, algorithms, evaluations, and future directions. arXiv preprint arXiv:2011.06801 (2020)
12. Briot, J.-P., Hadjeres, G., Pachet, F.-D.: Deep learning techniques for music generation-a survey. arXiv preprint arXiv:1709.01620 (2017)
13. Huang, C.-Z.A., et al.: Music transformer. arXiv preprint arXiv:1809.04281 (2018)
14. Hawthorne, C., et al.: Enabling factorized piano music modeling and generation with the maestro dataset. arXiv preprint arXiv:1810.12247 (2018)
15. Roberts, A., Engel, J., Raffel, C., Hawthorne, C., Eck, D.: A hierarchical latent vector model for learning long-term structure in music (2019)
16. Hershey, S., et al.: CNN architectures for large-scale audio classification. In: 2017 IEEE International Conference on Acoustics, Speech and Signal Processing (ICASSP), pp. 131–135. IEEE (2017)
17. Cramer, J., Wu, H.-H., Salamon, J., Bello, J.P.: Look, listen and learn more: design choices for deep audio embeddings. In: IEEE International Conference on Acoustics, Speech and Signal Processing (ICASSP), Brighton, UK, pp. 3852–3856, May 2019

18. Arandjelovic, R., Zisserman, A.: Look, listen and learn. In: Proceedings of the IEEE International Conference on Computer Vision, pp. 609–617 (2017)

19. Wilkinghoff, K.: On open-set classification with l3-net embeddings for machine listening applications. In: 2020 28th European Signal Processing Conference (EUSIPCO), pp. 800–804. IEEE (2021)

20. Koh, E., Dubnov, S.: Comparison and analysis of deep audio embeddings for music emotion recognition. arXiv preprint arXiv:2104.06517 (2021)

21. Pons, J., Serra, X.: musiCNN: pre-trained convolutional neural networks for music audio tagging. In: Late-Breaking/Demo Session in 20th International Society for Music Information Retrieval Conference (LBD-ISMIR2019) (2019)

22. Kim, J., Urbano, J., Liem, C., Hanjalic, A.: Are nearby neighbors relatives? Testing deep music embeddings. Front. Appl. Math. Stat. **5**, 53 (2019)

23. Kim, J., Won, M., Serra, X., Liem, C.C.S.: Transfer learning of artist group factors to musical genre classification. In: Companion Proceedings of the Web Conference 2018, pp. 1929–1934 (2018)

24. Choi, K., Fazekas, G., Sandler, M., Cho, K.: Transfer learning for music classification and regression tasks. In: The 18th International Society of Music Information Retrieval (ISMIR) Conference: Suzhou, p. 2017. International Society of Music Information Retrieval, China (2017)

25. Pons, J., Nieto, O., Prockup, M., Schmidt, E.M., Ehmann, A.F., Serra, X.: End-to-end learning for music audio tagging at scale. In: 19th International Society for Music Information Retrieval Conference (ISMIR 2018) (2018)

26. Simonyan, K., Zisserman, A.: Very deep convolutional networks for large-scale image recognition. arXiv preprint arXiv:1409.1556 (2014)

27. Abu-El-Haija, S., et al.: Youtube-8m: a large-scale video classification benchmark. arXiv preprint arXiv:1609.08675 (2016)

28. Gemmeke, J.F., et al.: Audio set: an ontology and human-labeled dataset for audio events. In: 2017 IEEE International Conference on Acoustics, Speech and Signal Processing (ICASSP), pp. 776–780. IEEE (2017)

29. McFee, B., et al.: librosa: Audio and music signal analysis in Python. In: Proceedings of the 14th Python in Science Conference, vol. 8, pp. 18–25. Citeseer (2015)

30. Tzanetakis, G., Cook, P.: Musical genre classification of audio signals. IEEE Trans. Speech Audio Process. **10**(5), 293–302 (2002)

31. Won, M., Spijkervet, J., Choi, K.: Music classification: beyond supervised learning, towards real-world applications. https://music-classification.github.io/tutorial (2021)

32. Sturm, B.L.: The gtzan dataset: its contents, its faults, their effects on evaluation, and its future use. arXiv preprint arXiv:1306.1461 (2013)

33. Tzanetakis, G.: Gtzan music/speech collection. Online (1999)

34. Bosch, J.J., Janer, J., Fuhrmann, F., Herrera, P.: A comparison of sound segregation techniques for predominant instrument recognition in musical audio signals. In: ISMIR, pp. 559–564. Citeseer (2012)

35. Cella, C.E., Ghisi, D., Lostanlen, V., Lévy, F., Fineberg, J., Maresz, Y.: Orchideasol: a dataset of extended instrumental techniques for computer-aided orchestration. arXiv preprint arXiv:2007.00763 (2020)

36. Cella, C., Dzwonczyk, L., Saldarriaga-Fuertes, A., Liu, H., Crayencour, H.-C.: A study on neural models for target-based computer-assisted musical orchestration. In: 2020 Joint Conference on AI Music Creativity (CSMC+ MuMe) (2020)

37. Wilkins, J., Seetharaman, P., Wahl, A., Pardo, B.: Vocalset: a singing voice dataset. In: ISMIR, pp. 468–474 (2018)

38. Cover, T., Hart, P.: Nearest neighbor pattern classification. IEEE Trans. Inf. Theory **13**(1), 21–27 (1967)
39. Cortes, C., Vapnik, V.: Support-vector networks. Mach. Learn. **20**(3), 273–297 (1995)
40. Suykens, J.A.K., Vandewalle, J.: Least squares support vector machine classifiers. Neural Process. Lett. **9**(3), 293–300 (1999)
41. Breiman, L., Friedman, J.H., Olshen, R.A., Stone, C.J.: Classification and regression trees, vol. 432, pp. 151–166. Wadsworth. International Group, Belmont (1984)
42. Breiman, L.: Random forests. Mach. Learn. **45**(1), 5–32 (2001)
43. Hinton, G.E.: Connectionist learning procedures. Mach. Learn. 555–610. Elsevier (1990)
44. Kingma, D.P., Ba, J.: Adam: a method for stochastic optimization. arXiv preprint arXiv:1412.6980 (2014)
45. Pedregosa, F., et al.: Scikit-learn: machine learning in python. J. Mach. Learn. Res. **12**, 2825–2830 (2011)
46. Chen, T., Guestrin, C.: XGBoost: a scalable tree boosting system. In: Proceedings of the 22nd ACM SIGKDD International Conference on Knowledge Discovery and Data Mining, pp. 785–794 (2016)

Analysis and Detection of DDoS Backscatter Using NetFlow Data, Hyperband-Optimised Deep Learning and Explainability Techniques

Marek Pawlicki[1,2]([✉]) [ID], Martin Zadnik[3] [ID], Rafał Kozik[1,2] [ID],
and Michał Choraś[1,2] [ID]

[1] ITTI Sp. z o.o., Poznań, Poland
[2] Bydgoszcz University of Science and Technology, Bydgoszcz, Poland
`marek.pawlicki@pbs.edu.pl`
[3] CESNET, Prague, Czech Republic

Abstract. The Denial of Service attacks are one of the most common attacks used to disrupt the services of public institutions. The criminal act of exhausting a network resource with the intent to obstruct the utility of a service is associated with hacktivism, blackmailing and extortion attempts. Intrusion Prevention Systems are an essential line of defence against this problem, strengthening public institutions, industrial and critical infrastructure alike. In the following work, an analysis of the detection of DDoS Backscatter with the use of neural networks is performed. To this end, a novel dataset is collected and described, on which a hyperband-optimized neural network is trained, and the decision process of the classifier is explained using LIME and SHAP.

Keywords: Cyber security · Machine learning · Explainable AI (xAI)

1 Introduction

Spoofing source IP addresses is a common technique for rendering mitigation of DoS attacks more difficult [13,19]. The addresses are usually spoofed at random [17]. The exceptions are reflection attacks, in which the source IP addresses are replaced by the victim IP address so that the reflected packets are forwarded to the victim [13]. In our work, we focus on attacks with randomly spoofed source IP addresses, namely, flooding attacks utilizing TCP and ICMP protocols. The randomness of IP spoofing makes it possible to observe artifacts of an attack even in the networks which do not carry the attack traffic, by monitoring the so-called backscattered packets (*backscatter*). These packets are received by the spoofed IP addresses from the victim of an attack. The backscattered packets are illustrated in Fig. 1. In this image, the attacker's packets are shown in red and are aimed at the victim with IP addresses B. The victim responds to these packets by sending a response (blue). However, response packets do not go back

L. Rutkowski et al. (Eds.): ICAISC 2022, LNAI 13588, pp. 82–92, 2023.
https://doi.org/10.1007/978-3-031-23492-7_8

to the attacker with IP address A, but to the devices with IP address C and D, which have been spoofed by the attacker. Therefore, it is possible to observe packets from the victim of the attack on devices C and D.

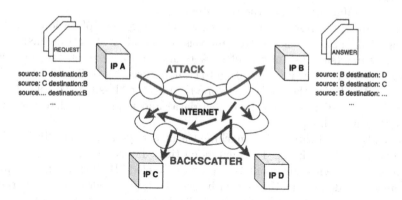

Fig. 1. An example of backscatter attack packets (blue). Attacker (A) performs an attack (red) on IP address B. The attacking machine spoofs its source address using IP addresses C and D. Therefore, the victim of an attack B sends a response (blue) to the attacker's request A to IP addresses C and D [5]. (Color figure online)

The observation of the backscatter traffic is performed by the so-called telescopes. Telescopes are made of a relatively large unused IP address block. There are no services nor clients, only a machine that receives all the packets for the whole assigned block. Therefore, the arrival of a packet to the address range of a telescope indicates an illegitimate activity. Typically, this is a backscatter or a scan. The detection of a backscatter at the telescope is relatively straightforward and is typically based on different quantitative thresholds that are able to recognize the backscatters from scans [5,17]. In this work, we experimentally evaluate if it is possible to detect a backscatter using data provided by the flow monitoring of a large productional backbone network. Our motivation is twofold; having the ability to observe backscatter without the need to dedicated large IP address space to the telescope since unassigned IP addresses are scarce resources in IPv4 world, and observing backscatter achieving high quality, i.e. the assumption is that the larger the IP address space, the better.

The methods for detecting DDoS attacks from backscattered packets are discussed in more detail in Sect. 2.

2   Related Work

The ground work on detection of backscatter was established by Moore [17]. Their work provided a theoretical background on the probability of observing backscatter at the network telescope of a specific size as well as proposed the first backscatter detector. The proposed detector applies thresholds over several traffic characteristics and it was constructed manually.

Blenn et al. performed another comprehensive study about DoS attacks using backscatter analysis ten years later. The study reports on evolution of DoS during five consecutive years on the TU Delft telescope. The detection of backscatter is based on several rules and thresholds, to differentiate between scans and backscatter traffic arriving at the telescope.

There are three research papers reporting on the use of various machine learning algorithms for backscatter analysis. Furutani et al. [11] defined twelve packet features (such as number of packets and average packet size) and used the Support Vector Machine algorithm to train backscatter traffic detector. They used HTTP traffic at the NICT telescope to train the classifier (thresholds over traffic characteristics were used to label data) and applied the detector on all the TCP traffic subsequently.

They also tested three other machine learning approaches, namely RBFNN, RAN and RAN-LHS, in [2]. They extended their dataset with UDP traffic and they used DNS (with thresholds) to provide labeled UDP traffic for training. They also extended the feature set with seven additional features (e.g., variance of interpacket intervals, average packets with unique source port, etc.).

Another work that served as an inspiration for our feature set is [22]. Skrjanc et al. investigated evolving Cauchy possibilistic clustering as a means of clustering similar traffic from NICT telescope. The authors labeled the resulting clusters using the well-known rules with thresholds, subsequently.

The main difference between the following work and the previous research is that in this work, the analysed network flow data was not collected from a telescope but from the whole backbone network. This introduces additional challenges to differentiating the backscatter traffic from the legitimate traffic and the specific events, such as misconfiguration. We build upon the previous works on telescopes by utilizing the existing CAIDA telescope to label our dataset.

3   Backscatter Detection Without Telescope

The current approach to detecting backscatter traffic is to capture packets at telescopes/honeypots [5] and differentiate between scans, attacks, misconfiguration and the DDoS backscatter. However, telescopes observe only a limited range of the IP address space. Moreover, telescopes are deployed using an unassigned IP address range. But, due to the depleted pool of IPv4 addresses, it is costly for organizations to sacrifice part of their address space to deploy telescopes to be able to observe the DDoS landscape. Our approach extends the existing work on detecting the backscatter traffic, to utilize a whole backbone network hosting several/16 prefixes as an observation point for the backscatter traffic. In such a case, it is not possible to use the raw packet capture as a source of data due to its volume; we use flow data collected from the edges of the backbone network. In such a setup, we face the additional challenge, in comparison to the telescopes, of spotting the backscatter in the vast amount of legitimate traffic, and, at the same time, not to misclassify the legitimate traffic for being a backscatter. Therefore, we apply machine learning to achieve a high success rate of classification between backscatter and non-backscatter traffic.

Our approach utilizes data inferred from the telescopes to annotate our dataset. The dataset serves for the training of a classifier which is subsequently applied on IP flow data. The telescope and its analysis engine recognize the backscatter traffic received from the attacked IP addresses. The output is a list of IP addresses that generated the backscatter traffic. This list serves as an annotation for our flow dataset. The feature vectors per each IP address are extracted from the annotated flow dataset. In comparison to the previous packet-based approaches, we are limited to the features that can be obtained from the flow data. The following list summarizes these features (Table 1).

Table 1. The list of all the features utilized in this work. The protocol column denotes if a feature is applicable for the given protocol. The Source column indicates if the feature was already used in the previous work, or if it is new. By the destination IP address we mean the receiver of the backscatter traffic, while the source is the victim of DDoS attack.

Features	Protocol	Source/inspiration
Number of bytes	TCP, ICMP	
Number of packets	TCP, ICMP	[11, 22]
Average bytes per packet	TCP, ICMP	[11, 22]
Std. deviation of bytes per packet	TCP, ICMP	[11, 22]
Number of flows	TCP, ICMP	
Average number of packets per flow	TCP, ICMP	
Std. deviation of packets per flow	TCP, ICMP	
Max. number of flows per minute	TCP, ICMP	[17]
Average number of flows per second	TCP, ICMP	
Number of unique destination IP addresses	TCP, ICMP	[11, 22]
Number of unique destination/24 networks	TCP, ICMP	
Number of unique destination ports	TCP	[11, 22]
Number of unique source ports	TCP	[11, 22]
Number of unique destination IP addresses normalized by the number of flows	TCP, ICMP	
Number of unique destination/24 networks normalized by the number of flows	TCP, ICMP	
Number of unique source ports normalized by the number of flows	TCP	
Number of unique source ports normalized by the number of flows	TCP	

For an IP address, both the ICMP and TCP features are calculated separately, e.g. packet count.

4   Evaluation

The evaluation empirically discovers whether the proposed approach is capable of recognizing the DDoS backscatter traffic from the vast amount of other network traffic flowing through the backbone network.

4.1   Deep Learning

Artificial Neural Networks (ANN), which sit at the heart of the Deep Learning revolution, are adaptable instruments capable of very accurate performance in specific tasks. With an ever-growing variety of applications, Deep Learning is successfully applied for data mining, classification, regression, clustering and time series analysis. This includes uses in Intrusion Detection Systems (IDS) [1,7, 10,12,21]. From a certain point of view, ANNs mimic the learning characteristics of biological neural networks, although heavily streamlined [16].

Hyperparameter Optimization. Apart from the parameters that are adjusted to the data in the training procedure, the ANN algorithms also have a set of parameters that cannot be inferred from the data. These are called 'hyper-parameters'. The proper setting of hyperparameters can influence the results of the ANN to a great extent [8]; thus, a myriad of hyperparameter tuning approaches have been formulated. This process governs the choice of parameters like the activation function, the learning rate, the optimizer, the batch size, the number of epochs and even the number of layers of the network along with the count of neurons on those layers. There are a number of approaches to hyperparameter optimization, starting from exhaustive searches of the parameter space with GridSearch [8], through Random Search [3] and Bayesian Optimization [23]. One of the most recent approaches is Hyperband Optimization [15]. Hyperband is an improvement to the successive halving algorithm, which discarded half of the worst performing hyperparameter setups out of a set on each iteration. Hyperband solves the resource allocation problem present in the successive halving algorithm (whether one should try a large number of setups over a short period of time, or try a smaller number of setups over a longer period of examination time). Hyperband performs a GridSearch on the number of hyperparameter setups for a fixed finite computation budget.

Employing the hyperband algorithm, the hyperparameters for the Deep Neural Network (DNN) used in this work were established as seen in Table 2. The hyperband completed 354 hyperparameter setup tests, choosing the finest combination of hyperparameters with regard to the number of layers (1–4), the number of neurons on the layers (minimum value set to four, maximum to 4096, with the step size of 32), the most suitable activation function (Rectified Linear Unit, Hyperbolic Tangent, Scaled Exponential Linear Unit or Sigmoid), the learning rate (0.1, 0.01, 0.001, 0.0001), the batch size (1, 2, 4, 8, 16, 32 and 64), and the number of Epochs (10, 30, 50, 100, 150).

Our dataset covers one week of network traffic captured at the edge of the CESNET backbone. The dataset is publicly available at [26].

Table 2. Best hyperparameter setup

Hyperparameter	Best value
layers	2
units	1668
activation	relu
optimizer	ADAM
learning_rate	0.0001
epochs	50
batch size	16

Dataset Balancing. The data imbalance problem affects some ML-based classifiers in the situation when the number of datapoints in one of the classes outweighs the number of samples in other classes. When trained on such data, some ML classifiers have a tendency to misclassify the minority samples as the majority samples. This poses an issue when the minority samples are the very reason of the deployment of ML algorithms, as in the case of anomaly detection in network traffic [14]. There is a number of established ways to handle the data imbalance problem in ML; in this study, random subsampling of the majority class was employed.

4.2 Detection Results

The DNN setup described in Table 2 was capable of detecting the DoS attacks with an accuracy (ACC) and balanced accuracy (BACC) of 0.986 and the Matthews Correlation Coefficient (MCC) of 0.971. The results, along with the precision, recall and f1-score for the particular classes are reported in Table 3.

Table 3. DoS detection results using DNN

	Precision	Recall	F1-score
False	0.99	0.98	0.99
True	0.99	0.99	0.99
			Value
ACC			0.99
BACC			0.99
MCC			0.97

4.3 Explaining the Attributes of a Deep Learning Based Intrusion Detection System

Explainability of Artificial Intelligence (xAI) can be defined as the pursuit of adequate intuitions on the behaviour of a black box AI model [4]. The authors of [9] notice that in this early stage of explainability and interpretability, it is not yet possible to formulate a homogeneous, formal definition of both of those aspects of AI. However, it is of paramount importance, as AI which directly impacts individuals is bound by legislation such as the GDPR [6]; this can include intrusion detection [18]. Apart from the legislative viewpoint, xAI can be used to provide crucial insights to security operators [25]. In this work, xAI is used to peek into the inner workings of DNN and see what the most relevant features to the classification of particular datapoints were.

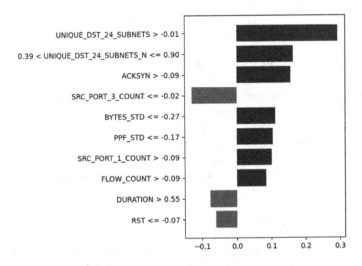

Fig. 2. Explanations for a sample marked as 'True' - LIME

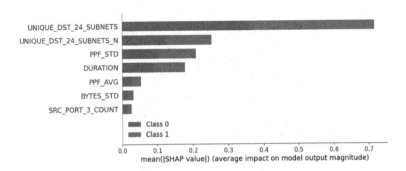

Fig. 3. Explanations for a sample marked as 'True' - SHAP

Local Interpretable Model-Agnostic Explanations (LIME). LIME is a model-agnostic Artificial Intelligence Explainability method. The premise of LIME is expressed by building a local linear model, following the assumption

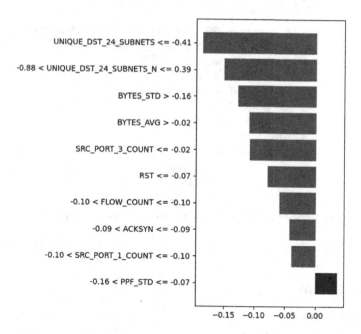

Fig. 4. Explanations for a sample marked as 'False' - LIME

that any complex model is linear at the local level [20]. LIME samples data-points in the neighbourhood of the explained instance and uses perturbations of the sampled data to train a linear model. In Fig. 2 and in Fig. 4, the corresponding feature contributions to the samples where symptoms of an attack were found (True label) and benign traffic (False label) are showcased. The x-axis displays the contributions themselves, while the numbers in the y-axis labels are the bounding values for the particular features. Green bars are contributing positively to the classification, the red bars are contributing negatively.

SHapley Additive exPlanations (SHAP). is an approach to explainability rooted in game theory. Using SHAP, one can figure out which features increase or decrease the probability of a particular classification, considering the interactions and redundancies among features. The method is validated in [24] in a controlled experiment, to provide better comprehension of the model behaviour. The plots in Fig. 3 and Fig. 5 showcase the features that contribute to the classification of the particular samples. The samples are the same as in the LIME explanations. The red and blue colors indicate that the feature will contribute to the classification regardless of the fact if the samples were to be classified as attacks or as benign traffic.

All explainability examples note the importance of the Unique_DST_24_ Subnets feature. The standard deviation of bytes per packet is of importance in all cases as well, as is the duration feature, along with the Packet Per Flow (PPF) standard deviation.

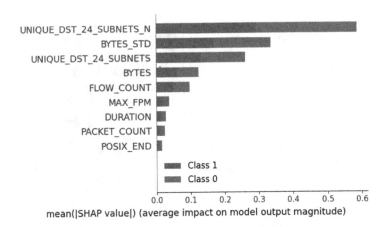

Fig. 5. Explanations for a sample marked as 'False' - SHAP

5 Conclusion

In this work, a novel cybersecurity dataset is presented; the collection process and the features available are described. The validity of the dataset is tested using a deep neural network which was optimized using the hyperband tuning method. The accuracy and balanced accuracy of the classifier reached almost 99%, and the recall for the attack class exceeded 99%. The classifications made with the use of the DNNs are explained using two xAI methods - LIME and SHAP.

Acknowledgement. This work is funded under the SPARTA project, which has received funding from the European Union's Horizon 2020 research and innovation programme under grant agreement No 830892.

References

1. Aldweesh, A., Derhab, A., Emam, A.Z.: Deep learning approaches for anomaly-based intrusion detection systems: a survey, taxonomy, and open issues. Knowl.-Based Syst. **189**, 105124 (2020)
2. Ali, S.H.A., Furutani, N., Ozawa, S., Nakazato, J., Ban, T., Shimamura, J.: Distributed denial of service (DDOS) backscatter detection system using resource allocating network with data selection. Memoirs Graduate Sch. Eng. Syst. Inform. Kobe Univ. (7), 8–13 (2015). https://doi.org/10.5047/gseku.e.2015.001
3. Bergstra, J., Bengio, Y.: Random search for hyper-parameter optimization. J. Mach. Learn. Res. **13**(2) (2012)
4. Bhatt, U., et al.: Explainable machine learning in deployment. arXiv preprint arXiv:1909.06342 (2019)
5. Blenn, N., Ghiëtte, V., Doerr, C.: Quantifying the spectrum of denial-of-service attacks through internet backscatter. In: Proceedings of the 12th International Conference on Availability, Reliability and Security. ACM, August 2017. https://doi.org/10.1145/3098954.3098985

6. Albert, C.: We are ready for machine learning explainability?, June 2019. https://towardsdatascience.com/we-are-ready-to-ml-explainability-2e7960cb950d. Accessed 31 Mar 2020
7. Chen, J.: Machine learning and cybersecurity: studying network behaviour to detect anomalies (2018)
8. Choraś, M., Pawlicki, M.: Intrusion detection approach based on optimised artificial neural network. Neurocomputing **452**, 705–715 (2021)
9. Dosilovic, F.K., Brcic, M., Hlupic, N.: Explainable artificial intelligence: a survey. In: 2018 41st International Convention on Information and Communication Technology, Electronics and Microelectronics (MIPRO). IEEE, May 2018. https://doi.org/10.23919/mipro.2018.8400040
10. Drewek-Ossowicka, A., Pietrołaj, M., Rumiński, J.: A survey of neural networks usage for intrusion detection systems. J. Ambient. Intell. Humaniz. Comput. **12**(1), 497–514 (2021)
11. Furutani, N., Ban, T., Nakazato, J., Shimamura, J., Kitazono, J., Ozawa, S.: Detection of DDOS backscatter based on traffic features of darknet TCP packets. In: 2014 Ninth Asia Joint Conference on Information Security. IEEE, September 2014). https://doi.org/10.1109/asiajcis.2014.23
12. Gamage, S., Samarabandu, J.: Deep learning methods in network intrusion detection: a survey and an objective comparison. J. Netw. Comput. Appl. **169**, 102767 (2020)
13. Jonker, M., King, A., Krupp, J., Rossow, C., Sperotto, A., Dainotti, A.: Millions of targets under attack. In: Proceedings of the 2017 Internet Measurement Conference. ACM, November 2017. https://doi.org/10.1145/3131365.3131383
14. Kozik, R., Pawlicki, M., Choraś, M.: Cost-sensitive distributed machine learning for netflow-based botnet activity detection. In: Security and Communication Networks 2018 (2018)
15. Li, L., Jamieson, K., DeSalvo, G., Rostamizadeh, A., Talwalkar, A.: Hyperband: a novel bandit-based approach to hyperparameter optimization. J. Mach. Learn. Res. **18**(1), 6765–6816 (2017)
16. Maimon, O., Rokach, L.: Data Mining and Knowledge Discovery Handbook, 2nd edn, January 2010
17. Moore, D., Shannon, C., Brown, D.J., Voelker, G.M., Savage, S.: Inferring internet denial-of-service activity. ACM Trans. Comput. Syst. **24**(2), 115–139 (2006). https://doi.org/10.1145/1132026.1132027
18. Pawlicka, A., Jaroszewska-Choras, D., Choras, M., Pawlicki, M.: Guidelines for stego/malware detection tools: achieving GDPR compliance. IEEE Technol. Soc. Mag. **39**(4), 60–70 (2020)
19. Syn flood attack. online. https://www.cloudflare.com/learning/ddos/syn-flood-ddos-attack/
20. Ribeiro, M.T., Singh, S., Guestrin, C.: "why should I trust you?": Explaining the predictions of any classifier. CoRR abs/1602.04938 (2016). http://arxiv.org/abs/1602.04938
21. Sani, Y., Mohamedou, A., Ali, K., Farjamfar, A., Azman, M., Shamsuddin, S.: An overview of neural networks use in anomaly intrusion detection systems. In: 2009 IEEE Student Conference on Research and Development (SCOReD), pp. 89–92, November 2009. https://doi.org/10.1109/SCORED.2009.5443289
22. Skrjanc, I., Ozawa, S., Dovzan, D., Tao, B., Nakazato, J., Shimamura, J.: Evolving Cauchy possibilistic clustering and its application to large-scale cyberattack monitoring. In: 2017 IEEE Symposium Series on Computational Intelligence (SSCI). IEEE, November 2017. https://doi.org/10.1109/ssci.2017.8285203

23. Snoek, J., Larochelle, H., Adams, R.P.: Practical Bayesian optimization of machine learning algorithms. In: Advances in Neural Information Processing System, vol. 25 (2012)
24. Štrumbelj, E., Kononenko, I.: Explaining prediction models and individual predictions with feature contributions. Knowl. Inf. Syst. **41**(3), 647–665 (2014)
25. Szczepański, M., Choraś, M., Pawlicki, M., Kozik, R.: Achieving explainability of intrusion detection system by hybrid oracle-explainer approach. In: 2020 International Joint Conference on Neural Networks (IJCNN), pp. 1–8. IEEE (2020)
26. Zadnik, M.: DDoS Backscatter Dataset, vol. 1 (2022). https://doi.org/10.17632/37zz4pvjzp.1

BiLSTM Deep Learning Model for Heart Problems Detection

Jakub Siłka(iD), Michał Wieczorek(iD), Martyna Kobielnik(iD),
and Marcin Woźniak(✉)(iD)

Faculty of Applied Mathematics, Silesian University of Technology, Kaszubska 23,
44-100 Gliwice, Poland
{martyna.kobielnik,marcin.wozniak}@polsl.pl

Abstract. Deep learning architectures find applications where analysis of complex data inputs is demanding and regular neural networks may have problems. There are many types of deep learning models, however the most important to fit architecture and training model to the input data. In this article we propose a model of deep learning based on architecture in which we use BiLSTM neural network. Proposed model is trained by using Adam algorithm. For the research experiment we have examined also other latest algorithms to select the best configuration of proposed model. Results show that our proposed BiLSTM deep learning neural network archived over 99% of accuracy.

Keywords: Deep learning · BiLSTM · Adam · Heart signal

1 Introduction

Computer simulations and decision models are very often powered by various aspects of Computational Intelligence. New ideas use variety of complex architectures to solve data analysis tasks. In recent years deep learning has been presented in many applications. There are different types of architectures which are mostly oriented particular applications. The type of architecture must be developed for the input data, to fit information and context of the data. We can read about many interesting models which use Long Short-Term Memory (LSTM) neurons, since this kind of neural unit simulates cognition processes and therefore improves classification of complex data structures.

LSTM units are mostly applied to numerical data, however we can also read about compositions with other types to process also complex data of various structure. Recent advances in machine learning show that bidirectional compositions of memory type units show excellent adaptation to data inputs which are oriented on time changing domain of inputs. This kind of complex structure however improve processing of inputs of various type which may show value fluctuation in time intervals. In [1] was presented how to use this model as sentiment analyzer for various types of texts that are posted in comments in the Internet. Similar proposal for text analysis from conversations, where developed in [2], where BiLSTM neurons were labelling information for further analysis. A

© The Author(s), under exclusive license to Springer Nature Switzerland AG 2023
L. Rutkowski et al. (Eds.): ICAISC 2022, LNAI 13588, pp. 93–104, 2023.
https://doi.org/10.1007/978-3-031-23492-7_9

model presented in [3] was developed to help in stock price prediction, however this application was built as composition of CNN with BiLSTM, since it was assumed that composed in this way architecture will be able to process images. Deep learning has also interesting applications in incomplete data analytic [4]. Model presented in [5] proposed recurrent neural network for technical purposes to optimize control of heating appliances. We can also find various applications of BiLSTM neurons in composition with other structures to operate on data inputs presents in a form complex sensor readings or knowledge graphs. In [6] was discussed how to apply a composition with CNN to search for optimal decision patterns in knowledge graph completion process, where a role of switching processor was given to particular attention mechanism. In [7] was presented how to use such compositions for automatic modulation of recognition target, while model discussed in [8] was using a concept of auto-encoder for training strategy to learn from soft-sensor data.

Models based on BiLSTM are also very important for the development in assisted environments or even life symptoms detection and analysis. In [9] was presented a very interesting model of human activity recognition in which sensor readings were analyzed by developed spatio-temporal deep learning model. A model of BiLSTM was applied in [10] to analyze life symptoms by double channel input to the network. Such constructions are also very efficient in hear signal reading analysis and detection. This type of signal is very well fitted to the nature of BiLSTM neurons since heart is giving a signal changing in the time interval. In [11] was presented devote proposition to use EEG signals as models of control for robotic arm, where complex CNN-BiLSTM network was analyzing inputs to improve the control of the arm. We can also read about using BiLSTM neurons as interpreters of the hear beat to analyze correctness of recorded electrocardiogram (ECG).

In this article we present a model of BiLSTM neurons applied in deep learning model to detect potential malfunctions of hear beats from recorded electrocardiogram (ECG) signals. Our research was started in [12] where was presented how to improve training algorithms for probabilistic neural networks. That idea was developed to preserve generalization of neural networks without complex signal processing. In this article we want to discuss our proposed novel approach to heart signal analysis. Proposed deep learning architecture is composed in the way to adapt time interval input signal. Proposed by us architecture is using simplified concept of auto-encoder in a form of connected bidirectional layers of the network, where BiLSTM units are implemented to analyze the signal. We show abilities of deep learning to adopt to the signal just by using BiLSTM neurons with simple data normalization. In our research experiments we have analyzed various training approaches to improve final classification form training on applied data set. Results show that our proposed model is very efficient and results of classification are reaching 99% Accuracy, which is a very promising result for further development of this idea.

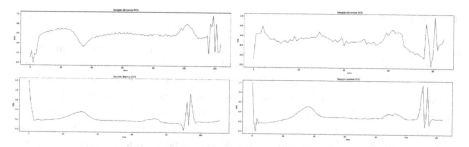

Fig. 1. Above we can see sample four ECG recordings. Two of them are correct and two are abnormal. It can be observed that the negative records are irregular and the positive ones, despite the imperfect overlap, can be observed to have distinguished characteristics.

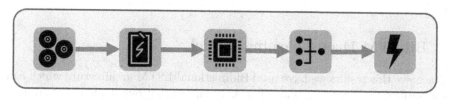

Fig. 2. Sample signal processing model. We assume that the data will be recorded by a set of sensors, then they are saved, processed and fed to the neural network inside the ECG and if a need is found, an impulse is sent from the ICD.

2 System

Our assumed system is based on smart-ICD in combination with ECG-holder, which ensures a much greater probability of detecting the right moment for the use of an implanted defibrillator or other necessary supports. Therefore, we have considered an innovative system for recognizing abnormal heart activity based on IoT and proposed deep learning model. A sample model is discussed in Fig. 2 with possible sequence of actions. We assume that first the data is collected from sensors located on the patient's chest, information is read and then written and transported to the ECG-holder which, using a deep neural network, decides what is the best action, and if necessary, will order ICD to cause a discharge or other device. The system is more extensive than the ICD itself, however, it has much better accuracy with which assesses the most appropriate method of patient assistance. Therefore, possible device application would be intended for people in the highest risk group. As the information itself is recorded, the next step would be to refine the entire system with all the data collected.

3   Dataset

As we are aware of the current global problem of heart disease and the enormous cost of treatment faced by millions of people every day, we recognized that there is a need for increased accuracy of ICD devices. We found relevant data on [13]. The data-set was created from 48 half-hour fragments that were recorded using a two-channel ECG. All these data were obtained from 47 patients studied by the BIH Arrhythmia Laboratory. Recorded data was digitized at 360 samples per second, for each of the channels all this was done with a resolution of eleven bits at 10mV. To make sure that none of the records was assigned to an incorrect abstraction class, each vector was checked by independent cardiologists. The data includes 187 values, each of the values ranges from 0 to 10. This set has been divided into five abstraction classes in turn: Normal beat (N), Supraventricular ectopic beat (S), Ventricular ectopic beat (V), Fusion beat (F), Unclassified beat (Q).

4   BiLSTM Deep Learning Model

To achieve this results we have used Bidirectional-LSTM architecture which can be described by Fig. 3, Fig. 4 and Fig. 5. To ensure that our signal is not entirely lost through the deep neural network model we have used a signal normalization layer in the middle of the network which allowed us to add more layers than was possible without the danger of instant loss of accuracy. This allowed our model to adapt to more complicated multi-dimensional mathematical functions than is possible with the more shallow neural network model.

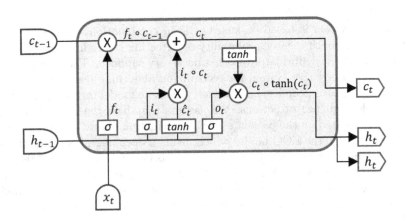

Fig. 3. LSTM gate unit in our neural network architecture.

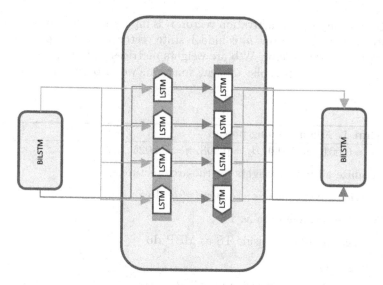

Fig. 4. A sample model of applied signal forwarding over our proposed BiLSTM architecture in which gated units are connected for double side communication to improve.

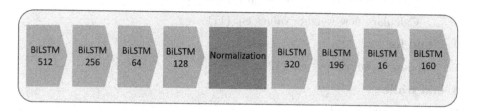

Fig. 5. Applied deep neural network architecture with normalization unit. Proposed concept serves as a simplified auto-encoder model in which normalization by using activation of the previous layer work independently on the data. It is implemented as operation in-between the recurrent layers of the network. It was applied the mean activation values between direct example close to 0 and standard deviation close to 1.

Applied LSTM layers work in a concept of a forget gate and memory recall. Mathematical model of this construction is defined as:

$$f_t = \sigma(W_f[h_{t-1}, x_t] + b_t) \tag{1}$$

$$i_t = \sigma(W_i[h_{t-1}, x_t] + b_i) \tag{2}$$

$$o_t = \sigma(W_o[h_{t-1}, x_t] + b_o) \tag{3}$$

$$\hat{c}_t = \tanh(W_c[h_{t-1}, x_t] + b_c) \tag{4}$$

$$c_t = f_t \circ c_{t-1} + i_t \circ \hat{c}_t \tag{5}$$

$$h_t = o_t \circ \tanh(c_t) \tag{6}$$

where x_t is the input, f_t is activation vector, i_t is input/update activation vector, o_t is output activation vector, h_t is hidden state vector, $\tilde{c}_t$ is cell input activation vector, c_t is cell state vector, W,b are weights matrices and bias, σ is a sigmoid function, tanh is a hyperbolic tangent function, we also assume $c_0 = 0$ and $h_0 = 0$.

Algorithm 1. Adam training process

Input: $\epsilon = 0.001$, $\beta_1 = 0.9$, $\beta_2 = 0.998$, $\eta = 0.0025$

1: Randomize all initial weights for the architecture,

2: **while** *global error value* $\varepsilon < error\_value$ **do**

3:     Reshuffle training data as TS,

4:     **for** each mini-batch inside TS as MBP **do**

5:         Step ++,

6:         Calculate gradient vector for MBP,

7:         Calculate values of momentum eq. (7) and oscillations eq. (8)

8:         Calculate correction values eq. (9) and eq. (10)

9:         Calculate new weights values eq. (11)

10:     **end for**

11:     Update *global error* ε eq. (12).

12: **end while**

Output: trained model

4.1   Adam

The training the model was most efficient by adaptive moment estimation algorithm called Adam. This algorithm is very fast and has low computational complexity. Adam is based on similar idea to RMSProp where first and second moments of gradients are used to update weights. First we calculate mean and variation

$$m_t = \beta_1 m_{t-1} + (1 - \beta_1)g_t, \tag{7}$$

$$v_t = \beta_2 v_{t-1} + (1 - \beta_2)g_t^2, \tag{8}$$

where hyper-parameters $\beta_1 = 0.9$, $\beta_2 = 0.998$ and g is gradient value of applied loss function. Next we calculate correlations of mean and variation

$$\hat{m}_t = \frac{m_t}{1 - \beta_1^t} \tag{9}$$

$$\hat{v}_t = \frac{v_t}{1 - \beta_2^t}.$$ (10)

which are used to update the weights

$$w_{t+1} = w_t - \frac{\eta}{\sqrt{\hat{v}_t} + \epsilon} \hat{m}_t,$$ (11)

where η is learning rate and ϵ is a constant small value. In our training we have used loss function in a form of Mean Squared Logarithmic Error

$$L(y, \hat{y}) = \frac{1}{N} \sum_{i=0}^{N} (log(y_i + 1) - log(\hat{y}_i + 1))^2$$ (12)

as it is very well fitted to the data type of our use. The training process is presented in Algorithm 1.

5 Numerical Experiments

To ensure we are using the best fitted algorithms we have conducted series of numerical experiments. Results can be seen in Fig. 6, Fig. 7 and Fig. 8. As we can see the system performs well on algorithms such as: Adam, NAdam, RMSprop and Adamax and is not suited well for the other four algorithms. What's interesting the loss function creates a logarithmic curve for all training algorithms which were giving promising results. For all the other the curve was almost linear. In terms of accuracy over time plots of Adadelta, Ftrl, Adagrad and SGD performed almost equally by classifying all values as the most numerous class thus giving us not useful results from training. After some deeper analysis of the plots mentioned before we can also spot that the smallest difference between

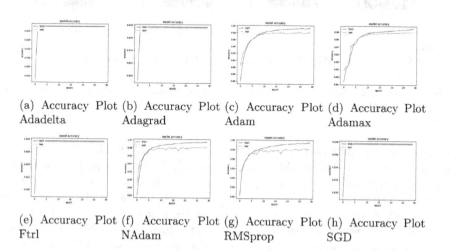

(a) Accuracy Plot Adadelta (b) Accuracy Plot Adagrad (c) Accuracy Plot Adam (d) Accuracy Plot Adamax

(e) Accuracy Plot Ftrl (f) Accuracy Plot NAdam (g) Accuracy Plot RMSprop (h) Accuracy Plot SGD

Fig. 6. Comparison of accuracy plots from examined models.

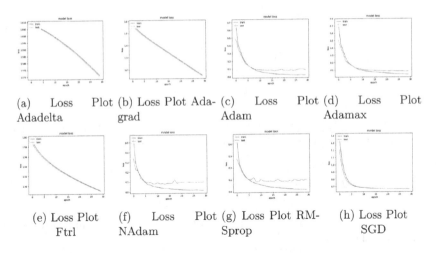

(a) Loss Plot Adadelta (b) Loss Plot Adagrad (c) Loss Plot Adam (d) Loss Plot Adamax

(e) Loss Plot Ftrl (f) Loss Plot NAdam (g) Loss Plot RMSprop (h) Loss Plot SGD

Fig. 7. Comparison of loss plots from examined models.

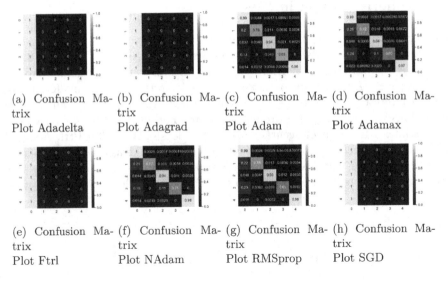

(a) Confusion Matrix Plot Adadelta (b) Confusion Matrix Plot Adagrad (c) Confusion Matrix Plot Adam (d) Confusion Matrix Plot Adamax

(e) Confusion Matrix Plot Ftrl (f) Confusion Matrix Plot NAdam (g) Confusion Matrix Plot RMSprop (h) Confusion Matrix Plot SGD

Fig. 8. Confusion Matrices from our experimental results comparisons among tested training algorithms.

training and testing sets. In the case of the Adamax algorithm loss decreases and accuracy increases. This could erroneously conclude that this is the best choice for our dataset. However from very important metric such as Confusion Matrix we can see that this algorithm may have some disadvantages. Here we can see the big fall of Adamax training algorithm for the benefit of Adam and NAdam methods. What we can see the first, third and the fifth classes are commonly well classified by all of them, however the other ones are much easier to miss-classify. With this task the best work was done by the Adam algorithm followed by the

Table 1. Comparison of results between different optimization algorithms

Algorithm	Accuracy	Precision	Recall	F1
Nadam	98.362273%	92.755022%	85.383819%	0.889169
Adam	98.21411%	88.697324%	88.432159%	0.885645
RMSprop	98.159116%	91.461394%	83.840187%	0.874851
Adamax	98.089926%	91.831089%	82.826612%	0.870967
Adadelta	89.321633%	22.330408%	25.0%	0.235899
Adagrad	89.321633%	22.330408%	25.0%	0.235899
Ftrl	89.321633%	22.330408%	25.0%	0.235899
SGD	89.321633%	22.330408%	25.0%	0.235899
Algorithm	Specificity	FDR	FPR	FNR
Nadam	91.810345%	99.789811%	8.189655%	0.649816%
Adam	84.540117%	99.562739%	15.459883%	0.618785%
RMSprop	89.808917%	99.73419%	10.191083%	0.667365%
Adamax	87.5%	99.684543%	12.5%	0.787662%
Adadelta	0.0%	100.0%	0.0%	2.977402%
Adagrad	0.0%	100.0%	0.0%	2.977402%
Ftrl	0.0%	100.0%	0.0%	2.977402%
SGD	0.0%	100.0%	0.0%	2.977402%

Nadam. However the best classification as the final accuracy of Nadam was a bit higher than the Adam's as it did a better job with the main 3 classes. Results of comparing tested algorithms for 30 epoch are presented in Table 1, while the final selected model training presented in next sections used 100 epochs. We can see that among results the best metrics in terms of accuracy are for Nadam. Very close came Adam algorithm which had smaller overall accuracy however performed much better in terms of classification consistency across all classes. Next are RMSprop and Adamax with the accuracy of 98.16% and 98.09%. And in the end we have Adadelta, Adagrad, Ftrl and SGD which did not learn any useful features on our architecture. As a result of our comparisons we have selected Adam as the best training model for our developed deep learning architecture as it gave us more stable results.

5.1 Results

As a result of our experiments we have select trained BiLSTM neural network with the final accuracy of 99.2%. The final plots and confusion matrix can be seen in Fig. 9. What we can see is that the final train accuracy reached almost 100% however the test one is reaching 99.2% at its best. Because of that we

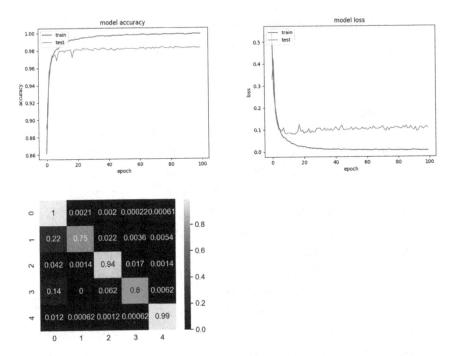

Fig. 9. Final results of training for our selected BiLSTM neural network architecture.

Table 2. Comparison of results between different approach

	Year	Type	Accuracy
Our model	2021	Bidirectional-LSTM	99.2%
Zhai et al. [14]	2018	2D-CNN	99.1%
Huang et al. [15]	2019	2-D Deep CNN	99.0%
Kiranyaz et al. [16]	2015	1-D CNN	99%
Oh et al. [17]	2018	CNN-LSTM	98.1%
Acharya et al. [18]	2017	9-layer CNN	94.03%

can conclude that there is probably some overfitting or the data features are not diverse enough to be correctly classified. Nevertheless we can still talk about high accuracy of prediction compared to other methods presented in Table 2. The network almost perfectly classifies the first class of abstraction with a small decrease in performance with the last fifth class. The third one has also a high accuracy and the second and fourth ones can be described as good enough with a high tendency to classify their examples as the members of the first class of abstraction. In Table we can see comparison of our proposed model to other solutions in literature. We can see that our proposed model is reaching best results among compared solution of machine learning, that gives good start for further development and research.

6 Final Remarks

Using our neural network model, we obtained 99.2% accuracy for validation data what is a considerable achievement as the classification method itself was very difficult without much interference with the input data to ensure the appropriate speed of operation. As a whole, we can say that in the future, when starting the classification of this set, it would be worth using various methods of balancing the input data, especially due to not perfect classification of the second class attracted by the first. In the future we will consider the use of dropout layers, data augmentation and more experiments with layer normalization to enhance the performance even more. What also could be beneficial is the use of some KFold derivatives to better fit the hyper-parameters such as batch size, learning rate, normalization momentum. We also hope that the field of science dealing with heart disease will develop significantly in the future as it is the greatest killer of our time.

References

1. Xu, G., Meng, Y., Qiu, X., Yu, Z., Wu, X.: "Sentiment analysis of comment texts based on bilstm," Ieee Access, vol. 7, pp. 51 522–51 532 (2019)
2. Pogiatzis, A., Samakovitis, G.: Using bilstm networks for context-aware deep sensitivity labelling on conversational data. Appl. Sci. **10**(24), 8924 (2020)
3. Lu, W., Li, J., Wang, J., Qin, L.: A cnn-bilstm-am method for stock price prediction. Neural Comput. Appl. **33**(10), 4741–4753 (2021)
4. Nowicki, R.K., Grzanek, K., Hayashi, Y.: "Rough support vector machine for classification with interval and incomplete data. J. Artif. Intell. Soft Comput. Res. **10** 2020
5. Niksa-Rynkiewicz, T., Szewczuk-Krypa, N., Witkowska, A., Cpałka, K., Zalasiński, M., Cader, A.: "Monitoring regenerative heat exchanger in steam power plant by making use of the recurrent neural network. J. Artif. Intell. Soft Comput. Res. **11** (2021)
6. Jagvaral, B., Lee, W.-K., Roh, J.-S., Kim, M.-S., Park, Y.-T.: Path-based reasoning approach for knowledge graph completion using cnn-bilstm with attention mechanism. Expert Syst. Appl. **142**, 112960 (2020)
7. Liu, K., Gao, W., Huang, Q.: Automatic modulation recognition based on a dcn-bilstm network. Sensors **21**(5), 1577 (2021)
8. Xie, W., Wang, J., Xing, C., Guo, S., Guo, M., Zhu, L.: Variational autoencoder bidirectional long and short-term memory neural network soft-sensor model based on batch training strategy. IEEE Trans. Industr. Inf. **17**(8), 5325–5334 (2020)
9. Nafea, O., Abdul, W., Muhammad, G., Alsulaiman, M.: Sensor-based human activity recognition with spatio-temporal deep learning. Sensors **21**(6), 2141 (2021)
10. Zhao, C., Huang, X., Li, Y., Yousaf Iqbal, M.: "A double-channel hybrid deep neural network based on cnn and bilstm for remaining useful life prediction." Sensors **20**(24) 7109 (2020)

11. Jeong, J.-H., Shim, K.-H., Kim, D.-J., Lee, S.-W.: Brain-controlled robotic arm system based on multi-directional cnn-bilstm network using eeg signals. IEEE Trans. Neural Syst. Rehabil. Eng. **28**(5), 1226–1238 (2020)

12. Beritelli, F., Capizzi, G., Sciuto, G.L., Napoli, C., Woźniak, M.: A novel training method to preserve generalization of rbpnn classifiers applied to ecg signals diagnosis. Neural Netw. **108**, 331–338 (2018)

13. Moody, G.B., Mark, R.G.: The impact of the mit-bih arrhythmia database. IEEE Eng. Med. Biol. Mag. **20**(3), 45–50 (2001)

14. Zhai, X., Tin, C.: "Automated ecg classification using dual heartbeat coupling based on convolutional neural network." IEEE Access. **6** 27 465–27 472 (2018)

15. Huang, J., Chen, B., Yao, B., He, W.: "Ecg arrhythmia classification using stft-based spectrogram and convolutional neural network." IEEE Access **7** 92 871–92 880 (2019)

16. Kiranyaz, S., Ince, T., Gabbouj, M.: Real-time patient-specific ecg classification by 1-d convolutional neural networks. IEEE Trans. Biomed. Eng. **63**(3), 664–675 (2015)

17. Oh, S.L., Ng, E.Y., San Tan, R., Acharya, U.R.: "Automated diagnosis of arrhythmia using combination of cnn and lstm techniques with variable length heart beats." Comput. Biol. Med. **102** 278–287 (2018)

18. Acharya, U.R.: "A deep convolutional neural network model to classify heartbeats." Comput. Biol. Med. **89** 389–396 (2017)

Short Texts Representations for Legal Domain Classification

Tomasz Zymkowski[1], Julian Szymański[1(✉)], Andrzej Sobecki[1],
Paweł Drozda[2], Konrad Szałapak[3], Kajetan Komar-Komarowski[3],
and Rafał Scherer[4]

[1] Faculty of Electronics, Telecommunications and Informatics,
Gdansk University of Technology, Gdansk, Poland
{tomasz.zymkowski,julian.szymanski,andrzej.sobecki}@eti.pg.edu.pl
[2] University of Warmia and Mazury, Olsztyn, Poland
pdrozda@matman.uwm.edu.pl
[3] Lex Secure 24H Opieka Prawna, Sopot, Poland
{ks,kkk}@lexsecure.com
[4] Department of Intelligent Computer Systems,
Częstochowa University of Technology, Częstochowa, Poland
rafal.scherer@pcz.pl

Abstract. This work presents the results of comparison text representations used for short text classification with SVM and neural network when challenged with imbalanced data. We analyze both direct and indirect methods for selecting the proper category and improve them with various representation techniques. As a baseline, we set up a BOW method and then use more sophisticated approaches: word embeddings and transformer-based. The study were done on a dataset from a legal domain where the task was to select the topic of the discussion with the layer. The experiments indicate that fine-tuned pre-trained BERT model for this task gives the best results.

Keywords: Text representation · Short text classification · Transformer · BERT

1 Introduction and Problem Statement

The recent development of advanced deep neural network architectures proves its suitability for the natural language processing domain. More sophisticated language models allow continuous improvements of tasks related to language analysis. One of them is text classification, where the aim is to assign a piece of text to a predefined category. Besides the classification algorithm, often the essential step of this task is to create a text representation that allows processing it using machines. Traditional approaches employ the so-called Vector Space Model that treats the set of documents as points in the high dimensional space of the features. These features are usually single words that aims to represent the text as a set of separate utterances that are related to the document with some weights, e.g.: $tf \times idf$ is used that is based on statistical word frequencies analysis [14].

In our research, we aim to evaluate how much information can be introduced to the model using different representation methods and evaluate them using a classification task. We compare the application of the typical BOW approach that forms the baseline for our research with word representations based on embeddings such as word2vec and fastText and document vectors constructed with transformer models. The experiments were conducted on our dataset dedicated to the classification of short texts in the legal domain. The goal was to assign a topic of the legal opinion based on a short sample of the text. This allows directing the client to the most suitable specialist in an automatic way.

The paper is constructed as follows. Section 2 provides a more detailed description of embedded text representations with particular emphasis on transformer methods used in NLP domain. Then we describe the setup of our experiment in Sect. 3 and provide the achieved results in Sect. 4. The paper is concluded with the discussion and description of directions for further improvements.

2   Neural Text Representations

One of the extensions of the BOW model is to code words as vectors instead of using a single numerical identifier. In such a case, words are encoded using a set of the numbers stored as a vector that allows to capture more sophisticated similarities and thus introduce elementary semantics into representation. One of the most popular approaches to building word vectors is to statistically analyze large text collections and, based on word concurrences in the context, modify the weights of the neural network that builds such a vector – word embedding. There are a number of methods to create the word embeddings, most well known are Word2vec [15], Glove [17], fastText [4]. For a large number of NLP tasks, it is sufficient to use already pretrained word embeddings that may be considered as a kind of transfer learning [19] within the representation level. The next step in text processing is to represent the whole document using word vectors. In the vector space model (VSM), a document is represented as a single point in the high dimensional space of words. Using word embeddings a tensor of vectors is formed that represent a document. To map it into VSM, usually, a simple average of word embeddings that occurs in the document is performed. It should be noticed that this approach works well for small texts, as while the number of averaged vectors increases, the resulted document representations turn out very similar and thus are hardly distinguishable. The detailed description and comparison of word embeddings used for text representation can be found in [24]. As word embeddings face several issues when creating text representation, and more efficient approaches should be used. The extensive review of neural approaches for text representation can be found in [3,30].

The transformer [27] architecture is dedicated for sequential data, in NLP tasks it employs analysis of word co-occurrences in the large text copra. The neural network is built with the encoder-decoder blocks with attention modules used to combine the information from the encoder with the results of the decoder operations. In machine translation tasks this approach is used to map words between languages [29].

At the model input the sequence of words is converted to embedding vectors. As there is no information about the position of a particular word in the sequence the positional encoding [25] is used. The approach employs an additional set of embeddings that contain information about the position of each token in the sentence. Thus the input representation of each token is the sum of word and position embeddings.

The typical architecture of a transformer is built from encoder-decoder blocks repeated six times. Each encoder block contains self-attention and feed-forward layers separated by a nonlinear transformation in the form of the ReLU function. It process the information provided by the attention mechanisms and transform them into a form used by successive layers of encoders or decoders. Between these layers, there are residual connections and normalizing layers. The decoder block has three main layers: masked self-attention, encoder-decoder attention and a feed-forward layer that scales the output vector from the decoder part to the number of vector dimensions of each word. Analogically, as in the encoder part, there are also residual connections. In the last decoder block a softmax layer produces the probability distribution of the particular word occurrence.

The introduction of the transformer architecture push forward the NLP domain. This resulted in the invention of further improvements to the Transformer network, creating new architectures for solving sophisticated lexical tasks.

The initial goal of the transformer architecture was to solve the machine translation task; hence it contains both encoder and decoder parts. Based on this approach, BERT [7] architecture was proposed to create a language representation model. To complete this task, it only needs a part of the encoder from the transformer network that encodes semantic and syntactic information into embeddings [10].

The BERT architecture employs two training techniques: masking [8] and *Next Sentence Prediction* (NSP) [22]. The first one extends training capabilities and thus improve predictions with the usage of the information stored in the embeddings so the model is capable to use more information from the input. The second training technique allows the encoder to predict the entire sentence taking into account the previous one.

The initial BERT architecture is a basis for many new variants [26], introducing improvements for particular tasks or adapting them to a particular language. Some of the important architectures are:

RoBERTa - Robustly Optimized BERT Pretraining Approach [12]. Extends the original BERT by training it longer and on larger data.

ELECTRA - Efficiently Learning an Encoder that Classifies Token Replacement Accurately [5] is a model that uses *replaced token detection technique*, where tokens are replaced with alternative samples. This approach provided more efficient results as the model is trained using all entry tokens, and not only the masked part of them.

XLNet - Generalized Autoregressive Pretraining for Language Understanding [31] is a combination of a large two-way transformer with auto-encoding noise reduction used in BERT. In BERT tokens were predicted in a specific order which limits capture long-term relationships between words. XLNet employs the permutation technique to capture a bidirectional context where the tokens are predicted randomly.

DistilBERT - Distilled version of BERT [21] is a light version of the BERT created to solve the problems with limited computational resources while using the original network for real-life tasks. The approach uses knowledge distillation [9] where, after the training of the larger model the knowledge is transferred to the smaller model. The method allows to obtain results similar to the original solution, despite a significant reduction in the size of the model and the processing time.

ALBERT - A lite version of BERT [11] was proposed to improve the learning process as well as the results achieved by the BERT architecture by applying *cross-layer parameter sharing* and *factorized embedding parameterization* techniques.

SBERT - Sentence BERT [18] – Architecture dedicated to simpler and less computationally complex comparisons of two sequences. It is a Siamese network composed of the BERT model, overlaid with a *pooling* layer, which builds a constant-size representation for input sentences of different lengths. This approach allows to compute the representation of each sentence separately, and then compute the similarity between any two sequences. The model performs significantly better than the standard BERT architecture in the semantic textual similarity task [13].

HerBERT [20] - BERT has been constructed for many languages, HerBERT using extensions introduced by ROBERTa is a network dedicated to Polish language, trained on the corpus composed of many sources, not only Wikipedia.

3   Experiment Setup

The goal of our research was to compare different text representation methods using the classification task. In our study, we test two groups of approaches. The first one creates a vector representation of a given text and then uses a classifier to predict the proper label. Here two processing steps are used, thus we call this an indirect approach as opposed to the direct one, which uses one transformer model both for text representation and classification tasks.

The approaches are evaluated on a dataset built from legal documents written in the Polish language and manually assigned to a particular category.

In this section, we describe in detail the data along with the models used in the classification task in our experiments.

3.1 Dataset Overview

Data used in this experiment has been obtained from legal opinions made for clients' inquiries, collected for the last seven years. Each of them was initially assigned to a category from a set of predefined 11 classes referring to separate branches of law.

Due to the fixed format of the documents, it was possible automatically extract client queries that describe the topic of the legal opinion. For documents with a larger number of queries, we concatenated them into one long string. Making this reduction would later allow the model to make a classification based on users' queries and therefore provide a high level of efficiency in the processing of each legal case.

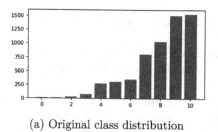

(a) Original class distribution

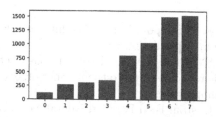

(b) Class distribution after merging smallest four categories

Fig. 1. Comparison of class distributions

Raw datasets usually require initial preprocessing: stop words removal, stemming etc. In our case, the distribution of classes was highly imbalanced ranging from 5 to over 1500 samples. To mitigate this issue it was decided to merge four of the smallest categories into one, finally obtaining 8 classes (1). The results of performing this adaptation are shown in Fig. 1a and Fig. 1b where class distributions are shown before and after concatenation of the smallest categories respectively.

After the above mentioned adaptations and preprocessing the resulting dataset consists of 5901 records of a key-value pairs stored as $(query, category)$. The detailed information of the dataset size is shown in Table 1.

3.2 Indirect Predictions

Indirect models are built from two separate steps, which we will refer to as a representation and classifier. The former maps text into an n-dimensional vector creating a numerical representation. The latter assigns to this vector a corresponding label and predicts the law category.

As this paper focuses on the evaluation of the information that is brought to the model using different representations, we compare representations obtained from BoW, Word2Vec, fastText, and BERT models and classify them with a

Table 1. Number of samples of each category

#	Class	Number of samples
1	Civil law	1530
2	Medical law	1504
3	Administrative law	1028
4	Labor law	804
5	Pharmaceutical law	349
6	Criminal law	302
7	Tax law	270
8	Rest	114

single C-Support Vector Machine and Feed Forward Neural Network to make the final prediction.

SVM has been used with its basic setup except gamma parameter set to $\frac{1}{n}$ (where n was equal to the number of features) and was proceeded with a feature scaling phase.

Custom Feed Forward Neural Network with one dense layer of size 128 activated with the ReLU function and an output layer containing eight neurons with softmax activation. Batch sizes and the number of epochs were selected for each model individually on the validation set to achieve its best performance.

Bag of Words. The first representation model was built on a full corpus as we use it as a baseline. Representation here is a high dimensional, sparse vector of length equal to the number of unique words in a whole corpus. To reduce the noise we proceed with additional steps of stop words removal and stemming.

Word2Vec. In this model, we use pretrained word vectors. We use 300-dimensional word vectors from the repository of Natural Language Processing resources for the Polish language [6]. The Polish word2vec vectors were trained on a corpus of 1.5 billion tokens from articles, books, and Wikipedia. For each sentence in the dataset, the words were processed to the basic form, embedded, and later aggregated by taking the average of all word vectors. In the case of an unrecognized word a zero vector was used.

FastText. Another pre-trained model producing 300-dimensional vectors comes from the repository of [1]. Since it operates on character n-grams instead of full words it does not require a stemming phase. Splitting sentences into words, embedding and aggregating steps were performed as in the Word2Vec approach.

SentenceBERT. The most advanced model was developed for the Polish language, leveraging a modification of the pretrained BERT network to derive semantically meaningful sentence embeddings [2]. Whole sentences were fed into this model without the need for word splitting or aggregating its embeddings. Resulting representation produces a 768-dimensional vector.

3.3 Direct Prediction

Direct models should be considered more of an end-to-end solution as they predict for a selected text the target class without preprocessing. This approach has become much more popular after the introduction of the transformer architecture. In the experiment we use the same pretrained SentenceBERT model as described in the previous section and finetune it for our classification task. It was achieved by adding a linear layer on top of the pooled output and then training it on our custom data. The tokenizer input sequences were either truncated or padded to fit the model architecture. Hyper-parameters were selected empirically on the validation set with a batch size set to 16 and the number of epochs to four.

4   Results

Each model has been tested with 10-fold stratified cross-validation and evaluated with a balanced accuracy metric defined as the average of recall obtained on each class. To compare effectiveness of text representation both SVM and neural network classifiers were used.

Results of indirect models are presented in Table 2. All three of more advanced methods proved to be more accurate than the BoW baseline by both classifiers. Moreover, it has been confirmed that more complex representation models allows to provide more information to the classifier. All the approaches reach better results using a custom neural network classifier improving their results by a few percentage points. The SentenceBERT model delivered the most information to the classifier as it achieved much higher results compared to others. However, it did not come close to the fine-tuned direct model which achieved the highest score of 0.68.

Table 2. Accuracy of classification using indirect representation methods

Model	SVM	Neural network
BoW	0.457	0.493
Word2Vec	0.497	0.521
FastText	0.501	0.513
SentenceBERT	0.561	0.571

5   Discussion and Future Works

In our research, we study the influence of different text representation models on the classification task. As can be expected, the usage of more sophisticated approaches for text representation allows us to increase the quality of the prediction. The best results have been achieved while using the end-to-end model that beats indirect approaches. It can also be observed usage of the neural network as a classifier allows achieving better results than using SVM.

The study allowed us to select the best method for short text classification that was used in our system, aiming to automatically select the specialist for a case reported by the client. The system has been implemented in the legal domain, but by providing the descriptive data, it could be easily adapted to other applications, e.g. in the medical area.

It seems the results may be slightly improved, using a bit more sophisticated models. In the future, we plan to test the usage of an additional convolutional layer in the representation model that may extract some additional features [28] useful for results improvement. It could also be possible that we are near the highest possible results that can be extracted statistically from this data. In this case, introducing more sophisticated models would not result in significant improvements. The solution would be adding more representative data as well as the further study on representation methods. One of the approaches we plan to test is extending the representations based on raw text and using external repositories as a reference for detected named entities. Here we plan to employ methods inspired by the Wikifiction approach. This method maps the text onto Wikipedia and uses additional information provided by this repository. This allows performing classification tasks based on more abstract features instead of ones created from words directly occurring in the text [16,23].

Acknowledgments. The work was supported by founds of the project A semi-autonomous system for generating legal advice and opinions based on automatic query analysis using the transformer-type deep neural network architecture with multitasking learning, POIR.01.01.01–00-1965/20.

References

1. Pre-trained word vectors of 30+ languages. Github (2017). https://github.com/Kyubyong/wordvectors/ (Accessed 10 April 2022)
2. Topic modelling with sentence bert (2022). https://voicelab.ai/topic-modelling-with-sentence-bert. (Accessed 1 May 2022)
3. Babić, K., Martinčić-Ipšić, S., Meštrović, A.: Survey of neural text representation models. Information 11(11), 511 (2020)
4. Bojanowski, P., Grave, E., Joulin, A., Mikolov, T.: Enriching word vectors with subword information. Trans. Ass. Comput. Linguist. 5, 135–146 (2017)
5. Clark, K., Luong, M.T., Le, Q.V., Manning, C.D.: Electra: Pre-training text encoders as discriminators rather than generators. arXiv preprint arXiv:2003.10555 (2020)

6. Dadas, S.: A repository of polish NLP resources. Github (2019). https://github.com/sdadas/polish-nlp-resources/ (Accessed 10 April 2022)
7. Devlin, J., Chang, M.W., Lee, K., Toutanova, K.: Bert: Pre-training of deep bidirectional transformers for language understanding (2019). https://arxiv.org/abs/1810.04805
8. Ghojogh, B., Ghodsi, A.: Attention mechanism, transformers, bert, and gpt: Tutorial and survey (2020)
9. Gou, J., Yu, B., Maybank, S.J., Tao, D.: Knowledge distillation: a survey. Int. J. Comput. Vis. **129**(6), 1789–1819 (2021)
10. Horan, C.: Google's bert - nlp and transformer architecture that are reshaping ai landcape (2021). https://neptune.ai/blog/bert-and-the-transformer-architecture-reshaping-the-ai-landscape
11. Lan, Z., Chen, M., Goodman, S., Gimpel, K., Sharma, P., Soricut, R.: Albert: A lite bert for self-supervised learning of language representations. arXiv preprint arXiv:1909.11942 (2019)
12. Liu, Y., et al.: Roberta: A robustly optimized bert pretraining approach. arXiv preprint arXiv:1907.11692 (2019)
13. Majumder, G., Pakray, P., Gelbukh, A., Pinto, D.: Semantic textual similarity methods, tools, and applications: a survey. Computación y Sistemas **20**(4), 647–665 (2016)
14. Manning, C., Raghavan, P., Schutze, H.: Term weighting, and the vector space model. In: Introduction to Information Retrieval, pp. 109–133 (2008)
15. Mikolov, T., Chen, K., Corrado, G., Dean, J.: Efficient estimation of word representations in vector space (2013). https://arxiv.org/abs/1301.3781
16. Olewniczak, Szymon, Szymański, Julian: Fast approximate string search for wikification. In: Paszynski, M., Kranzlmüller, D., Krzhizhanovskaya, V.V., Dongarra, J.J., Sloot, P.M.A. (eds.) ICCS 2021. LNCS, vol. 12744, pp. 347–361. Springer, Cham (2021). https://doi.org/10.1007/978-3-030-77967-2_29
17. Pennington, J., Socher, R., Manning, C.D.: Glove: Global vectors for word representation. In: Proceedings of the 2014 conference on empirical methods in natural language processing (EMNLP), pp. 1532–1543 (2014)
18. Reimers, N., Gurevych, I.: Sentence-bert: Sentence embeddings using siamese bert-networks. arXiv preprint arXiv:1908.10084 (2019)
19. Ruder, S., Peters, M.E., Swayamdipta, S., Wolf, T.: Transfer learning in natural language processing. In: Proceedings of the 2019 conference of the North American chapter of the association for computational linguistics: Tutorials. pp. 15–18 (2019)
20. Rybak, P., Mroczkowski, R., Tracz, J., Gawlik, I.: Klej: comprehensive benchmark for polish language understanding. arXiv preprint arXiv:2005.00630 (2020)
21. Sanh, V., Debut, L., Chaumond, J., Wolf, T.: Distilbert, a distilled version of bert: smaller, faster, cheaper and lighter. arXiv preprint arXiv:1910.01108 (2019)
22. Sun, Y., Zheng, Y., Hao, C., Qiu, H.: Nsp-bert: A prompt-based zero-shot learner through an original pre-training task-next sentence prediction. arXiv preprint arXiv:2109.03564 (2021)
23. Szymanski, J., Naruszewicz, M.: Review on wikification methods. AI Commun. **32**(3), 235–251 (2019). https://doi.org/10.3233/AIC-190581
24. Szymański, J., Kawalec, N.: An analysis of neural word representations for wikipedia articles classification. Cybern. Syst. **50**, 176–196 (2019)
25. Takase, S., Okazaki, N.: Positional encoding to control output sequence length. arXiv preprint arXiv:1904.07418 (2019)
26. Team, D.: Bert variants and their differences - 360digitmg (2021). https://360digitmg.com/bert-variants-and-their-differences (Accessed 19 April 2022)

27. Vaswani, A., et al.: Attention is all you need (2017). https://arxiv.org/abs/1706.03762
28. Wang, J., Wang, Z., Zhang, D., Yan, J.: Combining knowledge with deep convolutional neural networks for short text classification. In: IJCAI (2017)
29. Wang, Q., et al.: Learning deep transformer models for machine translation. arXiv preprint arXiv:1906.01787 (2019)
30. Wawrzyński, A., Szymański, J.: Study of statistical text representation methods for performance improvement of a hierarchical attention network. Appl. Sci. 11(13), 6113 (2021)
31. Yang, Z., Dai, Z., Yang, Y., Carbonell, J., Salakhutdinov, R.R., Le, Q.V.: Xlnet: Generalized autoregressive pretraining for language understanding. In: Advances in Neural Information Processing Systems, vol. 32 (2019)

Synthetic Slowness Shear Well-Log Prediction Using Supervised Machine Learning Models

Hugo Tamoto[1]([✉]) [iD], Rodrigo Colnago Contreras[2] [iD],
Franciso Lledo dos Santos[3] [iD], Monique Simplicio Viana[4] [iD],
Rafael dos Santos Gioria[1] [iD], and Cleyton de Carvalho Carneiro[1] [iD]

[1] University of São Paulo, Polytechnic School, Department of Mining and Petroleum
Engineering, São Paulo, SP 11013-560, Brazil
{hugo.tamoto,rafaelgioria,cleytoncarneiro}@usp.br
[2] São Paulo State University, Institute of Biosciences, Letters and Exact Sciences,
São José do Rio Preto, São Paulo, SP 15054-000, Brazil
contreras@usp.br
[3] Mato Grosso State University, Faculty or Architecture and Engineering, Cáceres,
MT 78217-900, Brazil
franciscolledo@unemat.br
[4] Federal University of São Carlos, Computing Department, São Carlos, SP
13565-905, Brazil
moniquesimplicioviana@estudante.ufscar.br

Abstract. The shear slowness well-log is a fundamental feature used in reservoir modeling, geomechanics, elastic properties, and borehole stability. This data is indirectly measured by well-logs and assists the geological, petrophysical, and geophysical subsurface characterization. However, the acquisition of shear slowness is not a standard procedure in the well-logging program, especially in mature fields that have a limited logging scope. In this research, we propose to develop machine learning models to create synthetic shear slowness well-logs to fill this gap. We used standard well-log features such as natural gamma-ray, density log, neutron porosity, resistivity logs, and compressional slowness as input data to train the models, and successfully predicted a synthetic shear slowness well-log. Additionally, we created five supervised models using Neural Networks, AdaBoost, XGBoost, and CatBoost algorithms. Among all models created, the neural network algorithm provided the most optimized model, using multilayer perceptron architecture reaching impressive scores as R^2 of 0.9306, adjusted R^2 of 0.9304, and MSE less than 0.0694.

Keywords: Synthetic well-logs · Machine learning · Regression models · Forecasting Time-series

1 Introduction

The high costs associated with the exploration and production in the oil and gas industry activities and the constant demand to reduce operational time impact

L. Rutkowski et al. (Eds.): ICAISC 2022, LNAI 13588, pp. 115–130, 2023.
https://doi.org/10.1007/978-3-031-23492-7_11

the geological and geophysical subsurface characterization of reservoirs [3]. This set of factors constantly prevents the complete acquisition of wireline logging data [17], although they are essential for the geological and geophysical modeling of oil and natural gas fields [7].

Moreover, there is often the need to reduce the scope of data acquisition of geophysical well-logs [3,17] and in particular for the acoustic velocities. Typically, the usual utilization for acoustic well-logs is in the pore pressure prediction, petrophysics, and geomechanics evaluation, in which the shear slowness data is an important input data [2,15,28].

The absence of a shear slowness dataset for all wells leads to a necessity in the use of the information of previous data acquired in older wells, to mitigate the lack of the information in new ones, especially in production and injection wells during the reservoir production phase, or the utilization of classical approach as the empirical correlations [4,5,12]. However, such correlations show limitations of generalization because they incorporate the particularities of the geological context or the geographic region that were developed for, and hence may not be successfully applied to all locations [18,19]. Therefore, an alternative solution to overcome this issue is to use previous information from the well-logging data and predict this feature for future/newest wells, generating synthetic well-logs. The use of machine learning models has highlighted the potential use of solving several problems of regression and classification in geoscience studies [17,21,23, 27] and is a powerful technique to estimate geological reservoirs. Recent examples of machine learning applications have been used to create synthetic geophysical well-logs, especially used for geological formation evaluation, petrophysical and geochemical characterization [17,26].

The main contribution of this study is to generate generalized synthetic shear slowness well-logs, using a supervised machine learning approach. To accomplish this task, it is necessary to use a database composed of similar well-logging features, often available in mature onshore fields. For this study, we used a Brazilian Northeast subsurface dataset. In this region, there is an increase in investments by major operators, but many fields are still profitable and currently with high production [9].

Finally, synthetic slowness shear well-log data obtained by supervised machine learning models, when estimated with precision and accuracy, may be used in subsurface characterization, aligned with the demand of technical needs and cost reduction in well-logging acquisition.

2   Material and Methods

The study was carried out using a public dataset available from the Brazilian National Agency of Petroleum, Natural Gas and Biofuels, ANP [11]. All wells belong to the Canto do Amaro onshore Field, located in the Potiguar Basin, in the Rio Grande do Norte State, Brazil. The development of this research follows to the workflow summarized below:

Initial preprocessing step:

1. Creation of the database and selection of wells and well-logging features.
2. Quality control of database well-logs.
3. Statistical analysis and data analysis, e.g., data standardization, exclusion of spurious values.
4. Supervised machine learning techniques applied to create the regression model and prediction of the shear wave slowness well-logs.
5. Error estimation used in machine learning metrics and comparison between synthetic and real well datasets.

After the preprocessing step, the database was divided into: (i) training set, (ii) validation set, composed of 30% of training data randomly selected, and (iii) test set. The supervised machine learning algorithms were applied to the training data, varying their hyperparameters to find the optimal configuration and generate the trained model, and for each model, we individually performed 10-fold cross-validation within the training data and, and finally compared machine learning algorithms with the real data set tested. This evaluation was performed using regression metrics and aid to validate the synthetic DTS well-logs results (Fig. 1).

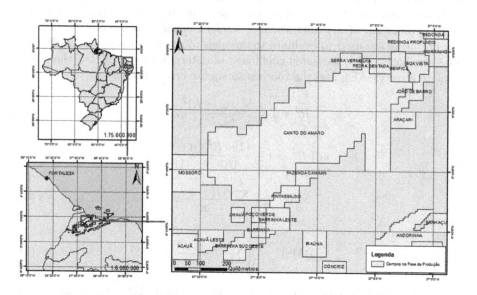

Fig. 1. Location map of the Canto do Amaro Field, Potiguar Basin, Rio de Grande do Norte State, Brazil [10].

2.1 Supervised Machine Learning Models

This research applied supervised regression algorithms used in highly relevant academic publications, such as artificial neural networks, boosting learners, and support vector machine. The neural networks model is based on neural structures. The neuron receives the input data and calculates the output through non-linear interactions assisted by activation functions. Among the several architectures used, the multilayer perceptron (MLP) showed recently successful applications to geological and petrophysical features [14,25,28]. This architecture uses multiple neurons connected in feedforward interactions, in which the training process updates the weights assigned to each neuron, using the backpropagation function to minimize errors in the created network [16]. The Extreme Gradient Boosting (XGBoost), CatBoost, and Adaptive Boosting (AdaBoost) are supervised ensemble models that use decision trees and can be applied to classification or regression problems [8,13]. The Boosting algorithms train the model sequentially and update each round of iteration. At the end of each round, the poorly classified cases are identified, and new interactions gain emphasis to be sequentially used and then provide feedback for the new training. Thus, these subsequent models seek to compensate for previous errors [22]. The Support Vector Machine (SVM) model is an algorithm based on linear relationships related to the principle of margin maximization [24] that minimizes the structural risk of classification and regression cases, aiming to improve the generalization of performance, enhancing the models' complexity. Additionally, the SVM traces a hyperplane that selectively separates the data sets to be predicted [1]. Finally, for the evaluation of machine learning models, usual metric scores for regression problems are used to evaluate the models' outputs, such as the coefficient of determination (R^2), adjusted coefficient of determination (adj R^2), the mean absolute error (MAE), and the root mean square error (RMSE).

$$R^2 = 1 - \frac{\sum_i (y_i - f_i)^2}{\sum_i (y_i - Y_i)^2}, \tag{1}$$

$$\text{adj } R^2 = 1 - \frac{(1 - R^2)(n - 1)}{(n - k - 1)}, \tag{2}$$

$$\text{MAE} = \frac{1}{n} \sum_{i=1}^{n} \frac{(f_i - y_i)}{n}, \tag{3}$$

$$\text{RMSE} = \sqrt{\sum_{i=1}^{n} \frac{(f_i - y_i)}{n}}, \tag{4}$$

where f_i: is the predicted value, y_i is the actual value; Y_i: is the average real value; n: is the number of points for each database; and k is the number of independent variables for each regressor.

2.2   Database Description

The public database is available in the ANP database, comprising 13 wells drilled at the Canto do Amaro onshore field, in the emerged portion of the Potiguar Basin, in the state of Rio Grande do Norte in northeastern Brazil (Fig. 1). This field was discovered in 1985 and is currently considered a mature producing field, with a development area of approximately $363\,\mathrm{km}^2$ [10].

The Geophysical well-logs are important sources of subsurface geological information, essential to any geological, petrophysical, and geophysical characterization of an oil field. They are acquired by wireline tools, which consist of sensors and receivers [7]. The 8 logs used in this work were Borehole Caliper (CALI), Natural Gamma-Ray (GR), Deep Resistivity (RT90), Shallow Resistivity (RT30) and Microspherical Resistivity (MSFL), Density Log (RHOB), Neutron Porosity (NPHI), compressional (DTC) and Shear (DTS) acoustic slowness well-logs. The borehole caliper consists of a profile capable of identifying the integrity of the open hole wall. The logging tool has arms (pads) that allow the identification of rough surfaces and washout zones when the well diameter is outside the nominal drilling of the well [7]. The natural gamma rays and neutron porosity logs explore the interaction of neutrons and the atom's collision, recording an energy spectrum emitted by sensors and receivers of the logging tools. The Gamma-ray logs are commonly obtained to identify the natural radioactivity in rocks and stratigraphic correlations [7,20]. Finally, the neutron porosity is a geophysical well-log that indicates the total porosity of the geological sections, used for petrophysical evaluation [7,20]. The deep, shallow, and microspherical resistivity well-logs measure the natural resistivity of rocks, used to estimate water and hydrocarbon saturation inside the reservoir zone [6]. The density log measures the Compton scattering when the geological formation is excited by electron collision/absorption while logging. The density is also an important property for the evaluation of porosity and geophysical correlations in reservoirs [7,20]. Finally, the sonic log measures the compressional and shear slowness of acoustic waves emitted by transmitters and acquired by receivers after propagation through rocks [7]. Additionally, compressional slowness is used for time-depth seismic tie, pore pressure calculation, and petrophysical evaluation, especially for total porosity [7,20]. Finally, the shear slowness is important for the calculation of elastic and mechanical properties [20]. The geophysical well-logs data are usually displayed in graphs that identify the variation of properties as a function of depth. Figure 2 shows an example applied to the database for well 1535, illustrating the spatial position of variables for the entire geological zone logged.

2.3 Pre-processing and Data Summary

The pre-processing stage consisted of creating a database with information about the 13 wells, shown in Table 1. The data set was randomly selected from the ANP collection. Firstly, we identified the features to be worked with, at this stage it was noted that in wells 1040, 1050, 1509, and 1512, shear wave data were not acquired. Thus, these data were not used in the creation of the database, because this is the variable to be predicted by the machine learning models assisted by the other geophysical well-logs.

Finally, we build the training and test sets described in Table 1. We used the Well 1535 as the test set because that was the most recent well logged among the others, and consequently Well 1356, 1438, 1456, 1462, 1475, 1505, 1519, and 1530

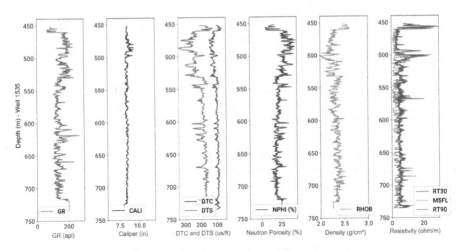

Fig. 2. Examples of well-logs feature composite applied to Well 1535.

Table 1. Well-logs data summary.

Well	CALI in	DTC us/ft	GR ᵒAPI	MSFL ohm/m	NPHI %	RHOB g/cm³	RT30 ohm/m	RT90 ohm/m	DTS us/ft	Data Set
1040	X	X	X	X	X	X	X	X		
1050	X	X	X	X	X	X	X	X		
1356	X	X	X	X	X	X	X	X	X	Training
1438	X	X	X	X	X	X	X	X	X	Training
1456	X	X	X	X	X	X	X	X	X	Training
146²	X	X	X	X	X	X	X	X	X	Training
1475	X	X	X	X	X	X	X	X	X	Training
1505	X	X	X	X	X	X	X	X	X	Training
1509	X	X	X	X	X	X	X	X		
151²	X	X	X	X	X	X	X	X	X	Training
1519	X	X	X	X	X	X	X	X	X	Training
1530	X	X	X	X	X	X	X	X	X	Training
1535	X	X	X	X	X	X	X	X	X	Testing

comprised the training set. We used this selection criterion based on the data acquisition, and our intent sought to build supervised machine learning models using the oldest information, and then predict for the recent well to simulate the potential use of applying machine learning techniques for the prediction of shear slowness, as close as to a real case scenario.

In addition to structuring the database, quality control of the features was carried out. This step consisted of statistical analysis and instances standardization, exclusion of null values, and spurious ones, caused by washouts and roughness of the borehole wall. Table 2 shows the statistical summary of the total of

26745 instances of the geophysical features applied to the training wells, and Table 3 presents the statistical summary for the test set (Well 1535) composed of 2832 instances.

Table 2. Statistic summary for features in the training set.

№ Instances 26475	CALI	DTC	GR	MSFL	NPHI	RHOB	RT30	RT90	DTS
Max	11,73	167	764	134	54.8	2.704	738	466	407
95%	9.2	134	173	11	40.5	2.508	21	24	292
Q3	8.85	108	143	5	28.2	2.373	6	7	220
Mean	8.69	101	119	5	24.7	2.316	7	8	203
Median	8.65	98	117	3	23.3	2.311	5	5	192
Q1	8.5	92	96	2	20.3	2.252	4	4	179
5%	8.28	77	55	1	12.4	2.165	3	3	145
Min	6.5	55	11	0	2.8	1.750	1	1	100
Std. Dev	315	16.2	39	6.86	7.87	102	12,1	13.5	42.4
Var	99	261	1.517	47.1	62	10	147	183	1.797

Table 3. Statistic summary for features of the testing set.

№ Instances 2832	CALI	DTC	GR	MSFL	NPHI	RHOB	RT30	RT90	DTS
Max	9.35	145.3	271.0	44	43.6	2.626	25.7	28.6	340
95%	8.67	130.8	202.0	7.6	36.1	2.456	9.5	10.7	272
Q3	8.55	109.4	176.0	4.8	25.8	2.380	5.8	6.3	226
Mean	8.45	101.4	15.0	4.0	22.7	2.329	5.5	5.9	199
Median	8.45	96.7	142.0	3.5	21.3	2.323	5.0	5.3	185
Q1	8.34	91.8	125.0	2.0	18.5	2.264	4.3	4.5	172
5%	8.25	86.5	110.0	0.9	15.2	2.169	3.0	3.1	161
Min	8.4	65.4	65.0	0.4	5.0	1.990	2.1	2.1	121
Std. Dev	161	13.8	32.2	3.56	6.27	93	2.63	3.2	37.1
Var	26	192.0	1.038	12.6	39.3	9	6.91	9.13	1.373

The comparison between the target feature (DTS) between the training and testing set is illustrated in the histogram in Fig. 3, in which it is possible to identify a similar distribution between these sets.

The following analysis conducted was to observe the correlations among the features, we used the Spearman correlation method for the data set and observed these relations with the DTS features illustrated by the correlation matrix in Fig. 4. The DTC, NPHI, and GR features show strong positive correlations with DTS as warm colors. In addition, the variables RT30, RT90, RHOB, MSFL, and CALI presented negative correlations with DTS, represented by cold colors.

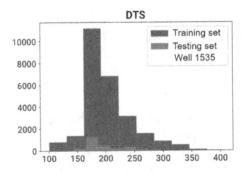

Fig. 3. Distribution of the shear slowness (DTS) instances for training and testing sets.

Finally, the standardization of all features was individually conducted for training and test sets, this transformation generated new data with zero mean (0) and standard deviation of one (1) value.

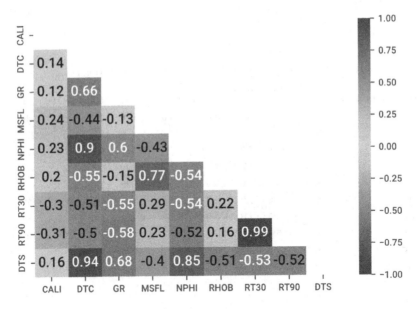

Fig. 4. Correlation Matrix of all well-log features.

3 Results and Discussions

3.1 Machine Learning DTS Synthetic Well-Logs

We conducted the generation of 5 supervised models created for the prediction of synthetic shear slowness well-logs. The models were trained using the training set, with an optimal hyper parametrization tested for each algorithm.

The scenario evaluated consisted of the use of 8 well-log features (CALI, DTC, GR, MSFL, NPHI, RHOB, RT30, and RT90) for training the models. Table 4 shows the optimal hyperparameter configuration found for the 5 machine learning models implemented. Initially, we applied the Gridsearch CV function, selecting a range of values for each hyperparameter of the supervised algorithms, then we chose the optimal combinations which returned the maximized R^2 values and minimized errors score. Additionally, to this analysis, the validation curve function was applied in both validation and test sets, and 10-fold cross-validation was used to evaluate the results of R^2 and RMSE.

Table 4. Summary of optimal hyperparameters configuration applied to scenario 1.

Model	Hyperparameter	Scenario
MLP	Hidden layers	10
	Hidden layers size	3
	Activation	ReLu
	Solver	Adam
	Max iterations	1000
	Learning rate	Adaptive
	Learning rate initial	1
	Others	* MLP default
SVM	kernel	RBF
	C (regularization parameter)	1
	Others	* SVM default
XGBoost	n_estimators (trees)	30
	Learning rate	0.1
	Max depth	3
	Others	* XGB default
AdaBoost	base_estimator	Decision Tree
	n_estimators (trees)	20
	Learning rate	0.12
	Loss	linear
	Others	* AdaBoost default
CatBoost	n_estimators (trees)	45
	Learning rate	0.12
	Max depth	4
	Others	* CatBoost default

Finally, for the model's evaluation, we calculated the regression metrics for the training and validation sets, using 10-fold cross-validation. The comparison between both sets for all models highly indicated agreement through the calculated metrics. In general, the results of the training and validation sets obtained vary between 0.89–0.91 (R^2) and small errors in the order of 0.08–0.1 (MSE) as displayed in Table 5.

Table 5. Score metrics achieved for training, validation and testing set for scenario evaluated.

	Scores metrics					
	Model	MLP	CatBoost	XGBoost	AdaBoost	SVM
Training set 10-fold cross-validation	R^2	8.927	9.104	9.068	8.860	9.173
	MSE	1.071	894	930	1.137	826
	RMSE	3.269	2.989	3.049	3.371	2.872
Validation set	R^2	8.959	9.035	8.997	8.833	9.103
	adj R^2	8.958	9.034	8.996	8.833	9.102
	MAE	1.818	1.800	1.865	2.122	1.611
	MSE	1.019	945	982	1.141	879
	RMSE	2.634	3.073	3.134	3.378	2.964
Testing set	R^2	9.306	9.254	9.228	9.051	8.874
	adj R^2	9.304	9.252	9.226	9.049	8.871
	MAE	1.710	1.763	1.883	2.127	2.036
	MSE	694	746	772	949	1.126
	RMSE	3.193	2.731	2.778	3.080	3.355

Following the analysis, we applied the models in the test set and evaluated the results using the same metrics used in the previous step, the results are summarized in Table 5. In general, the predicted well-logs obtained for the tested well achieved a variation of R^2 between 0.8874 and 0.9306. To directly compare all models the adj R^2 was calculated, considering the number of features and instances for each algorithm. The highest adj R^2 found was for the MLP (adj R^2 0.9304), followed by the Boosting models CatBoost (adj R^2 0.9252), XGBoost (adj R^2 0.9226) and AdaBoost (adj R^2 0.9049), respectively. Finally, the SVM presented the lowest result (adj R^2 0.8874) among all algorithms implemented. The analysis of the errors, in general, indicated that the lowest values of MAE and MSE are found for the MLP model and the highest for the SVM. Additionally, it was observed that Boosting models presented very similar results (Table 5), with the lowest found for CatBoost and the highest errors for AdaBoost. In addition to the calculated scores, the synthetic machine learning DTS well-logs results were compared to the real results logged for Well 1535, graphically displayed in Fig. 5. The gray curve shows the actual data obtained during the logging operation, and the blue curves are the synthetic results obtained for each of the regression models. In general, it is possible to observe that trends of the synthetic curves of all models follow the real data and that the curves are similar. Additionally, small variations are in between them, as exemplified in the 475–525m interval, where the curves diverge.

The visual interpretation of the models corroborates the analytical result found by the metrics previously calculated. Graphically it is observed that the

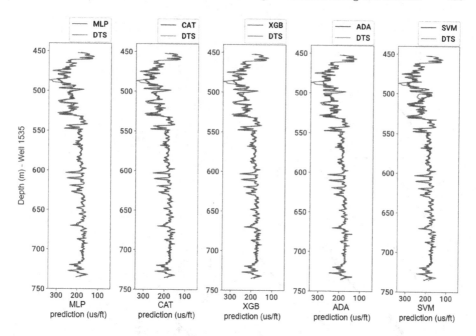

Fig. 5. Scenario evaluated. Machine learning synthetic DTS well-logs in comparison with real DTS slowness well-log applied to Well 1535; right to left. Multi-layer perceptron (MLP), CatBoost (CAT), XGBoost (XGB), AdaBoost (ADA), and Support Vector Machine (SVM) results, respectively.

MLP synthetic DTS well-logs are the most similar to the real data, followed by the boosting algorithms: CatBoost, XGBoost, and AdaBoost. Finally, among all models implemented the SVM synthetic well-log presented more differences from the real DTS well-log, which is evidenced around the intervals 500–510 m.

Figure 6 shows a pair-plots comparison between the data distribution of synthetic shear and real data well. The real data distribution presents a multimodal distribution with 2 peaks, at approximately 150 and 250 us/ft, and is visible a decreasing trend toward values greater than 275 us/ft. This trend was also observed in the format of the MLP, CatBoost, and XGBoost distributions. Among all distributions the MLP data is the most similar to the real result distribution, reflecting the higher value of the adjusted R^2 result calculated in Table 5.

Additionally, the results obtained using AdaBoost and SVM models presented differences in the distribution shape, when compared to the real distribution. For example, the result predicted by the AdaBoost indicates a third mode at 300 us/ft, while the SVM model shows only one peak as shown in Fig. 6. Finally, these differences indeed reflect the lower scores obtained in adjusted R^2 and the higher results from the RMSE of the supervised models.

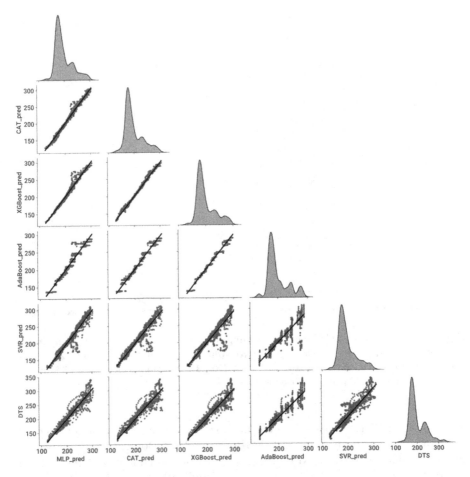

Fig. 6. Scenario 1 pair-plot with the synthetic shear well-log data predicted for Multi-layer Perceptron, CatBoost, XGBoost, AdaBoost, and Support Vector Machine algorithms applied in the Well 1535

3.2 Discussions

The results found in this research showed successfully that the implementation of supervised machine learning models may create trustable synthetic DTS well-logs, assisted by another well-log feature. The metrics obtained indicate highly reliable predictions, especially represented by high values of R^2 and adj R^2 (0.89-0.93), and low errors (MAE, MSE, and RMSE).

Among all models evaluated, the most interesting responses were obtained by the implementation of the neural network algorithms, using multi-layer perceptron architecture, with multiple layers and a few numbers of neutrons per layer in the scenario evaluated. In addition, the decision trees-based models of the Boosting family, especially the CatBoost and XGBoost algorithms provide equally good results for this scenario. On the other hand, the AdaBoost and SVM algorithm results presented the higher errors and dissimilarities with the real DTS well-log, which were evidenced graphically in Fig. 5 and Fig. 6.

Additionally, another point for discussion is the confidence level in the evaluation of the results obtained. The score metrics applied in similar publications using geological and petrophysical properties to regression problems are usually R^2 and RMSE-MSE scores. Table 6 shows the comparison between the results presented in this research and some examples found by other authors. Although there are considerable differences in previous research from the one presented here, such as the geological context, database features, and methodology; in general, it is possible to identify that R^2 results higher than 0.85 were considered with adequate quality for the synthetic well-logs. This is consistent with the results obtained in this study.

Table 6. R^2 and adjusted R^2 scores comparison from synthetical well-logs. Modified from [2].

Research	Algorithm	Synthetic Well-log	Testing Set (R^2)	Testing Set (Adj R^2)
Proposed	MLP	DTS slowness	$R^2 = 0.9306$	Adj $R^2 = 0.9304$
	CatBoost		$R^2 = 0.9254$	Adj $R^2 = 0.9252$
	XGBoost		$R^2 = 0.9228$	Adj $R^2 = 0.9226$
	AdaBoost		$R^2 = 0.9051$	Adj $R^2 = 0.9049$
	SVM		$R^2 = 0.8874$	Adj $R^2 = 0.8871$
[29]	Neural Networks	GR, Resistivity RHOB, NPHI	$0.85 \leq R^2 \leq 0.95$	Not Available
[30]	Fast Fuzzy	RHOB	$R^2 = 0.85$	Not Available
		DTC	$R^2 = 0.92$	Not Available
[31]	SVM Back-propagation neural network (BPNN)	Shear Velocity	$R^2 = 0.97$	Not Available
			$R^2 = 0.94$	Not Available
[32]	Neural Networks	Resistivity	$R^2 = 0.92$	Not Available
		RHOB	$R^2 = 0.97$	Not Available
[33]	Least Square SVM Cuckoo Optimization Algorithm (COA)	Shear Velocity	$0.868 \leq R^2 \leq 0.929$	Not Available
[2]	AdaBoost	Al-Ca-Fe-Mg-Na-Si-S-Ti	$0.843 \leq R^2 \leq 0.976$	Not Available

Furthermore, another evaluation that must be considered is the application of the synthetic well-log. In the results presented here, it was possible to correctly identify all trends in geological and petrophysical variations of the synthetic well-log created. It is noteworthy that the small local dissimilarities found do not affect the geological interpretation of the user, and therefore machine learning predictions can be successfully used. In addition, the creation of high confidence models opens the possibility to substitute the real data with synthetic ones. In this context, this application can be applied during the life cycle of a reservoir, particularly in cases where well-logs were not acquired due to operational problems or lack of technology available at the time of drilling the wells. Moreover, it is well known that subsurface data acquisition is highly expensive to the operators. Moreover, the use of predictive techniques used in this research can assist in the type of replacement, in mature fields, in which a large amount of data are already acquired, and consequently, a mature geological model and reservoir characterization are known. Finally, the substitution must be carefully evaluated for the activities of reservoir characterization and geomechanical modeling because errors associated with the machine learning predictions will certainly impact such analysis. Therefore, considerable limitations in the use of machine learning models are expected for exploratory fields, at exploratory frontiers, or even in the early stages of development. Last, it is expected that the learning curve of the models will evolve as the characterization of these reservoirs is updated and decrease the errors in the well-log prediction.

4   Conclusions

The implementation of supervised machine learning algorithms for the prediction of synthetic shear slowness at the Canto do Amaro oil field showed consistent results with high-reliability results for the test well.

The test performed indicated the potential use of optimization in the reduction of input features in the training of the models. The best result was obtained considering eight input features for all algorithms. In addition, the best models obtained are the MLP, CatBoost, XGBoost, AdaBoost, and SVM algorithms, respectively. In addition, it was observed that the strong dependence of the DTC variable impacted the trial conducted, and the most interesting results were obtained when this feature was used to build machine learning models.

Finally, it is concluded that all synthetic well-logs, independently of the algorithm, respected the trends of the DTS feature for the tested well, allowing a complete geological well-log interpretation, due to the high score metrics achieved in the R^2-adjusted R^2, and lower values in the errors scores.

Acknowledgments. This study was financed in part by the São Paulo Research Foundation (FAPESP), process #2022/05186-4, and by the *"Coordenação de Aperfeiçoamento de Pessoal de Nível Superior - Brasil"* (CAPES) - Finance Code 001.

References

1. Adankon, M.M., Cheriet, M., et al.: Support vector machine (2009)
2. Baouche, R., Sen, S., Ganguli, S.S., Feriel, H.A.: Petrophysical, geomechanical and depositional environment characterization of the triassic tagi reservoir from the hassi berkine south field, berkine basin, southeastern algeria. J. Natural Gas Sci. Eng. **92**, 104002 (2021)
3. Bruhn, C.H., Pinto, A.C., Johann, P.R., Branco, C., Salomão, M.C., Freire, E.B.: Campos and santos basins: 40 years of reservoir characterization and management of shallow-to ultra-deep water, post-and pre-salt reservoirs-historical overview and future challenges. In: OTC Brasil. OnePetro (2017)
4. Castagna, J., Batzle, M., Eastwood, R.: Relationships between compressional-wave in elastic silicate rocks and shear-wave velocities. Geophysics **50**(4), 571–581 (1985)

5. Eberhart-Phillips, D., Han, D.H., Zoback, M.: Empirical relationships among seismic velocity, effective pressure, porosity, and clay content in sandstone. Geophysics **54**(1), 82–89 (1989)
6. Ellis, Darwin V., Singer, Julian M.: Multi-array and triaxial induction devices. In: Ellis, Darwin V., Singer, Julian M. (eds.) Well Logging for Earth Scientists, pp. 179–212. Springer, Dordrecht (2007). https://doi.org/10.1007/978-1-4020-4602-5_8
7. Ellis, Darwin V., Singer, Julian M. (eds.): Well Logging for Earth Scientists. Springer, Dordrecht (2007). https://doi.org/10.1007/978-1-4020-4602-5
8. Friedman, J.H.: Greedy function approximation: a gradient boosting machine 1 function estimation 2 numerical optimization in function space. North **1**(3), 1–10 (1999)
9. Government, B.: Anp estabelece prazo para a petrobras finalizar a cessão de campos em desinvestimento. https://www.gov.br/anp/pt-br/canais_atendimento/imprensa/noticias-comunicados/anp-estabelece-prazo-para-a-petrobras-finalizar-a-cessao-de-campos-em-desinvestimento (2018)
10. Government, B.: Canto do amaro. https://www.gov.br/anp/pt-br/assuntos/exploracao-e-producao-de-oleo-e-gas/gestao-de-contratos-de-e-p/fase-de-producao/pd/canto_do_amaro.pdf(2018)
11. Government, B.: Acesso gratuito aos dados pÚblicos terrestres. https://reate.cprm.gov.br/anp/TERRESTRE (may 2021)
12. Johnston, J., Christensen, N.: Compressional to shear velocity ratios in sedimentary rocks. In: International journal of rock mechanics and mining sciences & geomechanics abstracts. vol. 30, pp. 751–754. Elsevier (1993)
13. Liu, L., Özsu, M.T.: Encyclopedia of database systems, vol. 6. Springer (2009). https://doi.org/10.1007/978-0-387-39940-9
14. Mahmoud, A.A., Elkatatny, S., Al Shehri, D.: Application of machine learning in evaluation of the static young's modulus for sandstone formations. Sustainability **12**(5), 1880 (2020)
15. Mancinelli, P., Scisciani, V.: Seismic velocity-depth relation in a siliciclastic turbiditic foreland basin: a case study from the central adriatic sea. Marine Petrol. Geol. **120**, 104554 (2020)
16. Miikkulainen, R.: Topology of a neural network (2010)
17. de Oliveira, L.A.B., de Carvalho Carneiro, C.: Synthetic geochemical well logs generation using ensemble machine learning techniques for the brazilian pre-salt reservoirs. J. Petrol. Sci. Eng. **196**, 108080 (2021)
18. Onalo, D., Adedigba, S., Khan, F., James, L.A., Butt, S.: Data driven model for sonic well log prediction. J. Petrol. Sci. Eng. **170**, 1022–1037 (2018)
19. Ramcharitar, K., Hosein, R.: Rock mechanical properties of shallow unconsolidated sandstone formations. In: SPE Trinidad and Tobago Section Energy Resources Conference. OnePetro (2016)
20. Rider, M., Kennedy, M.: The geological interpretation of well logs: Rider-french consulting limited (2011)
21. Rubo, R.A., de Carvalho Carneiro, C., Michelon, M.F., dos Santos Gioria, R.: Digital petrography: Mineralogy and porosity identification using machine learning algorithms in petrographic thin section images. J. Petrol. Sci. Eng. **183**, 106382 (2019)
22. Sammut, C., Webb, G.I.: Encyclopedia of machine learning. Springer Science & Business Media (2011). https://doi.org/10.1007/978-0-387-30164-8
23. Song, S., Hou, J., Dou, L., Song, Z., Sun, S.: Geologist-level wireline log shape identification with recurrent neural networks. Comput. Geosci. **134**, 104313 (2020)

24. Ukil, A.: Support vector machine. In: Intelligent Systems and Signal Processing in Power Engineering, pp. 161–226. Springer (2007)
25. Valentín, M.B., et al: Estimation of permeability and effective porosity logs using deep autoencoders in borehole image logs from the brazilian pre-salt carbonate. J. Petrol. Sci. Eng. **170** 315–330 (2018)
26. Wang, J., Cao, J., You, J., Cheng, M., Zhou, P.: A method for well log data generation based on a spatio-temporal neural network. J. Geophys. Eng. **18**(5), 700–711 (2021)
27. Yang, S., Wang, Y., Le Nir, I., He, A.: Ai-boosted geological facies analysis from high-resolution borehole images. In: SPWLA 61st Annual Logging Symposium. OnePetro (2020)
28. Yu, H., Chen, G., Gu, H.: A machine learning methodology for multivariate pore-pressure prediction. Comput. Geosci. **143**, 104548 (2020)

ARIMA for Short-Term and LSTM for Long-Term in Daily Bitcoin Price Prediction

Tran Kim Toai[1,3]([✉]), Roman Senkerik[2], Ivan Zelinka[3], Adam Ulrich[2], Vo Thi Xuan Hanh[1], and Vo Minh Huan[1]

[1] Ho Chi Minh University of Technology Education, No 1 Vo Van Ngan Street, Linh Chieu Ward, Thu Duc City, Vietnam
{toaitk,hanhvtx,huanvm}@hcmute.edu.vn

[2] Faculty of Applied Informatics, Tomas Bata University in Zlin, T. G. Masaryka 5555, 760 01 Zlin, Czech Republic
senkerik@utb.cz

[3] Department of Computer Science, Faculty of Electrical Engineering and Computer Science, VSB-Technical University of Ostrava, 17. Listopadu 2172/15, 708 00 Ostrava-Poruba, Czech Republic
ivan.zelinka@vsb.cz

Abstract. The goal of this paper is the insight into the forecasting of Bitcoin price using machine learning models like AutoRegressive Integrated Moving Average (ARIMA), Support vector machines (SVM), hybrid ARIMA-SVM, and Long short-term memory (LSTM). Depending on the different types of data and the period, various models are used for prediction. A single model may be the best fit in the short term but may not be the best in long-term series data. Thus, using only a single model may not be suitable for forecasting time series data that depends on data sampling length and prediction time, and the type of specific applications. As a result, the ARIMA model produces better error results with a short prediction period or a small data set. In contrast, the Hybrid ARIMA-SVM model will help improve the performance of the ARIMA model when predicting over a long period, specifically 7 and 30 days for Bitcoin price prediction used in this research paper. The paper aims to compare traditional models such as the ARIMA, the Hybrid ARIMA-SVM, and deep learning models such as LSTM on a specific cryptocurrency prediction task using different scenarios.

Keywords: ARIMA · SVM · LSTM · Hybrid models · Bitcoin prediction

Supported by grant of SGS No. SP2022/22, VSB-Technical University of Ostrava, Czech Republic, by Internal Grant Agency of Tomas Bata University under the project no. IGA/CebiaTech/2022/001, and further by the resources of A.I.Lab at the Faculty of Applied Informatics, Tomas Bata University in Zlin.

L. Rutkowski et al. (Eds.): ICAISC 2022, LNAI 13588, pp. 131–143, 2023.
https://doi.org/10.1007/978-3-031-23492-7_12

1   Introduction

Because of the innovative technology based on Blockchain, Bitcoin, a kind of decentralized digital currency, has attracted much attention from investors. The price of Bitcoin is extremely high volatility. The accurate prediction of Bitcoin price can not only provide strong decision support for investors but may also provide a reference for governments to create or change the policies [14].

Numerous studies have been investigated recently to predict the price of Bitcoin from time-series data either through some traditional machine learning methods or deep learning models, as evidenced in survey papers [10,14]. The traditional forecasting techniques are extensively used and easy to implement [15]; for example, the Authors of the study [5] have used Bayesian Neural Networks. Also, the recent approaches in deep learning have been utilized [8]. Different types of time-series data, such as financial and meteorological data, suffer significant fluctuation; numerical data often has a nonlinear relationship [1,11]. That is why nonlinear time series forecasting is a suitable object to predict the change in time series as a basis for risk management [6,16,17]. A popular model for nonlinear time series forecasting includes the models such as ARIMA, LSTM, SVM.

There are three different time types such as short-term period, mid-term period, and long-term period [4,9]. Short-term forecasting focuses on the period of less than one day whereas the mid-term and long-term consider the period more than one week. A model may provide a best fit in short term prediction task, but may not provide good results in long-term series data prediction task. Thus, using only a single model is not suitable for forecasting time series data that depends on how long the time data is sampled and predicted and the type of specific applications.

As mentioned above, many machine learning algorithms and models have been recently published for the time-series cryptocurrency data prediction. Some exploit the hybrid methods by combining more than one model or combining optimization algorithms and prediction models to improve accuracy. The Hybrid ARIMA-SVM model is considered a promising candidate, which is applied to various applications related to the investigated task [12,13,18].

With recent hardware advantages with high performance and low power computation, more advanced machine learning algorithms have been proposed, such as deep learning. The new models and architectures implementations are developed to forecast time series data. An important research question is an accuracy when comparing traditional forecasting techniques with advanced algorithms. According to the literature search, there is no specific designed method for forecasting Bitcoin time series data.

The main goal of this study is to compare the different models and analyze their performance in reducing the error rate. The secondary objective also defining the originality was to answer the research question of whether the newly developed algorithms are better than the traditional algorithms such as ARIMA. As stated in the literature, the ARIMA should be chosen with the non-stationary property of collected data. Similarly, the LSTM is an example of a deep learning method used due to the features of a long time of data. This paper also provides the procedure for processing data and training model for a Bitcoin time series data set. The study exploits the analysis to investigate the performance of traditional forecasting techniques and deep learning-based algorithms. Based on the current research directions and in contrast with research [7], this study aims to develop and deploy a new hybrid forecasting approach to simplify the model and improve forecasting accuracy.

2   Methods

This section contains a description of the models used for the cryptocurrency prediction task.

2.1   ARIMA Model

ARIMA stands for AutoRegressive Integrated Moving Average and it represents a generalization of simpler AutoRegressive Moving Average with the notion of integration. There are two linear time series models which are used widely such as Autoregression (AR) and Moving Average (MA). ARIMA mathematical model is a combination of AR (p), Integration, and MA (q) models [2].

The ARIMA model was proposed to include the case of non-stationarity. In the ARIMA model, the future value of a variable is supposed to be a combination of past values and past errors.

2.2   Support Vector Machines

Support Vector Machines (SVM) model is used to solve two common types of tasks which are classification and regression estimation problem. Support vector regression (SVR) is applied to solve the regression task. The SVR attempts to minimize the generalization error boundary to achieve generalized performance. The SVR creates a decision boundary that separates n-dimensional space into classes so that we can put new data points to the correct category in the future. The computation of SVR is based on the linear regression function in a high dimensional feature space where input data is mapped through a nonlinear function [2].

The major limitation of the ARIMA model it does not consider the factors of the input with non-linear patterns. Whereas, the SVR is a method designed to improve forecasting accuracy. The idea of the SVR algorithm is to find a hyperplane $f(x)$ with a certain deviation (ϵ) from the input training in the form of an Eq. (1):

$$f(x) = y = \omega \cdot x + b \tag{1}$$

The optimization problem in SVR is to find ω and b such that the margin reaches the maximum value at input training to the $f(x)$. The regression problem is transformed into an optimization function (2).

$$\min \frac{1}{2}\|\omega\|^2 + C \sum_{i=1}^{m} (\xi_i + \xi_i^*) \tag{2}$$

With constrain conditions of optimization function (3):

$$\begin{cases} y_i - \omega_1, x_i - b \le \varepsilon + \xi_i \\ \omega_1, x_i + b - y_i \le \varepsilon + \xi_i^* \\ \xi_i, \xi_i^* \ge 0, i = 1, \dots, m \end{cases} \tag{3}$$

where C is the parameter determining penalty degree and $C > 0$, ω is the weight, b is the parameter of mapping, ϵ is the loss function and $\varepsilon > 0$.

The above-defined hyperplane determination is assumed under ideal conditions when the input training has a margin of less than or equal to ϵ. Thus, in the case of data sets with confounding points, these points will not meet the above conditions, and the solution will not be found. For those cases, we need to use slack variables $\xi_i \ge 0$. The slack variables present the distance from the actual values to corresponding boundary values.

When the data problem is non-linear, we have to use the kernel that maps the data to a more dimensional space so that the data can be represented in a computational form. In more dimensional space, the calculation of each data point takes more memory and time. The kernel functions are implemented to make this calculation more manageable. The used SVR model utilizes the radial basis function kernel (RBF) in the form (4) [3].

$$K(x, y) = e^{-\gamma \|x - y\|^2} \tag{4}$$

To obtain the best result, two parameters, C and γ (gamma), are adjusted based on the data sets. The parameters with the less error are used for the best model. The SVR algorithm uses RBF kernel with many different parameters such as $C = 1, 10, 100$ and $\gamma = 0.1, 0.2..., 10$. To find the best parameter, we calculate the accuracy between the predicted data and test data according to many different C and γ parameters. When applying the algorithm to the experiment, the algorithm chooses the best C and γ parameters.

2.3   The Hybrid Model

The different prediction model hybridizations have been studied extensively in various types of research. A Hybrid prediction model capable of solving both linear and non-linear tasks is a good choice for weather or financial data prediction cases. The Hybrid model (Z_t) can be represented as in (5), where Y_t is the linear part, N_t is the non-linear part.

$$Z_t = Y_t + N_t \tag{5}$$

Both Y_t and N_t are predicted from the data set. Consequently, ε_t represents the error at time t obtained from the linear model (6).

$$\varepsilon_t = Z_t - \tilde{Y}_t \tag{6}$$

where $\tilde{Y}_t$ is the predicted data from the linear model at time t. These errors will be predicted from the non-linear model (SVR) and can be expressed as follows (7):

$$\varepsilon_t = f\left(\varepsilon_{t-1}, \varepsilon_{t-2}, \ldots, \varepsilon_{t-n}\right) + \Delta_t \tag{7}$$

where f is the non-linear function generated by the SVR model and Δ_t is the random error. Finally, the model is combined:

$$\tilde{Z}_t = \tilde{Y}_t + \tilde{N}_t \tag{8}$$

where $\tilde{N}_t$ is the predicted result from a non-linear model. In the used hybrid model, the linear part will be handled by the ARIMA model and the non-linear part by the SVR model. The ARIMA model is used to filter the linear patterns of the data set. The error terms of the ARIMA model are applied to SVR model in the hybrid model.

2.4   LSTM

Long Short-Term Memory (LSTM) represents an improvement of a recurrent neural network that allows learning long relationships. While recurrent neural networks (RNN) are limited when the long-term gradients can vanish or explode after propagating through multiple layers in a time-series model, LSTM is effective for learning long-term relationships. Unlike standard feed-forward neural networks, LSTM has feedback connections. LSTM can process a single data point and entire sequences of data. The long-term series represents the correlation between the outputs of the previous and later stages in a time-series model. The model of one layer in an RNN is represented by a processing block with the output of the Tanh function [3]. LSTM has a more complex model, but it is also based on RNN connections. RNN layers are referred to as RNN cells. Cells are connected in a temporal pattern to form recurrent neural networks that allow time series processing. LSTM cells are also connected similarly to RNN cells to create LSTM networks [3].

3   Simulation Results

This section describes both the data set used and the results of the individual models, including the necessary pre-processing and data analysis to estimate the coefficients of the models.

3.1   Data Set

This research uses daily data of the closing price of Bitcoin as the object of study. Bitcoin data was collected from January 1, 2015, to September 23, 2021, from Yahoo Finance[1]. The repository also allows exporting data to a .csv file as

---

[1] https://finance.yahoo.com/quote/BTC-USD/.

input for the prediction algorithm. Bitcoin cryptocurrency price data is extracted continuously by day. The experiment implementation used the first 1966 points (80%) to build the prediction model. The remaining 492 data points (20%) were used to predict and evaluate the accuracy, as shown in Fig. 1(a).

3.2   ARIMA Model

The model preprocesses the data by replacing the missing values with the value of the previous date. The logarithm function transforms the daily Bitcoin price data. In the case of ARIMA model, we take the first order of difference for log-transformation daily Bitcoin price to make the time series stationary. ADF Test (Augmented Dickey Fuller Test), also known as unit root test, helps check whether a time series of data is stationary or not by basing it on the p-value of the test method. If p-value < 0.05, the model is stationary; otherwise, the model is non-stationary.

To determine the parameter p, d, q values of the model, the presented research encompasses a two-step analysis. We analyze data based on an autocorrelation chart (ACF - AutoCorrelation Function) and Partial AutoCorrelation Function (PACF) or based on the p,d,q values at which the model's RMSE (Root Mean Square Error) reaches the minimum value. Then, the dataset is split into 2 sets of train and test. After training the ARIMA model, the experiment implementation predicts future Bitcoin price data in 1-day, 7-day, and 30-day intervals. Then, the model updates the actual value corresponding to the predicted value of each time interval into the training set so that the model continues to learn and predict. Finally, the model converts the data into its original form to evaluate the accuracy of the model.

To select parameters for the ARIMA model, we must first evaluate whether the Bitcoin price over time is stationary or non-stationary. The original data source for the Bitcoin price is shown in Fig. 1(a), whose variance and standard deviation vary over time. Figure 1 also shows characteristics that make time-series non-stationary.

The stationary nature of the data is a prerequisite for predictive modeling, especially when using an autoregressive time series model such as ARIMA. Table 1 shows the results of data stationary testing using the Augmented Dicky-Fuller test (ADF) and Phillips-Perron test (PP). The original and log-transformed data are both non-stationary. Still, they are stationary after the first difference for both the original and the data after the log-to-log conversion. Where p-values are shown in parentheses, a p-value less than 0.05 proves the data is stationary.

Figure 2 shows the ACF and PACF graph for Bitcoin price data. Often, non-stationary data cannot be predicted or modeled. Results obtained using non-stationary time series may be wrong because they may indicate a relationship between two variables for which neither of them exists. To get consistent and reliable results, non-stationary data must be converted to stationary data. Unlike non-stationary data series, which have variable variance and zero mean, stationary data series return to mean and have constant and time-independent variance. The autocorrelation function (ACF) shows that values tend to gradually decrease, which is an indication of the non-stationary nature of the data and this turns it into a stationary series.

Partial autocorrelation and analyzed autocorrelation did not yield accurate values for the parameters p and q after the time series had been made station-

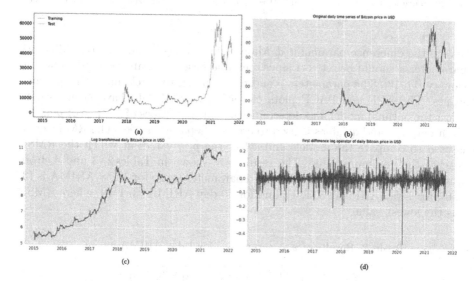

Fig. 1. (a) Train set and test set for predictive model in daily Bitcoin price (b) Original daily series data of Bitcoin price in USD (c) logarithmic transformed daily Bitcoin price (d) The first difference for log-transformation daily Bitcoin price.

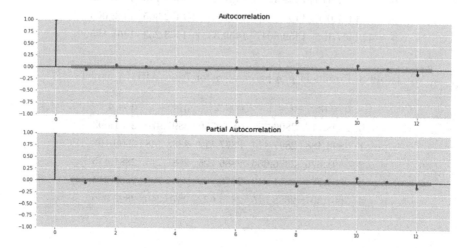

Fig. 2. Graph of autocorrelation function (ACF) and graph of partial autocorrelation function (PACF) for the first difference log transformation.

Table 1. Data stationary condition test

Data	Training sample	ADF test	PP test
Original data	1/1/2015–19/5/2020	−0.477 (0.896)	−0.197 (0.939)
1st difference	1/1/2015–19/5/2020	−7.521 (3.785e−11)	−52.22 (0.000)
Log transformed data	1/1/2015–19/5/2020	−0.622 (0.866)	−0.490 (0.894)
1st difference log transformation	1/1/2015–19/5/2020	−14.940 (1.325e−26)	−51.042 (0.000)

ary by the difference parameter d. Many research has found that the difference parameter d should be set to 1 since our time series data are required to be stationary [11–13]. The parameters p and q are, in this research, manually assigned values. The ACF and PACF plots of the differenced data have been used to select the optimal model and determine the optimal values of a range of p and q. Each pair of p and q values is used to create a separate model. The RMSE error is used to compare the models and select the p, q values at which the RMSE reaches the minimum value. The results are shown in Tables 2, 3 and 4, from where we can observe that the best p, q parameter setup for 1-day ARIMA is (3, 1, 0), for 7-day ARIMA (3, 1, 1), and for 30-day ARIMA (3, 1, 0), where RMSE has the lowest value.

Table 2. Results of p, q parameters for 1-day ARIMA model.

	q = 0	q = 1	q = 2	q = 3	q = 4
p = 0	1459.129	1457.637	1457.407	1457.096	1457.160
p = 1	1457.581	1457.750	1457.4087	1457.097	1457.1613
p = 2	1457.479	1458.427	1457.31	1461.193	1460.143
p = 3	1457.090	1458.245	1457.306	1462.365	1462.005
p = 4	1457.105	1459.8047	1459.3331	1463.253	1465.155

Table 3. Results of p, q parameters for 7-day ARIMA model

	q = 0	q = 1	q = 2	q = 3	q = 4
p = 0	2898.117	2896.455	2898.750	2897.970	2901.093
p = 1	2896.488	2897.474	2877.112	2888.148	2900.195
p = 2	2898.670	2876.656	2886.725	2885.715	2881.415
p = 3	2897.774	2871.819	2876.772	2886.725	2889.715
p = 4	2901.307	2888.515	2873.622	2876.260	2892.386

3.3 SVM Model

The SVM algorithm uses the Radial Basis Function (RBF) kernel with different C and *gamma* values. The SVM network model can determine the C and *gamma*

Table 4. Results of p, q parameters for 30-day ARIMA model

	q = 0	q = 1	q = 2	q = 3	q = 4
p = 0	6248.683	6240.709	6246.273	6236.782	6235.950
p = 1	6240.851	6243.860	6247.160	6252.136	6231.187
p = 2	6246.000	6295.073	6254.032	6198.412	6270.301
p = 3	6236.282	6181.004	6270.172	6272.666	6205.004
p = 4	6236.322	6327.818	6216.045	6214.917	6208.199

values by re-running it many times. Then, we can find the optimal value of the C and *gamma* values at which the RMSE error between the predicted data and actual data reaches the minimum. In addition to selecting the parameters, the kernel selection for the model is also essential. This research investigation utilizes values C and *gamma* for each prediction period of 1 day, 7 days, and 30 days as indicated in Table 5. C and *gamma* parameter values are designed by experiment. To find the best model, different values are assigned C and *gamma*. The parameter values with lower errors are used as the best accuracy model.

Table 5. Parameters of C and *gamma* for SVM model

Parameters	C	Gamma
1-day prediction	10000	0.0001
7-day prediction	10000	0.0001
30-day prediction	10000	0.0001

3.4   Hybrid ARIMA-SVM Model

The presented research uses the best parameter (p, d, q) corresponding to each period time in Tables 2, 3 and 4 to run ARIMA algorithm. The residuals will be used as input for the SVM model. The corresponding C and *gamma* parameters are shown in the Table 6.

Table 6. Parameters of C and *gamma* for Hybrid ARIMA-SVM model

Parameters	C	Gamma
1-day prediction	0.0001	0.01
7-day prediction	0.1	0.001
30-day prediction	100	0.001

3.5 LSTM Model

For the Bitcoin cryptocurrency price prediction model, the parameters for the LSTM network model have been set as follows. The number of input features is one corresponding to the daily Bitcoin price input. The number of hidden layers in the LSTM network structure is 100. There is one neuron output layer. Dropout is set to 0.2, which is a technique to remove connections to avoid overfitting problems in a multilayer neural network model. The Tanh activation function is applied in the hidden layer.

4 Results

This section contains the performance comparisons for four models, three case studies supported by graphical data, and the definitions of performance evaluation factors used to analyze the time-series data. The accuracy result of the algorithm has been calculated as follows. With a total of N given historical data in the dataset, y and $\hat{y}$ are the real data and corresponding predicted values. The formula of RMSE (Root Mean Square Error) and MAE (Mean Absolute Error) is given in (9) and (10). Table 7 shows the model's accuracy through the RMSE and MAE error metrics.

Considering the values of error metrics in Table 7, the LSTM model predicts with better accuracy than the ARIMA, SVM, and Hybrid ARIMA-SVM models when predicting for the long period of 7 days and 30 days. Meanwhile, the ARIMA model predicts in a 1-day period better than the LSTM model. At the same time, the 1-day prediction is not effective compared to the long period of 7 days and 30 days when applying the Hybrid ARIMA-SVM prediction model. The graph showing the actual value and the predicted value of the Bitcoin price when using the LSTM algorithm is shown in Fig. 3(a). All models are compared in Figs. 3(b)–3(d).

$$RMSE = \sqrt{MSE} = \sqrt{\frac{1}{N}\sum_{i=1}^{N}(y - \hat{y})^2} \tag{9}$$

$$MAE = \frac{\sum_{i=1}^{N} abs(y - \hat{y})}{N} \tag{10}$$

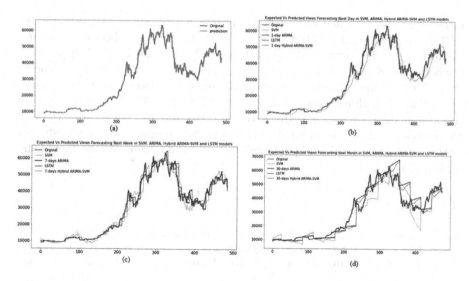

Fig. 3. (a) Actual and predicted data of the LSTM model. Actual and predicted data of LSTM and ARIMA, SVM, Hybrid ARIMA-SVM (b) 1-day prediction (c) 7-day prediction (d) 30-day prediction.

Table 7. Comparison of results of prediction models

Prediction period	Prediction model	MAE	RMSE
long term	LSTM	919.4612	1469.3486
1-day	ARIMA	899.6506	1457.0917
1-day	SVM	2699.6564	3907.7522
1-day	Hybrid ARIMA-SVM	907.2164	1459.6408
7-day	ARIMA	1816.1934	2871.8153
7-day	SVM	2773.8227	3853.02596
7-day	Hybrid ARIMA-SVM	1808.3425	2846.6203
30-day	ARIMA	4176.0200	6181.004
30-day	SVM	5053.8817	6834.2842
30-day	Hybrid ARIMA-SVM	3999.1995	5814.5864

5   Conclusion

In this study, daily Bitcoin price prediction is compared between the traditional algorithm model such as ARIMA, SVM, hybrid ARIMA-SVM, and the model based on a deep learning network (LSTM). The results given in Table 7 and supported by Figs. 3(b)–3(d) show that LSTM is a very efficient technique with a lower error rate so that it can be used more often for forecasting than other models. It is commonly known that LSTM can be implemented with a deep learning approach to get more efficient prediction results because of its pattern recognition property that works efficiently over a long period. Accordingly, the ARIMA model produces better error results with a short prediction period or a small data set. In contrast, the Hybrid ARIMA-SVM model will help improve the performance of the ARIMA model when predicting over a long period, specifically 7 and 30 days. However, when predicting over a long period of time, it is found that algorithms based on deep learning, such as LSTM, are better than traditional algorithms such as the ARIMA model and also the Hybrid ARIMA-SVM model. But of course, for the price of higher computational demands. For further study, other real-world data will be evaluated to compare the results and confirm conclusions.

References

1. Abu Bakar, N., Rosbi, S.: Autoregressive integrated moving average (ARIMA) model for forecasting cryptocurrency exchange rate in high volatility environment: a new insight of bitcoin transaction. Int. J. Adv. Eng. Res. Sci. 4(11), 130–137 (2017)
2. Adhikari, R., Agrawal, R.K.: An introductory study on time series modeling and forecasting. arXiv preprint arXiv:1302.6613 (2013)
3. Elsaraiti, M., Merabet, A.: A comparative analysis of the ARIMA and LSTM predictive models and their effectiveness for predicting wind speed. Energies 14(20), 6782 (2021)
4. Guo, H., Pedrycz, W., Liu, X.: Hidden Markov models based approaches to long-term prediction for granular time series. IEEE Trans. Fuzzy Syst. 26(5), 2807–2817 (2018)
5. Jang, H., Lee, J.: An empirical study on modeling and prediction of bitcoin prices with Bayesian neural networks based on blockchain information. IEEE Access 6, 5427–5437 (2017)
6. Jay, P., Kalariya, V., Parmar, P., Tanwar, S., Kumar, N., Alazab, M.: Stochastic neural networks for cryptocurrency price prediction. IEEE access 8, 82804–82818 (2020)
7. Karakoyun, E.S., Cibikdiken, A.: Comparison of ARIMA time series model and LSTM deep learning algorithm for bitcoin price forecasting. In: The 13th Multidisciplinary Academic Conference in Prague, vol. 2018, pp. 171–180 (2018)
8. Liu, M., Li, G., Li, J., Zhu, X., Yao, Y.: Forecasting the price of bitcoin using deep learning. Finance Res. Lett. 40, 101755 (2021)
9. Liu, Z., Zhu, Z., Gao, J., Xu, C.: Forecast methods for time series data: a survey. IEEE Access 9, 91896–91912 (2021)
10. Mahalakshmi, G., Sridevi, S., Rajaram, S.: A survey on forecasting of time series data. In: 2016 International Conference on Computing Technologies and Intelligent Data Engineering (ICCTIDE 2016), pp. 1–8. IEEE (2016)
11. Munim, Z.H., Shakil, M.H., Alon, I.: Next-day bitcoin price forecast. J. Risk Financ. Manag. 12(2), 103 (2019)
12. Ordóñez, C., Lasheras, F.S., Roca-Pardiñas, J., de Cos Juez, F.J.: A hybrid ARIMA-SVM model for the study of the remaining useful life of aircraft engines. J. Comput. Appl. Math. 346, 184–191 (2019)
13. Ruiz-Aguilar, J.J., Moscoso-López, J.A., Urda, D., González-Enrique, J., Turias, I.: A clustering-based hybrid support vector regression model to predict container volume at seaport sanitary facilities. Appl. Sci. 10(23), 8326 (2020)

14. Saad, M., Choi, J., Nyang, D., Kim, J., Mohaisen, A.: Toward characterizing blockchain-based cryptocurrencies for highly accurate predictions. IEEE Syst. J. **14**(1), 321–332 (2019)

15. Saad, M., Mohaisen, A.: Towards characterizing blockchain-based cryptocurrencies for highly-accurate predictions. In: IEEE INFOCOM 2018 - IEEE Conference on Computer Communications Workshops (INFOCOM WKSHPS), pp. 704–709 (2018). https://doi.org/10.1109/INFCOMW.2018.8406859

16. Sivaram, M., et al.: An optimal least square support vector machine based earnings prediction of blockchain financial products. IEEE Access **8**, 120321–120330 (2020)

17. Sujatha, R., Mareeswari, V., Chatterjee, J.M., Abd Allah, A.M., Hassanien, A.E.: A Bayesian regularized neural network for analyzing bitcoin trends. IEEE Access **9**, 37989–38000 (2021)

18. Syah, R., Davarpanah, A., Elveny, M., Karmaker, A.K., Nasution, M.K., Hossain, M., et al.: Forecasting daily electricity price by hybrid model of fractional wavelet transform, feature selection, support vector machine and optimization algorithm. Electronics **10**(18), 2214 (2021)

Multi-objective Bayesian Optimization for Neural Architecture Search

Petra Vidnerová$^{(\boxtimes)}$ ⬤ and Jan Kalina ⬤

Institute of Computer Science, The Czech Academy of Sciences,
Prague, Czech Republic
petra@cs.cas.cz

Abstract. A novel multi-objective algorithm denoted as MO-BayONet is proposed for the Neural Architecture Search (NAS) in this paper. The method based on Bayesian optimization encodes the candidate architectures directly as lists of layers and constructs an extra feature vector for the corresponding surrogate model. The general method allows to accompany the search for the optimal network by additional criteria besides the network performance. The NAS method is applied to combine classification accuracy with network size on two benchmark datasets here. The results indicate that MO-BayONet is able to outperform an available genetic algorithm based approach.

Keywords: Bayesian optimization · Multi-objective optimization · Neural architecture search · Number of parameters

1 Introduction

Deep neural networks (DNNs) are nowadays used in a huge variety of applications in various fields including (but not limited to) image analysis, signal processing, or natural language processing [9,14]. Their growing popularity increases the necessity to have tools for finding the most suitable neural architecture for a given task. The search for the optimal model for a given task should not take into account only its performance, but also other aspects related to its complexity. For example, the network size, computational costs or time, or energy consumption for the computation may also be relevant. This is especially true when the computation is performed on mobile phones or other devices with high energetic demands.

The most common approaches for choosing the optimal architecture for a particular task are based on solid expert experience, various rules of thumbs, or brute force. Automatic procedures for finding the optimal architecture are highly desirable. The neural architecture search (NAS) has already become an established research field with a variety of available algorithms that differ in the search space coding (how they perform coding of architectures) and in their optimization tools. The majority of NAS algorithms exploit evolutionary optimization or Bayesian optimization [4].

The work is supported by the project GA 22-02067S ("AppNeCo: Approximate Neurocomputing") of the Czech Science Foundation.

L. Rutkowski et al. (Eds.): ICAISC 2022, LNAI 13588, pp. 144–153, 2023.
https://doi.org/10.1007/978-3-031-23492-7_13

Computational demands represent the typical bottleneck of NAS approaches. The objective function being optimized reflects the performance of the resulting network and thus requires evaluation of this network including its learning phase. The Bayesian optimization [2] is particularly aimed for costly (in terms of time) objective functions. It benefits from two key advantages over the classical methods: efficiency, i.e. very low number of objective function evaluations due to surrogate modeling, and no need for analytical knowledge of the objective function. The latter property holds also for evolutionary approaches. Therefore, we decided to utilize Bayesian optimization for multi-objective NAS in this paper.

Our contribution is an extension of the multi-objective Bayesian optimization procedure MOBopt [7]. To allow it to work with candidate solutions representing architectures of deep neural networks, we use direct encoding of candidate architectures. To explain this, a candidate solution is not represented by a vector of numeric values as usual, but is represented directly by a list of layers, each layer encoded by a tuple of its characteristics (a number of neurons or filters, an activation function type, a dropout value, etc.). For such representation we construct the operations *crossover* and *mutation* needed during the surrogate function optimization phase. In addition, each candidate solution is accompanied by a feature vector. These vectors are of a fixed length, contain only numeric values, and describe the main characteristics of a given network. These feature vectors are used for the purpose of the surrogate model.

The multi-objectivity used in the proposed approach is motivated by the need for smaller and energy efficient architectures. Such need is motivated by the fact that many applications are run on mobile devices with limited memory and powered by a battery. The proposed approach allows to accompany the search for the optimal architecture with other criteria, which may be especially useful for controlling the network size (or complexity in general) of the final network. The method presented here allows to consider the trade-off between accuracy and network size in a unique way; the presented approach can be interpreted as a regularization, which is tailor-made for the NAS context, or dimensionality reduction of the considered parametric space.

Section 2 recalls available results on using Bayesian optimization in the context of NAS. Our proposed NAS framework is described in Sect. 3. Section 4 presents the results of our experiments and Sect. 5 brings conclusion. The full algorithm used for the computations of this paper is made publicly available at GitHub [21].

2 Related Work

Although NAS algorithms have been studied since 1990s, the field attracted an enormous interest in the last decade due to an easy access to sufficiently efficient hardware [12,16]. Available algorithms can be classified into categories according to the way of the search space encoding and the optimization algorithm used. The majority of NAS algorithms are based on evolutionary algorithms [19,22]; Bayesian optimization is also successful for the NAS task [11],

but still remains underutilized in such context. This is mainly because typical
Bayesian optimization toolboxes are based on Gaussian processes and focus on
low-dimensional continuous optimization problems, as claimed in the comprehensive survey [4] of NAS techniques for deep learning. A recent application
of Bayesian optimization on NAS can be found in [24], where a path encoding [23] is used as a representation of neural network architecture and Bayesian
optimization is enhanced with a neural predictor.

Multi-objective Bayesian optimization remains only rarely used for NAS,
although multi-objective problems were characterized as a promising research
direction in [4]. The first application of multi-objective Bayesian optimization
to the NAS problem was presented in [5]. The work considered two objectives,
namely performance and on-device inference time (latency) of the networks;
highly accurate networks naturally tend to have a high latency. Nevertheless, the
experiments did not bring comparisons with results of available NAS approaches
there. A very recent publicly available implementation of the multi-objective
Bayesian optimization was presented in [7], where it was applied to optimization
problems not related to neural networks.

3   Multi-objective NAS Framework

An approach to solving NAS based on multi-objective Bayesian optimization,
denoted here as MO-BayONet (Multi-Objective BAYesian Optimization for
NETwork architecture), is proposed in this section. First, the NAS problem
is introduced in Sect. 3.1. In Sect. 3.2, our coding used for the network architectures is described. Section 3.3 lists the set of features used for the surrogate
modeling. Section 3.4 introduces the crossover and mutation operators needed in
the main procedure that is described in Sect. 3.5.

3.1   Problem Definition

NAS can be defined as a global multi-objective optimization problem of finding
the architecture that satisfies our requirements in the best way. The requirements
are expressed by means of objective functions

$$\{O_1, \ldots, O_m\}, \tag{1}$$

where m is a number of objectives, typically two or three. Objectives O_i are
usually computationally costly (requiring the network training) and black-box
functions (without knowledge about derivatives or other characteristics).

The aim of NAS to find the architecture A that minimizes O_i across all i
may be formally expressed as the task to solve

$$\min_{A \in \mathcal{A}} O_i(A), \tag{2}$$

where $\mathcal{A}$ is a space of all possible solutions (*search space*). Solving (2) is usually
not possible, as the individual objectives O_i are conflicting in the sense that

decreasing one implies increasing another. Thus, the desired solution has to find a compromise among all the objectives.

Therefore, the result of multi-objective optimization is not a single solution but rather a set of solutions (in our case architectures) $\{A^\star\}$. Such set is known as the Pareto set, which contains all solutions that are Pareto optimal. The solution A is called Pareto optimal, if it is not dominated by another solution. Solution A_1 is said to dominate A_2 if $O_i(A_1) \leq O_i(A_2)$ for all i and at the same time $O_i(A_1) < O_i(A_2)$ for at least one i. The corresponding objective values (points in the objective space) are known as the Pareto front (Pareto frontier).

3.2   Search Space

In this work, the family of feed-forward convolutional networks represents the search space. The networks consist of convolutional and dense parts, where the convolutional part is a list of convolutional and pooling layers and the dense part consists only of dense layers. The first formal study of encodings for neural architecture was presented in [23]; the paper states the encoding is typically performed by means of directed acyclic graphs, and claims that there remains a need for new non-trivial encodings.

An encoding of candidate architectures is proposed here that allows a fast network construction needed during the evaluation of the objective function, and also allows a straightforward design of the crossover and mutation operators. A candidate solution A is retained within the procedure as a list of layer codes, where a layer code is a tuple of layer characteristics in the form

$$A = [[(t_1, n_1, k_1, a_1), \dots, (t_N, n_N, k_N, a_N)],$$
$$[(m_1, a_1, d_1), \dots, (m_M, a_M, d_M)]]. \tag{3}$$

Here, t represents the layer type, n is the number of filters, k is the filter size, a is the activation function type, N is the number of layers in the convolutional part; m is the number of neurons, d is the dropout parameter, and M is the number of dense layers. The input and output layers are not a part of the encoding, since they are defined by the problem at hand. An illustrative example of the encoding is shown in Fig. 1.

3.3   Feature Space

The encoding proposed in the previous subsection cannot be used as input for the purpose of surrogate modeling, and therefore we construct a feature vector for each candidate solution. Such vector contains the main characteristics of the network, has a fixed length, and contains only numerical values. The particular features used in this work are listed in Table 1.

3.4   Mutation and Crossover Operators

Since the proposed approach uses the NSGA2 [3] algorithm for optimization of surrogate functions, we have to implement the *crossover* and *mutation* operators.

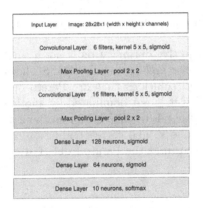
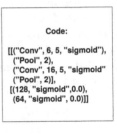

Fig. 1. An illustrative example of the encoding of a particular feed-forward convolutional neural network.

Table 1. List of the features used for the surrogate modeling.

N_A	The number of network parameters
N	The number of convolutional layers
P	The number of pooling layers
$\bar{k}$	The mean size of the convolutional filter
a_{at}^C	Relative numbers of individual activations in convolutional part
M	The number of dense layers
a_{at}^D	Relative numbers of individual activations in dense part
d	The minimal, maximal, and mean dropout values

The aim of crossover is to produce two new candidate solutions by a combination of two existing ones, and mutation produces a new candidate solution by applying small random modifications of an existing one.

Crossover combines two parent architecture codes and produces two offspring codes. It is inspired by the one-point crossover used in genetic algorithms. Only the whole layers are interchanged and the crossover is applied separately to the convolutional part and the dense part.

To explain crossover on an example, let us have two parent architectures

$$A_1 = [C_1, D_1], \qquad A_2 = [C_2, D_2]. \qquad (4)$$

The two offsprings are constructed as

$$\begin{aligned} A_{o1} &= [C_{o1}, D_{o1}] & C_{o1}, C_{o2} &= \text{crossover}(C_1, C_2) \\ A_{o2} &= [C_{o2}, D_{o2}] & D_{o1}, D_{o2} &= \text{crossover}(D_1, D_2), \end{aligned} \qquad (5)$$

where the crossover applied to two parents X_{p1} and X_{p2}

$$X_{p1} = (B_1^{p1}, B_2^{p1}, \ldots, B_k^{p1}) \quad \text{and} \quad X_{p2} = (B_1^{p2}, B_2^{p2}, \ldots, B_l^{p2}) \qquad (6)$$

produces the offsprings

$$X_{o1} = (B_1^{p1}, \ldots, B_{cp1}^{p1}, B_{cp2+1}^{p2}, \ldots, B_l^{p2})$$
$$X_{o2} = (B_1^{p2}, \ldots, B_{cp2}^{p2}, B_{cp1+1}^{p1}, \ldots, B_k^{p1}) \qquad (7)$$

with $cp_1 \in \{1, \ldots, k-1\}$ and $cp_2 \in \{1, \ldots, l-1\}$.

Mutation randomly chooses one of these operations: deleting a randomly chosen layer, adding a randomly chosen layer, or mutating a randomly chosen layer. Mutating a chosen layer includes a random change of its characteristics.

3.5 Multi-objective Bayesian Optimization

Bayesian optimization can be described as an efficient and effective global optimization tool for functions with expensive evaluations [1]. It is particularly popular for tuning hyper-parameters of machine learning methods [20]. The Bayesian approach is built upon the idea of constructing probabilistic models for the objective functions, called *surrogate functions*, that are searched efficiently (instead of searching the true objectives) before the candidate samples are chosen for the evaluation of the true objective function [2]. Typically, Gaussian Processes (GP) [18] are used as surrogate models.

The core cycle of the optimization algorithm is depicted in Fig. 2. The algorithm starts with a set of N_{init} initial points (N_{init} being a small number) that are generated randomly and are evaluated by the objective functions. During the run, it is necessary to store the database of candidate solutions evaluated so far together with the corresponding values of the objective functions $D = \{A_i, O_1(A_i), \ldots, O_m(A_i)\}_{i=1}^t$. The dataset D is then used to train Gaussian Processes (GP) as surrogate models of the objective functions. For a given A, GP_i represents an estimate of the mean and variance of the Gaussian distribution that describes $O_i(A)$. Then, a standard optimization algorithm may be applied. Since we deal with a multi-objective case, we use the NSGA2 algorithm, which is an established multi-objective optimization method able to obtain qualitatively good Pareto fronts [3].

After obtaining an approximation of the Pareto front and the Pareto set, it is possible to use one of two approaches for choosing a new point to be evaluated by the true objectives. The first one selects a point close to $\{A^\star\}$. For the evaluation of distances between candidate solutions, the feature vectors are used. The second approach selects a point in the same way, but then applies mutation on it. The trade-off between these two approaches represents the trade-off between exploitation and exploration. Each new point is then evaluated by the true objective functions and is added to the database D. With the update of the database, the Gaussian Processes are retrained.

Our optimization algorithm is based on the multi-objective Bayesian optimization algorithm MOBopt [7] that is publicly available on GitHub; it uses NSGA2 [3] for optimization of surrogate functions and keeps the Pareto set of candidate solutions. We implemented the whole procedure in Python using libraries Tensorflow [8], and DEAP [6].

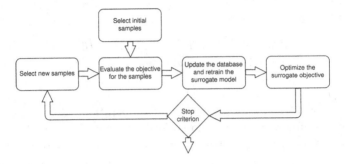

Fig. 2. The flow of the Bayesian optimization used within the proposed MO-BayONet procedure.

4   Experimental Results

In order to illustrate the performance of the novel method MO-BayONet, experiments on three popular benchmark datasets are performed. These datasets are MNIST, fashionMNIST and CIFAR10 [13,15,25]. The classification problem into one of 10 given groups is solved. We optimized an architecture of a feed-forward convolutional network and used two objectives.

The first objective is the network performance evaluated as the **classification accuracy**

$$O_1(A) = \frac{1}{K} \sum_{k=1}^{K} L_{acc}(A) \tag{8}$$

in K-fold cross-validation, where K is the number of folds (here $K = 3$) and L_{acc} is classification accuracy. The network was trained using the Adam optimizer [8] for 10 epochs and using categorical cross-entropy as a loss function.

The second objective, aiming to represent the energetic demands of the network and hardware, is considered as the **number of parameters** of the network. We denote it as

$$O_2(A) = N_A, \tag{9}$$

where N_A is the total number of the network parameters. No surrogate model was used for this objective and it was always evaluated directly, since it is not computationally expensive.

The algorithm was run for 100 iterations. The resulting networks were trained on the whole training set for 20 epochs for 5 trials. For each trial, the classification accuracy of the network was evaluated on the testing set. The mean and the standard deviation of test set accuracies were computed.

For comparison purposes, two architectures are used. One is a baseline solution proposed by a human expert (from tutorials on MNIST) and the other is a network found by a genetic algorithm for NAS (GA-NAS) [22]. The genetic algorithm was run for comparable number of objective function evaluations as was required by our algorithm.

Table 2. Results of the baseline solution (expert-designed network), GA-NAS solution (found by NAS based on genetic algorithms), and MO-BayONet solution (our proposed approach): classification accuracies (i.e. their averages and standard deviations, where applicable) and network sizes (numbers of parameters).

Task	Baseline	GA-NAS	MO-BayONet		
MNIST database					
Clas. accuracy	98.97	99.19	99.34	99.27	99.13
Std. deviation	–	0.26	0.09	0.06	0.05
Network size	600K	690K	233K	199K	39K
fashion-MNIST database					
Clas. accuracy	91.64	93.13	93.05	92.75	91.78
Std. deviation	–	0.20	0.19	0.16	0.14
Network size	356K	769K	2382K	360K	92K
CIRAR10 database					
Clas. accuracy	70.45	72.8	74.29	76.45	76.46
Std. deviation	–	0.59	0.49	0.51	0.17
Network size	122K	154K	81K	196K	455K

The obtained classification accuracies and sizes of the resulting networks are shown in Table 2. We can see that MO-BayONet is able to find competitive solutions in terms of classification accuracy, while using smaller network sizes.

A scatter plot of the objectives (classification accuracies and network sizes) of all obtained solutions is presented in Fig. 3. All resulting solutions (obtained approximations of Pareto sets) from 10 runs of MO-BayONet are shown. The solutions that are closer to the right bottom corner than the baseline and the GA-NAS solution outperform these two solutions.

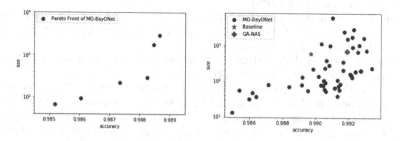

Fig. 3. Left: an example of the resulting Pareto front for MNIST dataset in terms of size and cross-validation classification accuracy. Right: the resulting networks for MNIST in terms of size and classification accuracy on the test set: baseline (expert-designed network); GA-NAS (result of NAS based on a genetic algorithm); the novel MO-Bayonet (Pareto sets from 10 computations).

5 Conclusion

A novel NAS approach based on Bayesian multi-objective optimization is proposed in this paper. In contrary to the available Bayesian optimization methods for NAS, the presented approach has the following unique features:

– It is **multi-objective**, and
– It uses a **direct encoding** of candidate solutions.

The proposed MO-BayONet approach yields promising results on 2 benchmark datasets, while the computational demands are comparable to available approaches. We made the software implementation of the novel approach publicly available on GitHub [21]. It does not use parallelization, but the parallelization is possible. Since the objective function contains cross-validation, network training and evaluation on individual folds can be done in parallel; inspiring ideas on a possible parallelization of Bayesian optimization can be found in [10].

There are several open directions left for the future work. As the method is general, it can be extended to more complex networks (arbitrary hierarchical structures) in a straightforward way. A more elaborated study of suitable features for surrogate models would also be desirable. As this paper was created in a broader framework of approximate neurocomputing research, other complexity objectives are planned to be investigated as well. It would be particularly useful to include objectives evaluating the energetic demands of the computations [17]; possible applications on mobile phones would allow energy savings and the work could thus contribute to the development of "green machine learning" [26].

References

1. Archetti, F., Candelieri, A.: Bayesian Optimization and Data Science. Springer, Cham (2019). https://doi.org/10.1007/978-3-030-24494-1
2. Brochu, E., Cora, V.M., de Freitas, N.: A tutorial on Bayesian optimization of expensive cost functions, with application to active user modeling and hierarchical reinforcement learning (2010)
3. Deb, K., Pratap, A., Agarwal, S., Meyarivan, T.: A fast and elitist multiobjective genetic algorithm: NSGA-II. IEEE Trans. Evol. Comput. **6**(2), 182–197 (2002). https://doi.org/10.1109/4235.996017
4. Elsken, T., Metzen, J.H., Hutter, F.: Neural architecture search: a survey. J. Mach. Learn. Res. **20**(1), 1997–2017 (2019)
5. Eriksson, D., et al.: Latency-aware neural architecture search with multi-objective Bayesian optimization. CoRR abs/2106.11890 (2021). arxiv.org/abs/2106.11890
6. Fortin, F.A., De Rainville, F.M., Gardner, M.A., Parizeau, M., Gagné, C.: DEAP: evolutionary algorithms made easy. J. Mach. Learn. Res. **13**, 2171–2175 (2012)
7. Galuzio, P.P., de Vasconcelos Segundo, E.H., dos Santos Coelho, L., Mariani, V.C.: MOBOpt - multi-objective Bayesian optimization. SoftwareX **12**, 100520 (2020). https://doi.org/10.1016/j.softx.2020.100520. http://www.sciencedirect.com/science/article/pii/S2352711020300911
8. Goodfellow, I., et al.: TensorFlow: large-scale machine learning on heterogeneous systems (2015). Software available from tensorflow.org. https://www.tensorflow.org/

9. Goodfellow, I., Bengio, Y., Courville, A.: Deep Learning. MIT Press (2016). http://www.deeplearningbook.org

10. Kandasamy, K., Krishnamurthy, A., Schneider, J., Póczos, B.: Parallelised Bayesian optimisation via Thompson sampling. In: AISTATS. Proceedings of Machine Learning Research, vol. 84, pp. 133–142. PMLR (2018)

11. Kandasamy, K., Neiswanger, W., Schneider, J., Póczos, B., Xing, E.P.: Neural architecture search with Bayesian optimisation and optimal transport. In: Proceedings of the 32nd International Conference on Neural Information Processing Systems, NIPS 2018, Red Hook, NY, USA, pp. 2020–2029. Curran Associates Inc. (2018)

12. Kitano, H.: Designing neural networks using genetic algorithms with graph generation system. Complex Syst. **4**, 461–476 (1990)

13. Krizhevsky, A., Nair, V., Hinton, G.: The CIFAR-10 dataset. http://www.cs.toronto.edu/kriz/cifar.html

14. Lecun, Y., Bengio, Y., Hinton, G.: Deep learning. Nature **521**(7553), 436–444 (2015). https://doi.org/10.1038/nature14539

15. LeCun, Y., Cortes, C.: The MNIST database of handwritten digits (2012). http://research.microsoft.com/apps/pubs/default.aspx?id=204699

16. Miikkulainen, R., et al.: Evolving deep neural networks. CoRR abs/1703.00548 (2017). http://arxiv.org/abs/1703.00548

17. Mrazek, V., Sarwar, S.S., Sekanina, L., Vasicek, Z., Roy, K.: Design of power-efficient approximate multipliers for approximate artificial neural networks. In: 2016 IEEE/ACM International Conference on Computer-Aided Design (ICCAD), pp. 1–7 (2016). https://doi.org/10.1145/2966986.2967021

18. Rasmussen, C.E., Nickisch, H.: Gaussian processes for machine learning (GPML) toolbox. J. Mach. Learn. Res. **11**, 3011–3015 (2010)

19. Real, E., Aggarwal, A., Huang, Y., Le, Q.: Regularized evolution for image classifier architecture search. In: Proceedings of the AAAI Conference on Artificial Intelligence, vol. 33, February 2018. https://doi.org/10.1609/aaai.v33i01.33014780

20. Snoek, J., Larochelle, H., Adams, R.P.: Practical Bayesian optimization of machine learning algorithms. In: Proceedings of the 25th International Conference on Neural Information Processing Systems, NIPS 2012, Red Hook, NY, USA, vol. 2, pp. 2951–2959. Curran Associates Inc. (2012)

21. Vidnerová, P., Kalina, J.: Bayonet (2022). https://github.com/PetraVidnerova/BayONet

22. Vidnerova, P., Neruda, R.: Evolving keras architectures for sensor data analysis. In: 2017 Federated Conference on Computer Science and Information Systems (FedCSIS), pp. 109–112, September 2017. https://doi.org/10.15439/2017F241

23. White, C., Neiswanger, W., Nolen, S., Savani, Y.: A study on encodings for neural architecture search. In: Advances in Neural Information Processing Systems (2020)

24. White, C., Neiswanger, W., Savani, Y.: BANANAS: Bayesian optimization with neural architectures for neural architecture search. In: AAAI Conference on Artificial Intelligence (AAAI-2021) (2021)

25. Xiao, H., Rasul, K., Vollgraf, R.: Fashion-MNIST: a novel image dataset for benchmarking machine learning algorithms (2017)

26. Xu, J., Zhou, W., Fu, Z., Zhou, H., Li, L.: A survey on green deep learning (2021)

Multilayer Perceptrons with Banach-Like Perceptrons Based on Semi-inner Products – About Approximation Completeness

Thomas Villmann$^{(\boxtimes)}$ ⑩ and Alexander Engelsberger

Saxon Institute for Computational Intelligence and Machine Learning (SICIM), University of Applied Sciences, Mittweida, Germany
thomas.villmann@hs-mittweida.de

Abstract. The paper reconsiders multilayer perceptron networks for the case where the Euclidean inner product is replaced by a semi-inner product. This would be of interest, if the dissimilarity measure between data is given by a general norm such that the Euclidean inner product is not longer consistent to that situation. We prove mathematically that the universal approximation completeness is guaranteed also for those networks where the used semi-inner products are related either to uniformly convex or to reflexive Banach-spaces. Most famous examples of uniformly convex Banach spaces are the spaces L_p and l_p for $1 < p < \infty$. The result is valid for all discriminatory activation functions including the sigmoid and the *ReLU* activation.

1 Introduction and Motivation

Various types of multilayer perceptrons (MLP) including deep networks belong nowadays certainly to the standard neural networks in machine learning for classification and regression tasks [1,8]. Biologically motivated by pyramid cells in brains the corresponding mathematical perceptron is the basis of those networks [24], see Fig. 1.

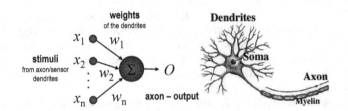

Fig. 1. Schematic illustration of a mathematical perceptron (left) according to a pyramid cell (right). The input vector $\mathbf{x} = (x_1, \ldots, x_n)^T$ is weighted by the weight vector $\mathbf{w} = (w_1, \ldots, w_n)^T$ to generate the output O.

A. Engelsberger—Supported by an ESF PhD grant.

L. Rutkowski et al. (Eds.): ICAISC 2022, LNAI 13588, pp. 154–169, 2023.
https://doi.org/10.1007/978-3-031-23492-7_14

The capability for these networks is justified by Cybenko's theorem with states the universal approximation capability for MLP's with sigmoidal activation functions [5]. One key ingredient in the proof of the respective theorem is the Hilbert-space-property needed to ensure the application of the Riesz-Representation-Theorem (RRT). This property is given for each perceptron in the network, because perceptrons generate their output based on the Euclidean inner product (EIP) between the input and the weight vector. Thus the data space is implicitly assumed to be a Hilbert space equipped with the Euclidean norm, which is generated by the standard inner product. However, depending on the task, other than the Euclidean metric might be more appropriate, e.g. l_p-norms (metrics) with $p \neq 2$ [18] or kernel metrics [28]. However, those metrics relate to so-called semi-inner products (SIP, [20]) which show weaker requirements than inner products. Hence, a consistent approach for a perceptron network should make use of SIPs instead of the EIP. Consequently the question arises whether those networks remain universal approximators. The paper tackles exactly this problem and will provide respective proofs.

The remainder of the paper is as follows: First we provide the basic mathematical concepts and definitions needed for the mathematical analysis of the problem. Thereafter, we recapitulate the proof of Cybenko's theorem regarding the approximation completeness to identify the keypoints of this proof in the light of the given problem. For this purpose, we analyze the class of discriminatory activation functions regarding the Euclidean inner product (or general inner products) and show that both sigmoidal and ReLU activation function belong to that class. In the next step we provide the results for SIP-based perceptrons, which we also denote as Banach-like-perceptrons (BlP). For this purpose, we show that the class of discriminatory functions with respect to a given SIP can be appropriately defined and, again sigmoidal and ReLU activation belong to this class. Further, we show which parts of the original Cybenko-theorem have to be modified. In particular, we identify those SIPs (and respective Banach-spaces), which can be equipped with an RRT compared to that valid for Hilbert spaces. The technical structure of this paper follows closely the mathematical description of MLP's given in [10].

2   The Standard Multilayer Perceptron Revisited

The mathematical modeling of standard perceptrons assumes stimulus vectors $\mathbf{x} \in \mathbb{R}^n$ and a weight vector $\mathbf{w} \in \mathbb{R}^n$ to generate the output according to

$$O(\mathbf{w}, \mathbf{x}) = f(\langle \mathbf{w}, \mathbf{x} \rangle + b) \tag{1}$$

where $b \in \mathbb{R}$ is the bias and f is the so-called activation function. The quantity $\langle \mathbf{w}, \mathbf{x} \rangle = \sum_{k=1}^{n} x_k \cdot w_k$ is the (real) Euclidean inner product, which is motivated biologically by the weighted sum of inputs, see Fig. 1. The activation function f usually is a monotonically increasing function. Common choices are the identity id $(z) = z$ (linear perceptron), the Heaviside function $H(z)$ (standard perceptron) or the sigmoid function

$$f_\theta(z) = \frac{1}{1 + \exp(\theta z)} \tag{2}$$

as smooth (differentiable) approximation of $H(z)$ as wellas the hyperbolic tangent. Nowadays, other activation functions became popular, rather motivated computationally than biologically [22]. Among them, the function

$$\text{ReLU}(z) = \max(0, z) \tag{3}$$

known as *Rectified Linear Unit* has gained great focus because of its easy computation and derivative [8].

MLPs are directed graphs with mathematical perceptrons as nodes organized in layers [13]. Only the first layer (input layer) receives direct data inputs. The last layer is denoted as output layer and delivers the network response $\mathbf{o}$ for a given data vector $\mathbf{x}$. The stimulus vectors of perceptrons in all layers except the input layer are output vectors of previous layers. Mathematically speaking, MLPs realize a mapping

$$F_{W,B} : \mathbb{R}^n \ni \mathbf{x} \longmapsto \mathbf{o} \in \mathbb{R}^m \tag{4}$$

if m output units are available and W is the set of all weights $\mathbf{w}$ and B is the set of all biases in the network. It was shown by CYBENKO that under certain conditions MLP's are universal approximators [5]. We will consider the proof of this theorem in detail after giving useful definitions and theorems from mathematical analysis needed for an adequate problem description in the proof of the Cybenko-theorem.

2.1   Basic Concepts, Mathematical Definitions and Theorems

Definition 1. The function σ is n-discriminatory with respect to the inner product $\langle \cdot, \cdot \rangle$ if for a measure $\mu \in \mathcal{M}(I_n)$ of the closed (compact) subset $I_n = [0, 1]^n \subset \mathbb{R}^n$ with the property

$$\int_{I_n} \sigma(\langle \mathbf{w}, \mathbf{x} \rangle + b)\, d\mu(\mathbf{x}) = 0$$

for all $\mathbf{w} \in \mathbb{R}^n$ and $b \in \mathbb{R}$ the implication $\mu \equiv 0$ follows. A function is said to be discriminatory with respect to the inner product $\langle \cdot, \cdot \rangle$ if it is n-discriminatory for all n.

A function σ is denoted as *sigmoidal* if

$$\sigma(z) \longrightarrow \begin{cases} 1 \text{ for } z \to \infty \\ 0 \text{ for } z \to -\infty \end{cases}$$

holds. Obviously, $f_\theta(z)$ from (2) is sigmoidal. Another example is

$$\lambda(z) = \begin{cases} 0 & \text{if } z < 0 \\ z & \text{if } z \in [0,1] \\ 1 & \text{if } z > 1 \end{cases} \tag{5}$$

denoted as interval-restricted linear function. Let the set $\Lambda_{[0,1]} = \{L(z) = \lambda(a \cdot z + b),\ a, b \in \mathbb{R}\}$ of *interval-restricted linear functions.*

Later, we will make use from the following lemma:

Lemma 2. *If the span* $\mathcal{S}(\Lambda_{[0,1]})$ *is dense in* $\mathcal{C}[0,1]$, *then the span* $\mathcal{S}(\Lambda_{[0,1]^n})$ *is dense in* $\mathcal{C}[0,1]^n$ *with* $\Lambda_{[0,1]^n} = \{L(z) = \lambda(\mathbf{a}^T\mathbf{b} + b),\ \mathbf{a}, \mathbf{b} \in \mathbb{R}^n\}$

The proof can be found in [14,17]

The following Lemma, proven in [5], relates sigmoidal functions to discriminatory functions:

Lemma 3. *Any bounded, measurable sigmoidal function is discriminatory with respect to the real inner product* $\langle \cdot, \cdot \rangle$ *and, hence, any continuous sigmoidal function is discriminatory.*

It turns out that also the function $\text{ReLU}(z)$ from (3) is discriminatory with respect to the inner product $\langle \cdot, \cdot \rangle$. In fact, we now prove the following lemma about the discriminatory property of the *ReLU*-activation with respect to a real inner product:

Lemma 4. *The* $\text{ReLU}(z)$ *from (3) is discriminatory with respect to the real inner product* $\langle \cdot, \cdot \rangle$ *for* $z(\mathbf{x}) = \langle \mathbf{w}, \mathbf{x} \rangle + b.$

Proof. We follow [10] and start with the case $n = 1$ (1-discriminatory), i.e. $\langle w, x \rangle = w \cdot x$ and $z(x) = w \cdot x + b$ for given w and b. For $w = 0$ we can rewrite an arbitrary $\lambda(z) \in \mathcal{S}(\Lambda_{[0,1]})$ into

$$\lambda(b) = \begin{cases} \text{ReLU}(\lambda(b)) & \text{if } \lambda(b) \geq 0 \\ -\text{ReLU}(-\lambda(b)) & \text{if } \lambda(b) \leq 0 \end{cases}$$

whereas for $w \neq 0$ we decompose $\lambda(z(x))$ into

$$\lambda(x) = \text{ReLU}\left(w \cdot x - \frac{b}{w}\right) - \text{ReLU}\left(w \cdot x + \frac{1-b}{w}\right) \tag{6}$$

using the linearity of the (inner) product $w \cdot x$. Applying this decomposition we prove immediately the assertion: Because $\lambda(z(x))$ is discriminatory according to the previous lemma we have that for the integral $I[\lambda] = \int \lambda(w \cdot x - b)\, d\mu(x)$ the equality $I[\lambda] = 0$ holds, which further implies that $\mu \equiv 0$ has to be valid. Hence, we get for the decomposition (6)

$$\begin{aligned} I[\lambda] &= \int \text{ReLU}\left(w \cdot x - \frac{b}{w}\right) d\mu(x) \\ &\quad - \int \text{ReLU}\left(w \cdot x + \frac{1-b}{w}\right) d\mu(x) \\ &\overset{\mu \equiv 0}{=} 0 - 0 \end{aligned}$$

which is the desired result.

For $n > 1$ we consider the span $\mathcal{S}(G)$ of the set $G = \{g(z(\mathbf{x})) \,|\, \text{nonlinear } g \in \mathcal{C}([0,1])\}$ of continuous functions depending on $\mathbf{x}$ with parameters $\mathbf{w}$ and b and keep in mind Lemma 2: Let $h(\mathbf{x}) \in \mathcal{S}(G)$, arbitrarily given. According to Kolmogorov's representation theorem [2,14] and [9] exist affine functions $g_k(z_\varepsilon(\mathbf{x})) \in \mathcal{C}([0,1])$ with $z_\varepsilon(\mathbf{x}) = \langle \mathbf{w}, \mathbf{x} \rangle + \frac{b}{N(\varepsilon)}$ such that

$$\left| h(\mathbf{x}) - g\left(\sum_{k=1}^{N(\varepsilon)} g_k(z_\varepsilon(\mathbf{x})) \right) \right| < \frac{\varepsilon}{2}$$

for arbitrarily chosen $\varepsilon > 0$ using the non-linearity of g. Because $z(\mathbf{x}) = \sum_{k,j} w_k x_j \langle \mathbf{e}_k, \mathbf{e}_j \rangle + b$ is an affine (linear) function in each variable x_j the introduced functions g_k are affine (linear) functions of x_j, i.e. we have $g_k(z_\varepsilon(\mathbf{x})) = \sum_{j=1}^{n} \hat{g}_k(z_\varepsilon(x_j))$ with $z_\varepsilon(x_j) = x_j \cdot w_j + b_j$. Each of the continuous functions $\hat{g}_k$ can be further approximated by

$$\left| \hat{g}_k(z_\varepsilon(x_j)) - \sum_{l=1}^{N_k(\varepsilon)} \lambda_{k,l}(z_\varepsilon(x_j)) \right| < \frac{\varepsilon}{2 \cdot N(\varepsilon) \cdot n}$$

with $\lambda_{k,l} \in \mathcal{S}(\Lambda_{[0,1]})$ which can be taken as combinations of ReLU-functions according to (6).

In consequence, we are able approximate each $h(\mathbf{x}) \in \mathcal{S}(G)$ with arbitrary precision which implies the n-discriminatory property using the first part of the proof. This completes the proof of the lemma.   □

Remark 5. We emphasize that for (6) the linearity of the inner product with respect to the first argument was used.

Definition 6. Let X be a vector space over $\mathbb{K} \in \{\mathbb{R}, \mathbb{C}\}$ and $\varphi : X \to \mathbb{K}$ be a functional. If both properties

- positive homogeneity: $\varphi(\lambda\mathbf{x}) = \lambda\varphi(\mathbf{x})$ for $\lambda \in \mathbb{R}_+$ and $\varphi(i\mathbf{x}) = i\varphi(\mathbf{x})$ is valid in the complex case
- subadditivity: $\varphi(\mathbf{x} + \mathbf{y}) \le \varphi(\mathbf{x}) + \varphi(\mathbf{y})$

hold, φ is denoted as *sublinear*.

We remark that every norm on a vector space X is sublinear. A central role in this paper plays the *Hahn-Banach-Theorem* which states the following [15,23]:

Theorem 7 (Hahn-Banach-Theorem). *Variant a): Let X be a vector space over $\mathbb{K} \in \{\mathbb{R}, \mathbb{C}\}$ and $Y \subseteq X$ a subspace. Let $\varphi : X \to \mathbb{R}$ be a sublinear functional and $f : Y \to \mathbb{K}$ be a linear functional with $\Re(f(\mathbf{y})) \le \varphi(\mathbf{y})$ for all $\mathbf{y} \in Y$. Then there exists a linear functional $F : X \to \mathbb{K}$ with $F|_Y = f$ and $\Re(F(\mathbf{x})) \le \varphi(\mathbf{x})$ is valid for all $\mathbf{x} \in X$.*

An alternative formulation is the variant [25,27] b): Let X be a normed space and Y is a subspace $Y \subset X$. Let be $f \in X^*$ with $f|_Y = 0$. The subspace Y is dense in X iff under these assumptions always follows $f(\mathbf{x}) = 0$ for all $\mathbf{x} \in X$.

The following theorem is known as the *Theorem of Dominated Convergence from Lebesgue* [15,23]:

Theorem 8 (Dominated-Convergence-Theorem). *Let X be a measure space, μ a Borel-measure on X and $g : X \longrightarrow \mathbb{R}$ absolute integrable, $g \in \mathcal{L}^1(X)$. Let further $\{f_k\}$ be a sequence of measurable functions $f_k : X \longrightarrow \mathbb{R}$ such that $|f_k(\mathbf{x})| \leq g(\mathbf{x})$ holds for all $\mathbf{x} \in X$, i.e. g dominates all f_k. If the sequence $\{f_n\}$ converges point-wise to a function f, i.e. $f_k(\mathbf{x}) \xrightarrow[k \to \infty]{pointwise} f(\mathbf{x})$ then f is absolute integrable, i.e. $f \in \mathcal{L}^1(X)$ with*

$$\lim_{k \to \infty} \int f_k(\mathbf{x}) \, d\mu(\mathbf{x}) = \int f(\mathbf{x}) \, d\mu(\mathbf{x}) \ .$$

2.2 Cybenko's Results for Standard MLP

The main statement regarding the universal approximation property of MLP's is given by the following theorem. For the sake of later considerations we also give the proof of the theorem as provided in [5]. We will later make use of that proof structure.

Theorem 9. *Let $I_n = [0,1]^n \subset \mathbb{R}^n$ be the closed hypercube equipped with the Euclidean metric., Let σ be a continuous discriminatory function with respect to the inner product $\langle \cdot, \cdot \rangle$. Further, let*

$$\Pi = \left\{ \pi(\mathbf{x}) \in \mathcal{C}(I_n) \,|\, \pi(\mathbf{x}) = \sum_{j=1}^{N} \alpha_j \cdot \sigma(\langle \mathbf{w}_j, \mathbf{x} \rangle + b_j) \right\} \tag{7}$$

be the set of continuous functions consisting of finite sums of perceptrons (1) with an activation function $f = \sigma$. Then the set $\mathcal{P} = \mathrm{span}(\Pi)$ of functions $\pi(\mathbf{x})$ is dense in the space $\mathcal{C}(I_n)$ of continuous functions over I_n.

Proof. The set $\mathcal{P}$ is dense in $\mathcal{C}(I_n)$ iff for any function $g(\mathbf{x}) \in \mathcal{C}(I_n)$ and $\varepsilon > 0$ exists a function $\pi(\mathbf{x}) \in \mathcal{P}$ with $|\pi(\mathbf{x}) - g(\mathbf{x})| < \varepsilon$ for all $\mathbf{x} \in I_n$. This statement is proven if we can show that for the closure $\overline{\mathcal{P}}$ of $\mathcal{P}$ the equality $\overline{\mathcal{P}} = \mathcal{C}(I_n)$ holds. We apply a proof by contradiction:

Obviously, $\mathcal{P}$ is a linear subspace of $\mathcal{C}(I_n)$. Thus, the closure $\overline{\mathcal{P}}$ is a closed subspace of $\mathcal{C}(I_n)$. We remark that I_n is equipped with the Euclidean norm such that it is a Banach-space or, more precisely, a Hilbert space. Now we suppose that $\overline{\mathcal{P}} \neq \mathcal{C}(I_n)$, i.e. $\mathcal{P}$ is not dense in $\mathcal{C}(I_n)$ and show that this assumption leads to a contradiction:

It follows from the assumed equality according to the Hahn-Banach-theorem that there is a bounded linear functional L on $\mathcal{C}(I_n)$ with $L(h) \neq 0$, i.e. it is not completely vanishing for $h \in \mathcal{C}(I_n)$ but $L(\mathcal{P}) = L(\overline{\mathcal{P}}) = 0$ is valid. We remark that L is continuous and we have $L \in \mathcal{C}^*(I_n)$ being the dual space of $\mathcal{C}(I_n)$.

According to the Hilbert-space property of I_n we can apply the Riesz-Representation-Theorem (RRT, [23]), which states that the functional L can be written in the form

$$L(h) = \int_{I_n} h(\mathbf{x}) \, d\mu(\mathbf{x}) \tag{8}$$

for some measure $\mu \in \mathcal{M}(I_n)$ and a continuous function $h \in \mathcal{C}(I_n)$. Yet, so far μ is unspecified.

Because for the continuous function $\sigma(\langle \mathbf{w}, \mathbf{x} \rangle + b) \in \overline{\mathcal{P}}$ is valid for all $\mathbf{w}$ and b we must have that

$$L(\sigma) = \int_{I_n} \sigma(\langle \mathbf{w}, \mathbf{x} \rangle + b) \, d\mu(\mathbf{x}) = 0$$

holds for all choices $\mathbf{w}$ and b according to $L(\overline{\mathcal{P}}) = 0$. Since σ is assumed to be discriminatory, the zero integral implies that $\mu \equiv 0$ has to be valid, which further implies, however, that $L(h) \equiv 0$ for any $h \in \mathcal{C}(I_n)$. This contradicts the assumption $\overline{\mathcal{P}} \neq \mathcal{C}(I_n)$. Hence, $\mathcal{G}$ is dense in $\mathcal{C}(I_n)$ which completes the proof. $\square$

According to this result and the Lemma 4 we can conclude that also the ReLU-activation ensures the universal approximation property.

Remark 10. In the proof of the Cybenko-theorem the Hilbert-space property of I_n was explicitly used which is guaranteed by the Euclidean metric/norm. Further, the Euclidean norm in I_n is consistent with the mathematical structure of the discriminatory functions $\sigma(\langle \mathbf{w}, \mathbf{x} \rangle + b)$ containing the Euclidean inner product in the argument.

Remark 11. We explicitly remark that the validity of the RRT provided by Eq. (8) is essential to complete the proof. The RRT, however, originally requires the Hilbert-space property.

3   Generalizations of Cybenko's Results for MLPs with Generalized Inner Products

In this chapter we generalize the Cybenko-Theorem 9. First, we make the easy step to kernel-based inner products replacing the inner product in perceptrons. Thereafter, we consider more general inner product variants, namely, semi-inner products and variants thereof.

3.1   Kernels for Hilbert-Spaces

Obviously, the proof of the Cybenko-theorem remains valid if we replace the Euclidean inner product $\langle \mathbf{w}, \mathbf{x} \rangle$ in the standard perceptron (1) by an arbitrary inner product and use the resulting norm as norm for the n-dimensional real space $\mathbb{R}^n$. We can continue this idea and, more generally, replace the inner product by a kernel κ, i.e. we consider

$$\kappa(\mathbf{w}, \mathbf{x}) = \langle \phi(\mathbf{w}), \phi(\mathbf{x}) \rangle$$

with $\phi(\mathbf{w}) \in \mathcal{H}$ where $\mathcal{H}$ is a reproducing kernel Hilbert space (RKHS) [26]. Then $\mathcal{I}_n = \phi(I_n)$ is compact in the Hilbert space $\mathcal{H}$ and the Cybenko's theorem is still applicable also for $\mathcal{I}_n$.

3.2   Semi-inner Products

In the second, more challenging case we want to exchange in the perceptron (1) the inner product $\langle \mathbf{w}, \mathbf{x} \rangle$ by a semi-inner product (SIP) $[\mathbf{w}, \mathbf{x}]$ [20].

Definition 12. A mapping $[\cdot, \cdot] : \mathcal{B} \times \mathcal{B} \to \mathbb{C}$ is called a semi-inner product (SIP) if the following relations are fulfilled:

1. linearity: $[\lambda \mathbf{x} + \mathbf{z}, \mathbf{y}] = \lambda [\mathbf{x}, \mathbf{y}] + [\mathbf{z}, \mathbf{y}]$ for $\lambda \in \mathbb{C}$
2. positiveness: $[\mathbf{x}, \mathbf{x}] > 0$ for $\mathbf{x} \neq \mathbf{0}$
3. Cauchy-Schwarz-inequality: $|[\mathbf{x}, \mathbf{y}]|^2 \leq [\mathbf{x}, \mathbf{x}][\mathbf{y}, \mathbf{y}]$

LUMER has shown that a SIP always generates a norm by $\|\mathbf{x}\| = \sqrt{[\mathbf{x}, \mathbf{x}]}$ as well as he has proofed that every Banach-space with norm $\|\mathbf{x}\|_{\mathcal{B}}$ is equipped with a SIP generating this norm [20]. Generally, there may exist several SIPs generating a given norm. Additional requirements are needed to ensure uniqueness. Further, given a norm, generally there is no constructive way to derive a respective SIP. Despite this impossibility, one can show that the homogeneity property $[\mathbf{x}, \lambda \mathbf{y}] = \overline{\lambda}[\mathbf{x}, \mathbf{y}]$ can be imposed without causing any significant restriction of the LUMER results [7].

Now we equip I_n with the norm $\|\mathbf{x}\| = \sqrt{[\mathbf{x}, \mathbf{x}]}$ denoted as $I_n^{\mathcal{B}} \subset \mathbb{R}_{\mathcal{B}}^n$. Thus $\mathbb{R}_{\mathcal{B}}^n$ becomes an n-dimensional real Banach-space. Considering now *Banach-like perceptrons (B-perceptron)* with output

$$O(\mathbf{w}, \mathbf{x}) = f([\mathbf{w}, \mathbf{x}] + b) \tag{9}$$

using real SIPs, we cannot simply apply the original Cybenko-theorem to show approximation completeness, because its proof requires the Hilbert-space property needed to apply the RRT. However, as mentioned before, $I_n^{\mathcal{B}}$ is not contained in a Hilbert space. Fortunately, there exist variants of the RRT which suppose weaker but special Banach-spaces instead of a Hilbert-space.

Before we will characterize those Banach-spaces, we have to extend the definition of a discriminatory functions:

Definition 13. The function σ is n-discriminatory with respect to the real-valued linear functional $l(\mathbf{w}, \mathbf{x})$ in $\mathbf{x}$, if for a measure $\mu \in \mathcal{M}(I_n)$ of the closed (compact) subset $I_n = [0, 1]^n \subset \mathbb{R}^n$ with the property

$$\int_{I_n} \sigma(l(\mathbf{w}, \mathbf{x}) + b) \, d\mu(\mathbf{x}) = 0$$

for all $\mathbf{w} \in \mathbb{R}^n$ and $b \in \mathbb{R}$ the implication $\mu \equiv 0$ follows. The function σ is said to be discriminatory with respect to the real-valued linear functional $l(\mathbf{w}, \mathbf{x})$ in $\mathbf{x}$, if it is n-discriminatory with respect to the real-valued linear functional $l(\mathbf{w}, \mathbf{x})$ for all n.

Lemma 14. *Any bounded, measurable sigmoidal function is discriminatory with respect to the real-valued linear functional $l(\mathbf{w}, \mathbf{x})$ and, hence, any continuous sigmoidal function is discriminatory.*

Proof. The proof we give here follows the argumentation in [5]. Doing so, we suppose a sigmoid function σ and a real-valued linear functional $l(\mathbf{w}, \mathbf{x})$ with $\int_{I_n} \sigma(l(\mathbf{w}, \mathbf{x}) + b)\, d\mu(\mathbf{x}) = 0$ for given signed measure μ. We have to show that $\mu \equiv 0$ follows.

For this purpose we consider the function

$$\sigma_\lambda(\mathbf{x}) = \sigma(\lambda \cdot (l(\mathbf{w}, \mathbf{x}) + b) + \varphi)$$

which converges point-wise and bounded to the function

$$\gamma(\mathbf{x}) = \begin{cases} 1 & \text{for } l(\mathbf{w}, \mathbf{x}) + b > 0 \\ 0 & \text{for } l(\mathbf{w}, \mathbf{x}) + b < 0 \\ \sigma(\varphi) & \text{for } l(\mathbf{w}, \mathbf{x}) + b = 0 \end{cases}$$

in the limit $\lambda \longrightarrow +\infty$, i.e. $\sigma_\lambda(\mathbf{x}) \xrightarrow[\lambda \to +\infty]{\text{pointwise}} \gamma(\mathbf{x})$. Hence, $|\sigma_\lambda(\mathbf{x})| \le \gamma(\mathbf{x})$ is valid. Applying the Dominant-Convergence-Theorem 8 we have

$$\int_{I_n} \gamma(\mathbf{x})\, d\mu(\mathbf{x}) = \lim_{\lambda \to +\infty} \int_{I_n} \sigma_\lambda(\mathbf{x})\, d\mu(\mathbf{x})$$

with $\int_{I_n} \sigma_\lambda(\mathbf{x})\, d\mu(\mathbf{x}) = 0$ according to the assumed discriminatory property of σ. Thus we can further calculate for an arbitrary choice of $\mathbf{w}$, b, and φ

$$\int_{I_n} \gamma(\mathbf{x})\, d\mu(\mathbf{x}) = \int_{X_{\mathbf{w},b}^+} 1\, d\mu(\mathbf{x}) + \int_{X_{\mathbf{w},b}^-} 0\, d\mu(\mathbf{x})$$

$$+ \int_{X_{\mathbf{w},b}^0} \sigma(\varphi)\, d\mu(\mathbf{x}) \tag{10}$$

$$= \mu\left(X_{\mathbf{w},b}^+\right) + \sigma(\varphi)\mu\left(X_{\mathbf{w},b}^0\right) \tag{11}$$

$$= 0$$

using the definition of $\gamma(\mathbf{x})$ in the first equation together with the half-planes $X_{\mathbf{w},b}^+ = \{\mathbf{x} \in I_n | l(\mathbf{w}, \mathbf{x}) + b > 0\}$ and $X_{\mathbf{w},b}^- = \{\mathbf{x} \in I_n | l(\mathbf{w}, \mathbf{x}) + b < 0\}$ whereas $X_{\mathbf{w},b}^0 = \{\mathbf{x} \in I_n | l(\mathbf{w}, \mathbf{x}) + b = 0\}$ is a hyperplane according to the linearity of $l(\mathbf{w}, \mathbf{x})$. For $\varphi \to +\infty$ we observe $\sigma \to 1$, because σ is sigmoid. Hence,

$$\mu\left(X_{\mathbf{w},b}^+\right) + \mu\left(X_{\mathbf{w},b}^0\right) = 0$$

must be valid in (11). Otherwise, if $\varphi \to -\infty$ we observe that $\sigma \to 0$ holds in (11) and, therefore, $\mu\left(X_{\mathbf{w},b}^+\right) = 0$ must be valid. Thus we have shown that the measures of all half-planes are zero. It remains to show that from this property it follows that the measure μ has to be zero. This would be trivial for positive measures, but this is not assumed here.

Thus, we now fix $\mathbf{w}$ and consider the linear functional

$$F(h) = \int_{I_n} h(l(\mathbf{w}, \mathbf{x})) \, d\mu(\mathbf{x})$$

for a bounded measurable function h. Hence, $F(h)$ is a bounded functional on $\mathcal{L}^\infty(\mathbb{R})$ because μ is a finite signed measure. We consider two choices for h: First we take the indicator function $\mathbf{1}_{[b,\infty)}$ obtaining

$$
\begin{aligned}
F(\mathbf{1}_{[b,\infty)}) &= \int_{I_n} \mathbf{1}_{[b,\infty)} (l(\mathbf{w}, \mathbf{x})) \, d\mu(\mathbf{x}) \\
&= \mu\left(X_{\mathbf{w},b}^+\right) + \mu\left(X_{\mathbf{w},b}^0\right) \\
&= 0
\end{aligned}
$$

for the functional. Second, we have the indicator function $\mathbf{1}_{(b,\infty)}$ obtaining

$$
\begin{aligned}
F(\mathbf{1}_{(b,\infty)}) &= \int_{I_n} \mathbf{1}_{(b,\infty)} (l(\mathbf{w}, \mathbf{x})) \, d\mu(\mathbf{x}) \\
&= \mu\left(X_{\mathbf{w},b}^+\right) \\
&= 0
\end{aligned}
$$

for the open interval (b, ∞). We can decompose indicator functions h_1 of arbitrary sets into sums of indicator functions of the above types. Due to the linearity of the functional F (linearity of the integral operator) all these integrals vanish and, hence, $F(h_1)$ vanishes for indicator functions. Yet, indicator functions are dense in $\mathcal{L}^\infty(\mathbb{R})$ and, therefore, $F(h) = 0$ for all $h \in \mathcal{L}^\infty(\mathbb{R})$ has to be valid.

In the last step of the proof we consider the functions $h_s(z) = \sin(z)$ and $h_c(z) = \cos(z)$, which are both in $\mathcal{L}^\infty(\mathbb{R})$. We take $z(\mathbf{x}) = l(\mathbf{w}, \mathbf{x})$ and calculate

$$
\begin{aligned}
F(h_c + i \cdot h_s) &= \int_{I_n} h_c(z(\mathbf{x})) + i \cdot h_s(z(\mathbf{x})) \, d\mu(\mathbf{x}) \\
&= \int_{I_n} \exp(i \cdot z(\mathbf{x})) \, d\mu(\mathbf{x})
\end{aligned}
$$

which is the Fourier-transform of the linear functional $l(\mathbf{w}, \mathbf{x})$ with an arbitrary chosen parameter $\mathbf{w}$. However, the Fourier-transform has to be zero in any case which is only possible for $\mu \equiv 0$, which completes the proof. $\square$

Lemma 15. *The ReLU (z) from (3) is discriminatory for $z(\mathbf{x}) = l(\mathbf{w}, \mathbf{x}) + b$, where $l(\mathbf{w}, \mathbf{x})$ is a real-valued linear functional in $\mathbf{x}$ and $\mathbf{w}$.*

Proof. The proof is in complete analogy to the proof for Lemma 4: Because in this proof only the linearity of the inner product was used as the essential property of the inner product, the argumentation remains valid also for linear functionals. $\square$

Now we start to characterize special Banach-spaces such that we can take them for a Cybenko-like theorem. In particular, we have to identify those Banach-spaces which preserve the possibility to apply an appropriate RRT as it was emphasized in Remark 11.

Theorem 16. *Let $\mathcal{B}$ be an uniformly convex Banach space with continuous SIP $[\cdot, \cdot]$. Then a RRT analogously to (8) is valid.*

Proof. The proof can be found in [7, Theorem 6].   □

The theorem can be extended to:

Theorem 17. *Let $\mathcal{B}$ be a reflexive Banach space. Then a RRT analogously to (8) is valid.*

Proof. Let $\mathcal{B}$ be a reflexive Banach space and $h \in \mathcal{B}^* = \mathcal{C}(\mathcal{B})$. Then exists a SIP $[\cdot, \cdot]$ and an element $\beta \in \mathcal{B}$ such that $\varphi(\mathbf{x}) = [\mathbf{x}, \beta]$ is a continuous linear functional [6]. Hence, the respective SIP determines a RRT analogously to (8). □

Both theorems are related according to the following lemma:

Lemma 18. *Every smooth (continuous) uniformly convex Banach space is also reflexive and strictly convex. The reverse direction is not valid. Hence, Theorem 16 is a special case of Theorem 17.*

Proof. The proof can be found in [6].   □

Now we are able to formulate a theorem which states the universal approximation property for perceptron networks consisting of Banach-like perceptrons.

Theorem 19 (Cybenko theorem for Banach-like perceptron networks). *Let σ be a continuous general discriminatory function with respect to the SIP $[\cdot, \cdot]$ for $I_n^{\mathcal{B}} \subset \mathbb{R}_{\mathcal{B}}^n$ equipped with the norm $\|\mathbf{x}\| = \sqrt{[\mathbf{x}, \mathbf{x}]}$ such that $\mathbb{R}_{\mathcal{B}}^n$ is a reflexive n-dimensional real Banach-space. Additionally, let*

$$\Pi_{\mathcal{B}}(\mathbf{x}) = \sum_{j=1}^{N} \alpha_j \cdot \sigma([\mathbf{w}_j, \mathbf{x}] + b_j) \tag{12}$$

be the finite sum of Banach-like perceptrons (9) with activation function $f = \sigma$. Then $\Pi_{\mathcal{B}}(\mathbf{x})$ is an universal approximator.

Proof. The proof is in complete analogy to the proof of the Cybenko-theorem. The application of the Hahn-Banach-theorem is not affected by the weaker assumption regarding the Banach-space. The existence of a respective RRT is guaranteed by the previous lemmata.   □

The most famous examples for (real) Banach-spaces are the spaces L^p and l^p. The latter one is equipped with the unique SIP

$$[\mathbf{w}, \mathbf{x}]_p = \frac{1}{\|\mathbf{x}\|_p^{p-2}} \sum_k w_k \cdot |x_k|^{p-1} \cdot \operatorname{sgn}(x_k) \tag{13}$$

with $1 \leq p < \infty$ [7]. Thus we can equip I_n^B with the SIP $[\mathbf{w}, \mathbf{x}]_p$. Further, the following lemma holds:

Lemma 20. *Both L^p and l^p are uniformly convex for $1 < p < \infty$.*

Proof. The proof can be found in [12]. □

Corollary 21. *The compact set I_n^B with the SIP $[\mathbf{w}, \mathbf{x}]_p$ from (13) is contained in the uniformly reflexive Banach space l_p for $1 < p < \infty$. Hence, a RRT analogously to (8) is valid.*

Proof. Just applying Theorem 17 gives the desired result. □

The last corollary leads to the following statement:

Lemma 22. *A MLP using Banach-like perceptrons with output*

$$O_p(\mathbf{w}, \mathbf{x}) = f\left([\mathbf{w}, \mathbf{x}]_p + b\right) \tag{14}$$

according to (9) generated by the SIP $[\mathbf{w}, \mathbf{x}]_p$ from (13) is an universal approximator in case of $1 < p < \infty$.

Proof. The previous corollary about uniform convexity of the l_p-space together with Lemma 18 guarantees that Theorem 19 is applicable. □

The particular B-perceptron (14) is denoted as B_p-perceptron.

ZHANG & ZHANG considered generalized SIPs (gSIP) [31] extending a first attempt by NATH [21]. They considered SIPs $[\mathbf{w}, \mathbf{x}]_\xi$ for a function $\xi : \mathbb{R}_+ \to \mathbb{R}_+$ fulfilling the requirements 1) and 2) of Definition 12. The Cauchy-Schwarz-inequality is replaced by

$$\left|[\mathbf{w}, \mathbf{x}]_\xi\right| \leq \xi\left([\mathbf{w}, \mathbf{w}]_\xi\right) \cdot \psi\left([\mathbf{x}, \mathbf{x}]_\xi\right)$$

for a conjugate function $\psi : \mathbb{R}_+ \to \mathbb{R}_+$, i.e. $\xi(t) \cdot \psi(t) = t$ has to be valid. According to statement in [31] a RRT is also valid for generalized SIP-spaces: For a RRT regarding those gSIPs it is assumed that $\xi(t)$ is a so-called gauge function, i.e. $\xi(0) = 0$ and $\lim_{t \to \infty} \xi(t) = \infty$. If $\xi(t)$ is surjective onto $\mathbb{R}_+$ and $\zeta(t) = \frac{\xi^{-1}(t)}{t}$ is a gauge function on $\mathbb{R}_+$ then a RRT can be formulated, because the resulting Banach-space is reflexive and strictly convex [31].

3.3   Kernels for Banach-Spaces

In the last step we extend Cybenko's theorem to the case of kernels regarding reproducing kernel Banach spaces (RKBS). As stated in [30, Theorem 4], a RKBS is always reflexive. Thus, we suppose a kernel $\kappa_\mathcal{B}$ corresponding to the kernel feature map $\phi_\mathcal{B} : I_n \rightarrow \mathcal{I}_n \subset \mathcal{B}$ with $\mathcal{B}$ being a RKBS [18]. From Theorem 17 we can conclude that Cybenko's theorem is applicable, accordingly.

4   Numerical Simulations and Conclusions

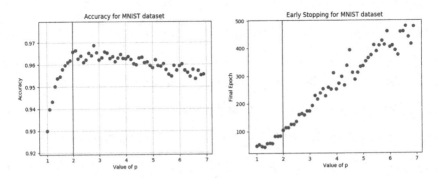

Fig. 2. left: Obtained accuracies of an MLP with B_p-perceptrons for the MNIST data set depending on the p-value for the SIP $[\mathbf{w}, \mathbf{x}]_p$. We observe a broad range of p-values delivering the same good accuracy. **right**: Investigation of the convergence behavior of MLPs with B_p-perceptrons for the MNIST data set depending on the p-value for the SIP $[\mathbf{w}, \mathbf{x}]_p$. A linear correlation between early stopping (number of learning epochs until convergence) and p-value is observable.

In the simulation part we trained MLPs using B-perceptrons with SIP $[\mathbf{w}, \mathbf{x}]_p$ from (13) for the two well-known data sets MNIST and CIFAR10 [16,19]. For the MNIST-problem, the gray-value images were vectorized and taken as an input for an MLP with only one hidden layer consisting of 32 B_p-perceptrons with sigmoid activation. For CIFAR10 we used a convolutional network with four convolutional layers and three max-pooling layers. The final dense layer was performed by 10 B_p-perceptrons with ReLU-activation. The convolutional layers were trained using the dense layer for $p = 2$. After this training, the convolutional layers were kept fix - only the dense layer was trained using different p-values.

Both networks were trained using cross-entropy loss for different p-values for the SIP $[\mathbf{w}, \mathbf{x}]_p$. The MNIST-results are depicted in Fig. 2.

The MLP is always capable to solve the classification problem appropriately. For large and small p-values, numerical instabilities and difficulties lead to a slightly decreased performance.

For the CIFAR10 data set the results are depicted in Fig. 3.

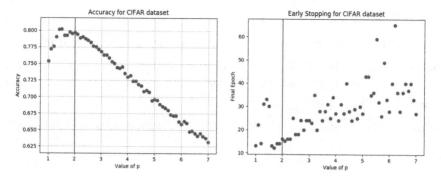

Fig. 3. left: Obtained accuracies of CNN-networks with final dense layers consisting of B_p-perceptrons for the CIFAR10 data set depending on the p-value for the SIP $[\mathbf{w}, \mathbf{x}]_p$. We observe a broad range of p-values delivering the same good accuracy. Particularly, p-values lower than one provide good performance. **right**: Investigation of the convergence behavior of CNN-networks with final dense layer using B_p-perceptrons for the CIFAR10 data set depending on the p-value for the SIP $[\mathbf{w}, \mathbf{x}]_p$. A rough linear correlation between early stopping (number of learning epochs until convergence) and p-value is observable.

Again, we can recognize a overall good performance for a wide range of p-values. The decrease of the performance for higher and very low p-values is again attributed to numerical difficulties. These can be observed also from the early-stopping analysis reflecting the somewhat instable convergence behavior.

In this paper we investigated the approximation completeness of multilayer perceptrons consisting of Banach-like perceptrons. These perceptrons use semi-inner products whereas usual perceptrons rely on the standard Euclidean inner product. Semi-inner products are related to Banach-spaces. We prove mathematically that for semi-inner products determining reflexive Banach-spaces the respective perceptron networks are approximation complete. The proof is valid for discriminatory activation functions which comprise both sigmoid and *ReLU*-functions. Numerical simulations accompany the theoretical considerations.

Future work will deal with indefinite inner products as well as will include the investigation of ResNets. Further, other more promising activation functions like *swish* (see [3,22,29]) should considered as well as networks with bounded width [11].

Appendix

In this appendix we give some useful definitions regarding SIPs and Banach spaces, which are used in the text as well as some basic statements and remarks.

Definition 23. A Banach space $\mathcal{B}$ is denoted as *strictly convex* iff for $\mathbf{x}, \mathbf{y} \neq 0$ with $\|\mathbf{x}\| + \|\mathbf{y}\| = \|\mathbf{x} + \mathbf{y}\|$ we can always conclude that $\mathbf{x} = \lambda \mathbf{y}$ for some $\lambda > 0$.

Lemma 24. *A Banach space $\mathcal{B}$ with SIP $[\cdot, \cdot]$ is strictly convex iff for $\mathbf{x}, \mathbf{y} \neq 0$ with $[\mathbf{x}, \mathbf{y}] = \|\mathbf{x}\| \cdot \|\mathbf{y}\|$ we can always conclude that $\mathbf{x} = \lambda \mathbf{y}$ for some $\lambda > 0$.*

Proof. The proof can be found in [7].   □

The following definition for the uniform convexity was introduced in [4]:

Definition 25. A Banach space $\mathcal{B}$ is denoted as *uniformly convex* iff for each $\varepsilon > 0$ exists a $\delta(\varepsilon) > 0$ such that if $\|\mathbf{x}\| = \|\mathbf{y}\| = 1$ with $\|\mathbf{x} - \mathbf{y}\| > \varepsilon$ then $\frac{\|(\mathbf{x}+\mathbf{y})\|}{2} < 1 - \delta(\varepsilon)$ is valid.

Definition 26. A Banach space $\mathcal{B}$ with SIP $[\cdot, \cdot]$ is denoted as *continuous* iff

$$\Re\{[\mathbf{x}, \mathbf{y} + \lambda\mathbf{x}]\} \xrightarrow[\lambda \to 0]{} \Re\{[\mathbf{x}, \mathbf{y}]\}$$

is valid for $\lambda \in \mathbb{R}$. The space is *uniformly continuous* iff this limit is approached uniformly.

Definition 27. A Banach space $\mathcal{B}$ is denoted as *reflexive* iff the mapping $J : \mathcal{B} \to \mathcal{B}^{**} = (\mathcal{B}^*)^*$ is surjective, where the star indicates the dual space.

Theorem 28. *Let $\mathcal{B}$ be a Banach space. Then a necessary and sufficient condition for $\mathcal{B}$ to be reflexive is that for every $f \in \mathcal{B}^*$ exists an SIP $[\cdot, \cdot]$ and an element $\mathbf{y} \in \mathcal{B}$ with $f(\mathbf{x}) = [\mathbf{x}, \mathbf{y}]$ for all $\mathbf{x} \in \mathcal{B}$. If $\mathcal{B}$ is strictly convex then $\mathbf{y}$ is unique.*

Proof. The proof can be found in [6, Theorem 2].   □

Definition 29. A Banach space $\mathcal{B}$ is denoted as *smooth* iff for each $\mathbf{x} \in \mathcal{B}$ with $\|\mathbf{x}\| = 1$ there exists a linear functional $f_{\mathbf{x}} \in \mathcal{B}^*$ with $f_{\mathbf{x}}(\mathbf{x}) = \|f_{\mathbf{x}}\|$. The existence of $f_{\mathbf{x}}$ is guaranteed by the Hahn-Banach-Theorem.

References

1. Bishop, C.: Pattern Recognition and Machine Learning. Springer, London (2006)
2. Braun, J., Griebel, M.: On a constructive proof of Kolmogorov's superposition theorem. Constr. Approx. **30**, 653–675 (2009). https://doi.org/10.1007/s00365-009-9054-2
3. Chieng, H., Wahid, N., Pauline, O., Perla, S.: Flatten-T Swish: a thresholded ReLU-Swish-like activation function for deep learning. Int. J. Adv. Intell. Inform. **4**(2), 76–86 (2018)
4. Clarkson, J.: Uniformly convex spaces. Trans. Am. Math. Soc. **40**, 396–414 (1936)
5. Cybenko, G.: Approximations by superpositions of a sigmoidal function. Math. Control Sig. Syst. **2**(4), 303–314 (1989). https://doi.org/10.1007/BF02551274
6. Faulkner, G.D.: Representation of linear functionals in a Banach space. Rocky Mt. J. Math. **7**(4), 789–792 (1977)
7. Giles, J.: Classes of semi-inner-product spaces. Trans. Am. Math. Soc. **129**, 436–446 (1967)
8. Goodfellow, I., Bengio, Y., Courville, A.: Deep Learning. MIT Press, Cambridge (2016)
9. Gorban, A.: Approximation of continuous functions of several variables by an arbitrary nonlinear continuous function of one variable, linear functions, and their superpositions. Appl. Math. Lett. **11**(3), 45–49 (1998)

10. Guilhoto, L.: An overview of artificial neural networks for mathematicians (2018). http://math.uchicago.edu/~may/REU2018/REUPapers/Guilhoto.pdf
11. Hanin, B.: Universal function approximation by deep neural networks with bounded width and ReLU activations. Mathematics **7**(992), 1–9 (2019)
12. Hanner, O.: On the uniform convexity of L^p and l^p. Ark. Mat. **3**(19), 239–244 (1956)
13. Hertz, J.A., Krogh, A., Palmer, R.G.: Introduction to the Theory of Neural Computation, Volume 1 of Santa Fe Institute Studies in the Sciences of Complexity: Lecture Notes. Addison-Wesley, Redwood City (1991)
14. Kolmogorov, A.: On the representation of continuous functions of several variables as superpositions of continuous functions of one variable and addition. Doklady Academ Nauk SSSR **114**(5), 953–956 (1957)
15. Kolmogorov, A., Fomin, S.: Reelle Funktionen und Funktionalanalysis. VEB Deutscher Verlag der Wissenschaften, Berlin (1975)
16. Krizhevsky, A., Sutskever, I., Hinton, G.: ImageNet classification with deep convolutional neural networks. In: Advances in Neural Information Processing Systems (NIPS), San Diego, vol. 25, pp. 1097–1105. Curran Associates Inc. (2012)
17. Kůrková, V.: Kolmogorov's theorem and multilayer neural networks. Neural Netw. **5**, 501–506 (1992)
18. Lange, M., Biehl, M., Villmann, T.: Non-Euclidean principal component analysis by Hebbian learning. Neurocomputing **147**, 107–119 (2015)
19. LeCun, Y., Cortes, C., Burges, C.: The MNIST database (1998)
20. Lumer, G.: Semi-inner-product spaces. Trans. Am. Math. Soc. **100**, 29–43 (1961)
21. Nath, B.: Topologies on generalized semi-inner product spaces. Composito Mathematica **23**(3), 309–316 (1971)
22. Ramachandran, P., Zoph, B., Le, Q.: Searching for activation functions. Technical report, Google Brain (2018). arXiv:1710.05941v1
23. Riesz, F., Nagy, B.Sz.: Vorlesungen über Functionalanalysis, 4th edn. Verlag Harri Deutsch, Frankfurt/M. (1982)
24. Rosenblatt, F.: The perceptron: a probabilistic model for information storage and organization in the brain. Psychol. Rev. **65**, 386–408 (1958)
25. Rudin, W.: Functional Analysis, 2nd edn. MacGraw-Hill Inc., New York (1991)
26. Steinwart, I., Christmann, A.: Support Vector Machines. Information Science and Statistics, Springer, Heidelberg (2008). https://doi.org/10.1007/978-0-387-77242-4
27. Triebel, H.: Analysis und mathematische Physik, 3rd revised edn. BSB B.G. Teubner Verlagsgesellschaft, Leipzig (1989)
28. Villmann, T., Haase, S., Kaden, M.: Kernelized vector quantization in gradient-descent learning. Neurocomputing **147**, 83–95 (2015)
29. Villmann, T., Ravichandran, J., Villmann, A., Nebel, D., Kaden, M.: Investigation of activation functions for generalized learning vector quantization. In: Vellido, A., Gibert, K., Angulo, C., Martín Guerrero, J.D. (eds.) WSOM 2019. AISC, vol. 976, pp. 179–188. Springer, Cham (2020). https://doi.org/10.1007/978-3-030-19642-4_18
30. Zhang, H., Xu, Y., Zhang, J.: Reproducing kernel banach spaces for machine learning. J. Mach. Learn. Res. **10**, 2741–2775 (2009)
31. Zhang, H., Zhang, J.: Generalized semi-inner products with applications to regularized learning. J. Math. Anal. Appl. **372**, 181–196 (2010)

Fuzzy Systems and Their Applications

Similarity Fuzzy Semantic Networks and Inference. An Application to Analysis of Radical Discourse in Twitter

Juan Luis Castro and Manuel Francisco(✉)

Department of Computer Science and Artificial Intelligence, University of Granada,
18071 Granada, Spain
{castro,francisco}@decsai.ugr.es
https://decsai.ugr.es/

Abstract. In this paper we introduce a new Knowledge Representation model, the Similarity Fuzzy Semantic Networks. It is an extension of Fuzzy Semantic Networks that incorporates *reasoning by similarity* through a Similarity Inference Rule. Moreover, we show as it can be effectively applied to a trending and complex problem like the analysis of radical discourse in Twitter.

Keywords: Fuzzy Semantic Networks · Similarity fuzzy reasoning · Social network analysis · Knowledge engineering · Semantic network

1 Introduction and Motivation

Semantic Networks are one of the first models proposed for Knowledge Representation, and they have been effectively applied over the years [5,10]. Later, graduations were introduced to obtain Fuzzy Semantic Networks, that have interesting and relevant applications [1,3]. Moreover, it is an effective approach to use reasoning by similarity in fuzzy systems [9]. Thus, it would be interesting to extend the Fuzzy Semantic Network model to include similarity reasoning.

In this paper, we propose a new model of knowledge representation which extend Fuzzy Semantic Network model, and incorporate an *inference by similarity* rule.

The rest of this paper is organised as follows. Sections 2 and 3 present a brief introduction to Semantic Networks and Fuzzy Semantic Networks models, respectively. Section 4 proposes our Similarity Fuzzy Semantic Network model, jointly with the similarity inference rule. Section 5 shows an inference strategy for an effective application of the model. Lastly, Sect. 6 applies it to a trending and complex problem: the analysis of radical discourse in social networks.

This work was financially supported by Junta de Andalucia, projects P18-FR-5020 and A-HUM-250-UGR18, and cofinanced by the European Social Fund (ESF). Manuel Francisco Aparicio was supported by the FPI 2017 predoctoral programme, from the Spanish Ministry of Economy and Competitiveness (MINECO), grant reference BES-2017-081202.

L. Rutkowski et al. (Eds.): ICAISC 2022, LNAI 13588, pp. 173–181, 2023.
https://doi.org/10.1007/978-3-031-23492-7_15

2 Semantic Networks

Semantic Networks represent knowledge with directed labelled graphs, where vertices represent concepts, which can be individuals or classes (sets of individuals), and labelled edges represent semantic relations between concepts, such that:

$$A \xrightarrow{relationS} B \tag{1}$$

represents the assertion " **A** *relationS* **B** ". Consequently, we can represent knowledge as "Bird has-part Wings", "Animal has-part legs" or "Bird is-an Animal".

We can distinguish between two types of semantic relations:

– Hierarchical semantic relations:
  • *instance-of* (an individual is an instance of a class)
  • *is-a* (a class is a subclass of another class)
– Domain-specific semantic relations, such as *is-an-opponent-of, owns...*

Hierarchical semantic relations are universal, in the sense that they are present in any semantic network, meanwhile each semantic network introduces its own domain-specific relations.

The main inference rule in a Semantic Network is *inference by inheritance*. It consists on deducing new assertions in accordance with the following scheme:

$$\frac{\begin{array}{l} \mathbf{A} \;\; is\text{-}a \;\; \mathbf{B} \;\; \vee \mathbf{A} \;\; instance\text{-}of \;\; \mathbf{B} \\ \mathbf{B} \;\; relationS \;\; \mathbf{C} \end{array}}{\mathbf{A} \;\; relationS \;\; \mathbf{C}} \tag{2}$$

3 Fuzzy Semantic Networks

It has been proposed to use graduations to obtain Fuzzy Semantic Networks [1,3]. These models represent knowledge as graded labelled directed graphs. Classes are now defined as fuzzy sets of individuals, and the degree of the relation *instance-of* is the membership function of the correspondent fuzzy set. Analogously, edges represent graded semantic relations:

– *instance-of*: α stands for an instance with grade α
– *is-a* : α stands for a class that inherits from other in grade α
– Domain-specific fuzzy semantic relations, such that each relation has a an associated degree in which the assertion meets.

In this way,

$$A \xrightarrow{relationS:\alpha} B \tag{3}$$

represents the fuzzy assertion

$$A \; relationS \; B \text{ in } \alpha \text{ degree.} \tag{4}$$

that can be abbreviated as

$$\mathbf{A}\ \textit{relationS:}\alpha\ \mathbf{B} \tag{5}$$

We can now define the *fuzzy inference by inheritance rule*. It consists on deducing new fuzzy assertions by the following scheme:

$$\frac{\mathbf{A}\ \textit{is-a:}\alpha\ \mathbf{B} \vee \mathbf{A}\ \textit{instance-of:}\alpha\ \mathbf{B}}{\mathbf{B}\ \textit{relationS:}\beta\ \mathbf{C}} \tag{6}$$
$$\mathbf{A}\ \textit{relationS:}t(\alpha,\beta)\ \mathbf{C}$$

being t a t-norm chosen to model the connective "and".

Obviously, the fuzzy inference by inheritance is a generalisation of the (non fuzzy) inference by inheritance: if we have crisp semantic relations in the premises ($\alpha = \beta = 1$), then we obtain the same crisp consequence ($t(1,1) = 1$).

3.1   Combining Inferences

After applying fuzzy inference by inheritance (or any other reasoning method), it is possible to obtain the same semantic relation between two given concepts but with different degrees. We can use an aggregation function [2] to combine both assertions in the following *combining inference rule*:

$$\frac{\mathbf{A}\ \text{relationS:}\alpha\ \mathbf{B}}{\mathbf{A}\ \text{relationS:}\beta\ \mathbf{B}} \tag{7}$$
$$\mathbf{A}\ \text{relationS:}g(\alpha,\beta)\ \mathbf{B}$$

were g is a previously chosen aggregation function.

4   Similarity Fuzzy Semantic Networks

In the same way that classes extend its semantic relations to its sub-classes and instances by inheritance, individual or classes may transmit properties, by similarity semantic relations, to similar individual or classes [6,7]. For example, if two persons have similar opinions about political topics, then it will be reasonable to think that the properties with political sense would affect one another.

In order to enrich the model of fuzzy semantic relation with this idea, we propose a new model for knowledge representation that we call *Similarity Fuzzy Semantic Networks*. It consist on fuzzy semantic networks with a specific family of semantic relations between classes or individuals, which we call *Similarity semantic relations*.

4.1   Similarity Semantic Relations

Similarity semantic relations are fuzzy semantic relations that represent that two individuals or two classes are similar in some sense or aspect:

$$\mathbf{A}\ \textit{is-similar-in-sense-D} : \alpha\ \mathbf{B}, \tag{8}$$

where D may be any topic or aspect, and it represents the assertion that concepts A and B are similar in the sense D in α degree.

We can have similarity relations between classes and also between individuals. Additionally, for every sense D, each concept will have a fuzzy neighbourhood of similar concepts in sense D.

On the other hand, we only might transmit by *similarity-in-sense-D* those semantic relations that are related to D. Thus, we introduce relations between *senses* and *semantic relations of the network*.

4.2   Meta-relations

Semantic relations of the network can be considered *second order concepts*, therefore it is possible to think in *second order relations* where relations between *semantic relations of the network* are established. We call them *meta-relations*.

Particularly, we introduce in our Similarity Fuzzy Semantic Networks model a meta-relation that will be used for the Similarity inference. It is a relation that goes from domain-specific semantic relations to *is-similar-in-sense-D* relations:

$$\textbf{relationS} \ \textit{is-related-to:}\gamma \ \textbf{senseD}, \tag{9}$$

representing the assertion that *relationS* is related to *senseD* and thus, it can be transmitted by *is-similar-in-sense-D*.

The similarity semantic relation specifies a correspondence between concepts in an specific aspect D, meanwhile *is-related-to* delimits the domain in which similarity relations apply. In fact, when using meta-relations, we are defining a new semantic network of a higher level in which concepts are similarity relations of the principal semantic network.

4.3   Similarity Inference

These new relations enable a new kind of reasoning based on similarity. New knowledge may be extracted upon propagation of semantic relations through the *is-similar-in-sense-D* by means of the *Similarity Inference Rule*:

$$\frac{\begin{array}{l} \textbf{A} \ \textit{is-similar-in-sense-D} : \alpha \ \textbf{B} \\ \textbf{B} \ \textit{relationS:}\beta \ \textbf{C} \\ \textbf{relationS} \ \textit{is-related-to:}\gamma \ \textbf{senseD} \end{array}}{\textbf{A} \ \textit{relationS:}(\gamma * t(\alpha, \beta)) \ \textbf{C}} \tag{10}$$

where t is a triangular norm (t-norm).

5   Inference Strategy

In the proposed similarity fuzzy semantic network, the properties of the concepts may be deduced by fuzzy inheritance and/or by similarity inference. Moreover, each reasoning process results in new knowledge that may lead to new inferences. Hence, we might to establish an inference strategy.

First of all, we may choose the prevalence between inheritance and similarity inference rules. Inheritance is a *depth reasoning*, while similarity can be considered a *breadth inference*, since it is based on the neighbourhood of similar concepts. Therefore, we can use the classical Z and N models of reasoning strategy:

- Z model: first similarity, then inheritance.
- N model: first inheritance, then similarity.

Lastly, we establish iterations or cycles, as it is usual when dealing with these kind of systems. In each step, we update the degree of every semantic relation by applying inheritance and similarity reasoning rules in the chosen order, and then applying the combining inference rule.

6   Application to Radical Discourse in Twitter

There are several cases in which it may be interesting to infer knowledge using similarity fuzzy semantic networks. In this case, we applied it to represent and infer new knowledge about radical discourse in Twitter.

Radical propaganda is disseminated through Social Networking Sites (SNS) such as Twitter, blogs and other platforms [8,12]. Recruitment and radicalisation of SNS users is due to diverse factors which radicals take advantage of [11]. Identifying these radical accounts and others that are susceptible of being radicalised are important tasks in order to deal with extremism.

We used Twitter API to obtain tweets about some specific topics that are frequently found in the radical discourse. The challenge that we are facing is to detect radical users in the social network, and its main handicap is that most of the users and tweets are not radical in any form.

Given a *twitter user* U, we consider a domain-specific fuzzy semantic relation *is-radical* to represent whether a user is radical or not:

$$U \xrightarrow{\text{is-radical:}\alpha} Yes \tag{11}$$

$$U \xrightarrow{\text{is-radical:}\beta} No \tag{12}$$

being α and β the degrees in which U is radical or not, respectively.

When two users, A and B, share opinions regarding the selected topics, we represent it by the similarity semantic relation *is-similar-in-sense-opinion-share*:

$$\mathbf{A} \; \textit{is-similar-in-sense-opinion-share} : w \; \mathbf{B} \tag{13}$$

where w stands for the degree in which they share opinions. This enables us to propagate knowledge from user A to B and vice versa. However, we still need to determine a way in which properties defined by the semantic fuzzy relation

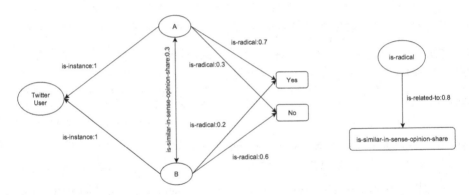

Fig. 1. Graphical representation of a similarity fuzzy semantic network for the radical discourse in Twitter.

is-radical can be propagated using the similarity relation. Let us define a meta-relation such that

$$\textbf{is-radical } \textit{is-related-to} : \gamma \textbf{ is-similar-in-sense-opinion-share} \qquad (14)$$

Figure 1 shows the graphic representation of the fuzzy semantic network.

Let us exemplify the results of the inference in this similarity fuzzy semantic network. We use the product t-norm and the sum aggregation.

First, we apply the similarity inference rule for every pair of similar users (users that share opinions about the selected topics):

$$
\begin{array}{l}
\textbf{A} \text{ is-similar-in-sense-opinion-share:}w \textbf{ B} \\
\textbf{B} \text{ is-radical:}p \textbf{ Yes} \\
\textbf{is-radical } \text{is-related-to:}\gamma \textbf{ is-similar-in-sense-opinion-share} \\
\hline
\textbf{A} \text{ is-radical:}\gamma * w * p \textbf{ Yes}
\end{array} \qquad (15)
$$

Then, using the combination inference rule, we obtain the degree in which every twitter user *is radical*:

$$
\begin{array}{l}
\textbf{A} \text{ is-radical:}p_1 \textbf{ Yes} \\
\textbf{A} \text{ is-radical:}p_2 \textbf{ Yes} \\
\hline
\textbf{A} \text{ is-radical:}p_1 + p_2 \textbf{ Yes}
\end{array} \qquad (16)
$$

For each cycle, the similarity inference is fired for every similar user to A, and since summation is an associative operator, the order in the combination inference rule is not relevant. Thus, we may conclude that, when the cycle i ends, it is possible to deduct that:

$$\textbf{A} \text{ is-radical:} \left(radical^{(i)}(A) \right) \textbf{ Yes} \qquad (17)$$

being

$$radical^{(i)}(A) = radical^{(i-1)}(A) + \gamma * \sum_{U | w_u \in neighbours(A)} w_u * radical^{(i-1)}(u) \quad (18)$$

where $neighbours(A)$ is the fuzzy set of twitter users similar to A in the sense that they share opinions about the selected topics:

$$neighbours(A) = \{U | w_u : \textbf{A} \text{ is-similar-in-sense-opinion-share} : w_u \textbf{ U}\} \quad (19)$$

and being

$$radical^{(0)}(A) = \alpha \quad (20)$$

where α is the initial degree (if any) for **A** is-radical : α **Yes**.

6.1   Determining Degrees for the Fuzzy Relations

The similarity inference process is conditioned to the initial degrees of at least one of the instances of Twitter users. Determining these values is not a trivial task, but it can be done in several manners.

is-radical is defined for a particular user and, initially, it can be calculated taking into account only the information available for such user (in this case, their tweets). It is possible to use an *oracle* that, given a tweet, returns a binary answer (yes or no) to the question *"is this tweet radical?"*. In our case, we used a human expert as an oracle, which answer this question for some tweets. The initial degree p would be the result of the aggregation of the answers. We used the *mean*, that result in the ratio between user's radical tweets an the total number of them.

is-similar-to-in-sense-opinion-share is defined between two users and it needs to be determined taking into account the information available for both of them. We used a predictor H using Twitter mechanics as proposed in [4]:

$$\forall u, v \in T, H(u, v) = cocopies(u, v) + cofavourites(u, v) + \|\{m : m \in M \\ \wedge author(u, m) \wedge \exists n \in M : [author(v, n) \\ \wedge (copy(m, n) \vee favourite(u, n))]\}\| \quad (21)$$

where:

- M is the set of all the tweets.
- $cocopies(u, v)$ stands for the number of retweets that both users have in common (which can be translated to the number of tweets that both users agree with).
- $cofavourites(u, v)$ stands for the number of favourites that both users have in common (analogously to *cocopies*).

Table 1. Results of the expert evaluation of the deductions made by the model. 3537 of the 4114 deductions were accepted, that yields 85.97% of accuracy.

Accepted deductions	3537
Rejected deductions	549
Undetermined deductions	28
Total deductions	4114

- *author*(*u, m*) checks if the tweet *m* belongs to the user *u*.
- *copy*(*m, n*) checks if the tweet *n* is a retweet of *m*.
- *favourite*(*u, n*) checks if the tweet *n* is marked as favourite by *u*.

After applying normalisation to H, we obtain a degree in which both users share opinions.

is-related-to is a context-dependant degree that should be decided after an analysis of the specific problem. It may be defined using statistical measures such as percentiles or centrality measures.

6.2   Real-World Experiments

We effectively conducted real-world experiments with a dataset that involve more than 430000 tweets authored by more than 30000 users using a human oracle to establish initial degrees for 778 tweets. Later, since our model implements an approximate reasoning, we evaluated the result of the inference process with the help of human experts to check for the soundness of these conclusions, and we obtained good results as shown in Table 1. We obtained 85.97% of accuracy, which is a better result than a baseline non-deductive model such as Support Vector Machines (SVM). Particularly, we trained a SVM model over the same dataset and we obtained a 68.97% accuracy in a cross-validation scheme.

7   Conclusions

Semantic Networks are widely used to represent Knowledge, and they have been specialised to Fuzzy Semantic Networks with useful applications. Throughout this paper, we extended these to provide them with similarity reasoning. In order to do so, we introduced a new family of semantic relations and a higher order meta-relation that allows to develop an *inference by similarity* rule, along with an inference strategy. We also showed how it can be applied to radical discourse in Twitter and how knowledge is inferred in a practical manner. This example illustrates that our proposal can be applied to complex problems and that it has great potential. We obtain effective and sound results that shows that deductions are precise in 85.97% of the cases, that is better than a baseline non-deductive machine learning model.

We intend to pursue further research in the future, both at a theoretically and at application level. In particular, we want to explore dissimilarities as a manner to complement similarity measures in order to better determine fuzzy memberships.

References

1. Alhiyafi, J., Atta-ur-Rahman, Alhaidari, F.A., Khan, M.A.: Automatic text categorization using fuzzy semantic network. In: Benavente-Peces, C., Slama, S., Zafar, B. (eds.) SEAHF 2019. SIST, vol. 150, pp. 24–34. Springer, Cham (2019). https://doi.org/10.1007/978-3-030-22964-1_3

2. Dujmović, J., Torra, V.: Aggregation functions in decision engineering: ten necessary properties and parameter-directedness. In: Kahraman, C., Cebi, S., Cevik Onar, S., Oztaysi, B., Tolga, A.C., Sari, I.U. (eds.) INFUS 2021. LNNS, vol. 307, pp. 173–181. Springer, Cham (2022). https://doi.org/10.1007/978-3-030-85626-7_21

3. Flores, D.L., Rodríguez-Díaz, A., Castro, J.R., Gaxiola, C.: TA-fuzzy semantic networks for interaction representation in social simulation. In: Castillo, O., Pedrycz, W., Kacprzyk, J. (eds.) Evolutionary Design of Intelligent Systems in Modeling, Simulation and Control. SCI, vol. 257, pp. 213–225. Studies in Computational Intelligence, Springer, Berlin, Heidelberg (2009). https://doi.org/10.1007/978-3-642-04514-1_12

4. Francisco, M., Castro, J.L.: A fuzzy model to enhance user profiles in microblogging sites using deep relations. Fuzzy Sets Syst. **401**, 133–149 (2020). https://doi.org/10.1016/j.fss.2020.05.006. https://www.sciencedirect.com/science/article/pii/S0165011419301782

5. Guo, L., Yan, F., Li, T., Yang, T., Lu, Y.: An automatic method for constructing machining process knowledge base from knowledge graph. Robot. Comput.-Integr. Manuf. **73**, 102222 (2022). https://doi.org/10.1016/j.rcim.2021.102222. https://www.sciencedirect.com/science/article/pii/S0736584521001058

6. Klawonn, F., Castro, J.L.: Similarity in fuzzy reasoning. Mathware Soft Comput. **3**, 197–228 (1995)

7. Luo, M., Zhao, R.: Fuzzy reasoning algorithms based on similarity. J. Intell. Fuzzy Syst. **34**, 213–219 (2018). https://doi.org/10.3233/JIFS-171140

8. Nouh, M., Jason Nurse, R., Goldsmith, M.: Understanding the radical mind: identifying signals to detect extremist content on Twitter, pp. 98–103 (2019). https://doi.org/10.1109/ISI.2019.8823548

9. Omri, M.N., Chouigui, N.: Measure of similarity between fuzzy concepts for identification of fuzzy user's requests in fuzzy semantic networks. Int. J. Uncertain. Fuzziness Knowl.-Based Syst. **9**, 743–748 (2001). https://doi.org/10.1016/S0218-4885(01)00119-8

10. Rahman, A.: Knowledge representation: a semantic network approach (2016). https://doi.org/10.4018/978-1-5225-0427-6

11. Spiller, P.Y.: Psicología y terrorismo: el terrorismo suicida. Estudio de variables que inciden en su aparición y desarrollo. Thesis, Universidad de Belgrano, Facultad de Humanidades (2005). http://repositorio.ub.edu.ar/handle/123456789/225. Accepted 23 July 2011

12. Ul Rehman, Z., et al.: Understanding the language of ISIS: an empirical approach to detect radical content on Twitter using machine learning. Comput. Mater. Continua **66**(2), 1075–1090 (2020). https://doi.org/10.32604/cmc.2020.012770

An Application of Information Granules to Detect Anomalies in COVID-19 Reports

Adam Kiersztyn[1]($\boxtimes$) ⓘ, Krystyna Kiersztyn[2] ⓘ, Rafał Łopucki[3] ⓘ,
and Patrycja Jedrzejewska-Rzezak[4] ⓘ

[1] Department of Computer Science, Lublin University of Technology, Lublin, Poland
adam.kiersztyn.pl@gmail.com, a.kiersztyn@pollub.pl
[2] Department of Mathematical Modelling, The John Paul II Catholic University
of Lublin, Lublin, Poland
krystyna.kiersztyn@gmail.com
[3] Centre for Interdisciplinary Research, The John Paul II Catholic University
of Lublin, Lublin, Poland
lopucki@kul.pl
[4] Department of Probability Theory and Statistics, The John Paul II Catholic
University of Lublin, Lublin, Poland
patrycja.jedrzejewska-rzezak@kul.pl

Abstract. The COVID-19 pandemic has affected almost every aspect of life. The patterns of interpersonal contacts, the ways of doing business and the methods of school education have changed. A significant part of worldwide business has migrated to the virtual world, and the global supply chains have been disrupted. On the other hand, this new situation created opportunities for a much faster development of some areas of business and science. For example, the observation and analysis of pandemic data has contributed to the development of new techniques for effective mathematical forecasting. It is worth noting that during a pandemic most political and economic decisions are based on official data on the number of new infections at the country level. Therefore, the quality of this data is very important for making difficult decisions, such as implementing new restrictions. In this study, we will focus on the problem of pandemic data quality and present a novel anomaly detection method based on information granules. In numerical experiments, data from several European countries were compared. The selection of data for analysis was based on the following information: the movement of people between countries, similar quality of medical care and the sanitary standards. An appropriate adaptation of the author's anomaly detection method based on information granules allowed to identify potential anomalies in daily COVID reports.

Keywords: Information granules · Granular computing · Anomaly detection · COVID-19 · Pandemic

The work was co-financed by the Lublin University of Technology Scientific Fund: FD-20/IT-3/002.

1   Introduction

To monitor and prevent the spread of disease, various public health surveillance systems are implemented at the national level [9]. Nowadays, these systems have found unprecedented use in monitoring the rapid spread of the SARS-CoV-2 coronavirus. Globally, confirmed cases of infection with this virus amount to hundreds of millions and nearly six million people have died [2]. In addition to health problems, the coronavirus pandemic has caused many and diverse social, economic and environmental consequences [8,17,18,20,21].

To limit the transmission of the virus, national governments have been forced to take a series of radical and unpopular decisions, such as outdoor and indoor masking orders, mobility restrictions or lockdowns [6]. Epidemic decisions were made under strong social pressure, and taking or not taking certain actions resulted in the death of thousands of people, a serious disturbance in business, or the risk of collapse of the national health care systems [7].

Crisis management decisions were usually made on the basis of the national system of data collection, modelling and prediction of the future course of the epidemic [9]. Meanwhile, due to the unprecedented scale and nature of the SARS-CoV-2 epidemic, the epidemiological data collection and reporting system was error prone [10]. Basic pandemic data collected at different levels by public health surveillance systems were usually: number of coronavirus tests performed, number of confirmed cases, number of deaths, number of convalescent patients, and number of hospitalisations. However, the number of uncontrolled variables affecting these indicators was so large that modelling and predicting the course of the epidemic turned out to be very difficult and in most cases ineffective [5,6,11,24]. At the same time, in outbreak forecasting, extreme errors can lead to suboptimal decision-making, such as unexpected shortages in or oversupply of resources [19].

Since single models used to describe or predict the development of an epidemic have often turned out to be ineffective, major public health government units, such as the Centre for Disease Control and Prevention (CDC), adopted ensemble forecasts for modelling and forecasting epidemic development [1]. This approach proved to be sufficiently accurate and precise at short-term prediction horizons, with a general increase in error at longer horizons [19]. However, the precision of ensemble forecasts and their prediction horizons depends on the precision of the individual models that are aggregated. Thus, the most accurate single models and methods of their optimal aggregation are still sought in order to better understand and predict the epidemic at the national and global level [3–6,22,23].

In this study, we propose modelling the course of COVID-19 epidemic using a novel approach based on information granules. This new method of fuzzy analysis has already been described and applied for various data [12,15,16]. In this paper, we focus on data from 37 European countries, because it is the continent where the highest number of confirmed cases of coronavirus infections in the world has been recorded (over 160 million) [2]. However, the novel method we present is universally applicable.

The work is organised as follows. Section 2 provides a theoretical description of the proposed anomaly detection method based on information granules. In the next Sect. 3, the results of numerical experiments are presented. Finally, Sect. 4 contains conclusions and future work directions.

2 Methodology

Data describing the dynamics of the COVID-19 pandemic are collected by many institutions and made public. This study will use data from the Our World in Data website (https://ourworldindata.org/coronavirus-source-data). The structure of the data allows the observation of many phenomena related to the pandemic in the following days, broken down into individual regions of the world and specific countries. These data are organised in such a way that it is possible to identify a number of phenomena, including: date; total_cases; new_cases; new_cases_smoothed (7-day smoothed); total_deaths – total deaths attributed to COVID-19; new_deaths – new deaths attributed to COVID-19; new_deaths_smoothed (7-day smoothed); total_cases_per_million – total confirmed cases of COVID-19 per 1,000,000 people; new_cases_per_million – new confirmed cases of COVID-19 per 1,000,000 people; total_deaths_per_million – total deaths attributed to COVID-19 per 1,000,000 people; new_deaths_per_million – new deaths attributed to COVID-19 per 1,000,000 people; icu_patients; hosp_patients; new_tests.

From the theoretical point of view, these data can be equated with time series describing particular phenomena for individual countries. As part of this study, interesting relationships between the individual ranks will be examined. In addition, in-depth analyses within individual time series will be carried out. Contrary to the classic approaches that are presented in the media, the analyses will be carried out using fuzzy techniques, in particular with the use of information granules.

At the outset, it is necessary to get acquainted with the basic properties of individual time series. If not necessary, considerations will be made on a general case. Let us denote by $X[n]$, $n \geq 1$ the time series describing one of the above-mentioned phenomena changing with time. It is obvious that, due to the dynamics of the pandemic itself, the series will show some seasonal fluctuations that can be identified with successive waves of the pandemic. Additionally, due to technical conditions, in many cases weekly fluctuations can be noted. Underestimated values are most often reported during the weekend and on Mondays. The largest increases in individual phenomena are recorded on Tuesdays and Wednesdays. Therefore, it is advisable to use smoothing within a week. You should compare smoothed values or data on a weekly basis.

Within individual time series, an important element is the study of the direction and pace of growth. The use of simple relative increment or absolute increment is unreliable. It is because the weekly fluctuations mentioned above should be taken into account. Data smoothing is based on the application of the transformation given by the formula

$$\tilde{X}[n] = \frac{1}{2k+1} \sum_{i=-k}^{k} X[n+k] \tag{1}$$

Typically, all analyses present data taking into account weekly smoothing with the parameter $k = 3$. This approach is absolutely justified, but other levels of smoothing can also be considered.

In traditional mainstream approaches, increases of a certain phenomenon are said to be when the values of the smoothed time series are increasing, or when there is a relation

$$X[n] - X[n-7] > 0. \tag{2}$$

Both approaches are correct, however, they are not free from fluctuations caused by various factors. It seems reasonable to consider certain states of the analysed phenomena, to which the membership function to this class should be properly defined. For the purposes of this study, we distinguish the following 7 classes of states of the phenomenon:

- series $X[n]$ increases strongly
- series $X[n]$ increases
- series $X[n]$ increases slightly
- series $X[n]$ is stable
- series $X[n]$ decreases slightly
- series $X[n]$ decreases
- series $X[n]$ decreases strongly.

Of course, for analytical and predictive reasons, it is not advisable to use the values of the $X[n]$ series for $n > n_0$ to describe the series at time n_0. However, one should be aware of weekly fluctuations and it is not reasonable to use unsmoothed values. Let us apply a slightly different smoothing given by the formula

$$\bar{\bar{X}}[n] = \frac{1}{k+1} \sum_{i=0}^{k} X[n-k] \tag{3}$$

Then, for such a modified time series, statistics describing the location and dispersion should be determined. For this purpose, the classical mean and standard deviation of the last N days given by the formulas can be used

$$\bar{X} = \frac{1}{N} \sum_{k=0}^{N-1} \bar{\bar{X}}[n-k] \tag{4}$$

$$\bar{S} = \sqrt{\frac{1}{N} \sum_{k=0}^{N-1} (\bar{\bar{X}}[n-k] - \bar{X})^2} \tag{5}$$

Instead of arithmetic mean and standard deviation, median and quadrant deviations can be used, which by their nature are less sensitive to outliers. Nevertheless, the use of series smoothing protects us against the occurrence of strong fluctuations.

Due to the high intuitiveness and ease of interpretation, the use of the trapezoidal membership function was proposed to describe the degree of membership to particular states. For example, "stable" state membership function is given by the formula

$$\mu(X[n], \text{stable}) = \begin{cases} \frac{X[n]-(\bar{X}-0.5S)}{0.25S}, & X[n] \in [\bar{X}-0.5S; \bar{X}-0.25S) \\ 1, & X[n] \in [\bar{X}-0.25S; \bar{X}+0.25S] \\ \frac{\bar{X}+0.5S-X[n]}{0.25S}, & X[n] \in (\bar{X}+0.25S; \bar{X}+0.5S) \\ 0, & X[n] \notin [\bar{X}-0.5S; \bar{X}+0.5S] \end{cases} \quad (6)$$

For the remaining states, membership functions are defined analogously. The transformation of any value of the $X[n]$ series is presented in Fig. 1.

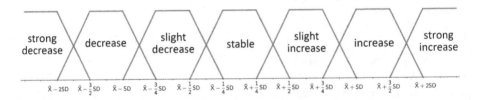

Fig. 1. Membership functions of descriptors describing the time series state

After applying such a transformation for M different time series describing various phenomena within one country, we obtain an information granule that can be identified with an element of space χ^M of a given feature, where

$$\chi = [\mu(\text{strong decrease}), \mu(\text{decrease}), \mu(\text{slight decrease}), \ldots].$$

In such a defined space of states, the process of detecting and classifying anomalies can be carried out [13,14]. In this way, it is possible to search for anomalies and define trends of individual features within one country. An interesting issue seems to be the confrontation of the state of the pandemic in different countries.

In this case, the above considerations should be slightly modified and the transformation into space χ^M should be made with the use of fuzzy statistical semantics determined for different countries. More precisely, when determining the mean and standard deviation, one should not go back in time and use properly smoothed (possibly shifted) data from different countries. It should be noted that countries in the same region should be used for comparison. Additionally, based on the conducted preliminary analyses, it is possible to perform a certain transformation of the data consisting in the transfer of data from certain countries. It can be noticed, based on previous waves, that successive peaks of the disease are shifting within a given region. In the case of Europe, it is most often from west to east.

Additionally, it should be noted that the comparison of individual features should not be made on the raw data. The population of the country should be taken into account. It is most convenient to operate on data representing the size of a given phenomenon per a given number of inhabitants.

3 Numerical Experiments

In the experimental section, all calculations are made of data per 100,000 inhabitants. This transformation does not disturb the directions or strength of increments, and on the other hand it allows for the comparison of values between different countries. In the case of limiting the analysed period of time to a selected time interval, a more in-depth analysis is possible.

To reduce the impact of weekly fluctuations, the smoothing given by formula (1) should be performed. The smoothing effect with the parameter values $k = 1, 2, 3$ is shown in Fig. 2.

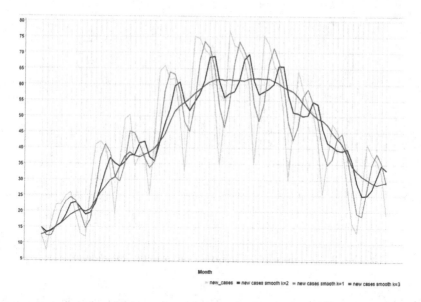

Month

new_cases ▪ new cases smooth k=2 ▪ new cases smooth k=1 ▪ new cases smooth k=3

Fig. 2. Smoothing the data describing the number of new cases

Comparing the influence of the parameter k on the smoothing level, it can be seen that the fluctuations disappear with the increase of this parameter. In further considerations, smoothing will be used with the parameter $k = 3$ corresponding to the weekly smoothing. Based on the smoothed data, information granules were determined for 5 series corresponding to the following features: new cases, new deaths, hosp patients, new test, and the quotient of new cases by new test.

In the process of determining the information granules, it was assumed that the value of parameter $N = 31$. This value corresponds to one month. For such information granules, it is possible to visualise them in the form presented in Fig. 3.

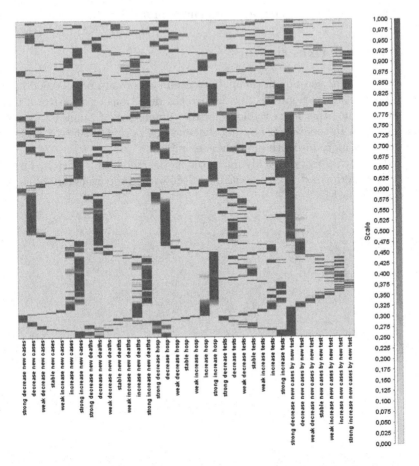

Fig. 3. Visualization of information granules for Poland

Analyzing the results shown in Fig. 3, it can be seen that there is a significant correlation between the number of new cases, the number of new deaths, and the number of hospitalized patients. In most cases, the membership degrees of each state are close to each other. The level of the quotient of new cases to new tests has an interesting tendency. The ratio strongly decreases when the number of new cases and the number of tests performed in successive waves of the pandemic increases. This may be due to the fact that a significant number of tests are positive, but the quotient decreases slightly in subsequent days. However, it should be noted that in the case of Poland, as well as other Eastern

European countries, a very high level of positive tests can be observed. This is motivated by the reluctance to test and only those who are sick and in mandatory quarantine are tested. The result is a high number of positive tests. Therefore, it can be assumed that many infected people are not included in the official counts.

An interesting issue seems to be the comparison of such defined information granules for different countries. It is worth noting that for each of the compared countries, the mean and standard deviation are determined independently, which are the basis for determining the degrees of membership in particular states. The summary of information granules describing the number of new cases for 3 selected countries is presented in Fig. 4.

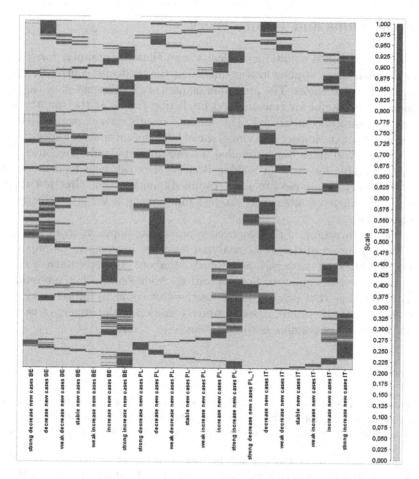

Fig. 4. Comparison of information granules for Belgium (BE), Poland (PL) and Italy (IT)

The results presented in Fig. 4 shows differences between the duration and specificity of subsequent pandemic waves in individual countries. The greatest differences can be seen in the occurrence of the 4th and 5th wave of the pandemic. The reasons for this may be found in the percentage of fully vaccinated persons in the compared countries. In the case of Poland, where this ratio is low, higher fluctuations can be observed.

Besides constructing information granules for each country, a specific case was independently considered, where information granules were created based on the mean and median determined using data from 36 countries in the European region. In this case, as before, data per 100,000 inhabitants were analyzed. It is a very important assumption, otherwise the time series for different countries should not be compared.

4   Conclusions and Future Work

Based on the obtained results, it is clearly seen that the granular fuzzy techniques can find more applications in modeling the epidemic course, detecting patterns and dependencies. The presented numerical experiments show that this approach can be useful for tracking and predicting trends at the country level and for cross-country comparisons. The obtained results even allow very detailed conclusions about the impact of regional social behavior on epidemic spread rates and hospitalization rates. Such detailed insights into epidemic phenomena are extremely valuable for decision making because they can help to make better choices. Granular techniques can also be applied to analyze the direction of pandemic spread, an issue intensively studied today using a variety of analytical techniques [23].

Among the directions of the proposed approach development should be mentioned the expansion of the set of analyzed features. It seems to be important to consider additional data, such as information about the percentage of people with a positive result going to hospitalization. Additionally, it is advisable to determine the time that passes from a positive test result to full recovery or to hospitalization. Furthermore, it seems worth considering the impact of vaccination on the period of possible hospitalization.

References

1. Centers for Disease Control and Prevention. https://www.cdc.gov/coronavirus/2019-ncov/science/forecasting/forecasting.html. Accessed 11 Jan 2022
2. World Health Organization Coronavirus. https://covid19.who.int/. Accessed 11 Jan 2022
3. Aldawish, I., Ibrahim, R.W.: A new mathematical model of multi-faced COVID-19 formulated by fractional derivative chains. Adv. Continuous Discrete Models **2022**(1), 1–10 (2022). https://doi.org/10.1186/s13662-022-03677-w
4. Askar, S., Ghosh, D., Santra, P., Elsadany, A.A., Mahapatra, G.: A fractional order SITR mathematical model for forecasting of transmission of COVID-19 of India with lockdown effect. Results Phys. **24**, 104067 (2021)

5. Ayinde, K., et al.: Modeling Covid-19 cases in west african countries: a comparative analysis of quartic curve estimation models and estimators. In: Azar, A.T., Hassanien, A.E. (eds.) Modeling, Control and Drug Development for COVID-19 Outbreak Prevention. SSDC, vol. 366, pp. 359–454. Springer, Cham (2022). https://doi.org/10.1007/978-3-030-72834-2_12

6. Chandra, R., Jain, A., Singh Chauhan, D.: Deep learning via LSTM models for COVID-19 infection forecasting in India. PLoS ONE **17**(1), e0262708 (2022)

7. Glanz, J., Robertson, C.: Lockdown delays cost at least 36,000 lives, data show. The New York Times, vol. 21 (2020)

8. Gonçalves, C.P., et al.: The impact of COVID-19 on the Brazilian power sector: operational, commercial, and regulatory aspects. IEEE Lat. Am. Trans. **20**(4), 529–536 (2022)

9. Groseclose, S.L., Buckeridge, D.L.: Public health surveillance systems: recent advances in their use and evaluation. Annu. Rev. Public Health **38**, 57–79 (2017)

10. Guharoy, R., Krenzelok, E.P.: Lessons from the mismanagement of the COVID-19 pandemic: a blueprint to reform CDC. Am. J. Health Syst. Pharm. **78**(18), 1739–1741 (2021)

11. Ioannidis, J.P., Cripps, S., Tanner, M.A.: Forecasting for COVID-19 has failed. Int. J. Forecast. (2020)

12. Kiersztyn, A., Karczmarek, P., Kiersztyn, K., Łopucki, R., Grzegórski, S., Pedrycz, W.: The concept of granular representation of the information potential of variables. In: 2021 IEEE International Conference on Fuzzy Systems (FUZZ-IEEE), pp. 1–6. IEEE (2021)

13. Kiersztyn, A., Karczmarek, P., Kiersztyn, K., Pedrycz, W.: The concept of detecting and classifying anomalies in large data sets on a basis of information granules. In: 2020 IEEE International Conference on Fuzzy Systems (FUZZ-IEEE), pp. 1–7. IEEE (2020)

14. Kiersztyn, A., Karczmarek, P., Kiersztyn, K., Pedrycz, W.: Detection and classification of anomalies in large data sets on the basis of information granules. IEEE Trans. Fuzzy Syst. (2021)

15. Kiersztyn, A., et al.: The use of information granules to detect anomalies in spatial behavior of animals. Ecol. Indic. **136**, 108583 (2022)

16. Kiersztyn, A., et al.: Classification of complex ecological objects with the use of information granules. In: 2021 IEEE International Conference on Fuzzy Systems (FUZZ-IEEE), pp. 1–6. IEEE (2021)

17. Łopucki, R., Kitowski, I., Perlińska-Teresiak, M., Klich, D.: How is wildlife affected by the COVID-19 pandemic? Lockdown effect on the road mortality of hedgehogs. Animals **11**(3), 868 (2021)

18. Mofijur, M., et al.: Impact of COVID-19 on the social, economic, environmental and energy domains: lessons learnt from a global pandemic. Sustain. Prod. Consum. **26**, 343–359 (2021)

19. Ray, E.L., et al.: Ensemble forecasts of coronavirus disease 2019 (COVID-19) in the US. MedRXiv (2020)

20. Selim, T., Eltarabily, M.G.: Impact of COVID-19 lockdown on small-scale farming in Northeastern Nile Delta of Egypt and learned lessons for water conservation potentials. Ain Shams Eng. J. **13**(4), 101649 (2022)

21. Sidorova, E.: Overcoming COVID-19 impact in the EU: supranational financial aspect. Mirovaia Ekon. Mezdunar. Otnosheiia **65**(1), 24–32 (2021)

22. Sinha, A.K., Namdev, N., Shende, P.: Mathematical modeling of the outbreak of COVID-19. Netw. Model. Anal. Health Inform. Bioinform. **11**(1), 1–19 (2022)

23. Xiang, L., Ma, S., Yu, L., Wang, W., Yin, Z.: Modeling the global dynamic contagion of COVID-19. Front. Public Health **9** (2021)
24. Yin, C., Zhao, W., Pereira, P.: Meteorological factors' effects on COVID-19 show seasonality and spatiality in Brazil. Environ. Res. 112690 (2022)

Type-2 Fuzzy Classifier with Smooth Type-Reduction

Katarzyna Nieszporek[1] , Giorgio De Magistris[2] , Christian Napoli[2] ,
and Janusz T. Starczewski[1(✉)]

[1] Department of Intelligent Computer Systems, Częstochowa University
of Technology, Częstochowa, Poland
{katarzyna.nieszporek,Janusz.Starczewski}@pcz.pl
[2] Department of Computer, Control and Management Engineering,
Sapienza University of Rome, Via Ariosto 25, 00185 Roma, Italy

Abstract. The defuzzification of a type-2 fuzzy set is a two-stage process consisting of firstly type-reduction, and a secondly defuzzification of the resultant type-1 set. All accurate type reduction methods used to build fuzzy classifiers are based on the recursive Karnik-Mendel algorithm, which is troublesome to obtain a feedforward type-2 fuzzy network structure. Moreover, the KM algorithm and its modifications complicate the learning process due to the non-differentiability of the maximum and minimum functions. Therefore, this paper proposes to use the smooth maximum function to develop a new structure of the fuzzy type-2 classifier.

Keywords: Smooth type reduction · Interval type-2 fuzzy logic systems

1 Introduction

In recent years, fuzzy logic methodology has shown to be very effective in solving complex nonlinear systems containing uncertainties that are otherwise challenging. However, it is also noted that fuzzy rules working in an uncertain or non-stationary environment require a higher order of fuzziness. This is due to the fact that type-1 fuzzy sets, whose membership grades are real numbers, could have limitations in minimizing the effect of uncertainty, whereas the membership grades of a type-2 fuzzy logic system are themselves fuzzy logic systems in $[0, 1]$. Describing a type-2 fuzzy set by a rectangular membership function sufficiently describes the uncertainty in modeling of most processes. However, for the output, there will need a type reduction to convert the output of the fuzzy inference engine into a type 1 fuzzy sets before defuzzification can be performed to obtain a crisp output. The center-of-sets iterative Karnik-Mendel (KM) approach to type reduction is of great interest. Over the years, modifications have been made to the basic KM algorithm, including Wu and Tan [17] who presented their concept using a genetic algorithm. In this paper, the smooth method is used to design efficient type reduction algorithms (Fig. 1).

L. Rutkowski et al. (Eds.): ICAISC 2022, LNAI 13588, pp. 193–202, 2023.
https://doi.org/10.1007/978-3-031-23492-7_17

Type-2 FLS

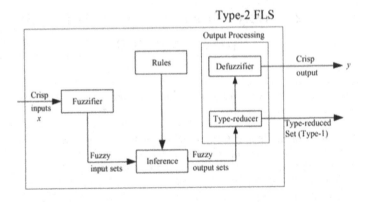

Fig. 1. Type-2 FLS (from Mendel [8])

2 An Overview

The Type-1 Fuzzy Set. Let X be a universe of discourse. A fuzzy set A on X is characterised by a membership function $\mu_A : X \to [0,1]$ and can be represented as follows:

$$A = \{(x, \mu_A(x)); \mu_A(x) \in [0,1] \forall x \in X\} \tag{1}$$

The Type-2 Fuzzy Set. Let $\tilde{P}(U)$, where $U = [0,1]$, be set of fuzzy sets in U. A type-2 fuzzy set $\tilde{A}$ in X is a fuzzy set whose membership grades are themselves fuzzy [23].

$$\tilde{A} = \{(x, \mu_{\tilde{A}}(x)); \mu_{\tilde{A}}(x) \in \tilde{P}(U) \forall x \in X\} \tag{2}$$

where $\mu_{\tilde{A}(x)}$ is a fuzzy set in U for all x, i.e. $\mu_{\tilde{A}(x)} : X \to \tilde{P}(U)$.
It implies that $\forall x \in X \; \exists J_x \subseteq U$ such that $\mu_{\tilde{A}}(x) : J_x \to U$ [4].

$$\mu_A(x)) = \{(u, \mu_{\tilde{A}}(x)(u)) | \mu_{\tilde{A}}(x)(u) \in U \forall u \in J_x \subseteq U\} \tag{3}$$

where X is called the primary domain, J_x the primary membership of x, U is known as the secondary domain and $\mu_{\tilde{A}}(x)$ is the secondary membership of x.

In this paper, an interval singleton type-2 fuzzy logic system type is used. This means that the fuzzifier converts the fuzzy logic system input signals into fuzzy singletons and then the inference engine adjusts the fuzzy singletons with the fuzzy rules in the rule base.

Considering a type-2 fuzzy system with K rules will be used with the following scheme [7]:

$$\tilde{R}^k : \text{IF } \tilde{A}' \text{ is } \tilde{A}_k \text{ THEN } \tilde{B}' \text{ is } \tilde{B}_k. \tag{4}$$

where $\tilde{A}'$, $\tilde{A}_k$, $\tilde{B}'$ and $\tilde{B}_k$ are type-2 fuzzy sets. In the interval case, they are subintervals of $[0,1]$ expressed by of upper and lower bounds, e.g. $\tilde{A}_k = \left[\underline{\mu}_{A_k}(x), \overline{\mu}_{A_k}(x)\right] \subseteq [0,1]$ for each $x \in X$.

The output needs a type reduction to convert into a type 1 fuzzy sets before defuzzification can be performed to obtain a crisp output. This is the main

1. Let the consequent values be aranged in the ascending order
   $y_1 < y_2 < \ldots < y_K$
2. calculate type-1 system output y_0 as an average of y_k weighted by mean membership grades, i.e., $\left(\underline{\mu}_k + \overline{\mu}_k\right)/2$,
3. set the initial values $y_{\min} = y_{\max} = y_0$,
4. for each $k = 1, 2, \ldots, K$, if $y_k > y_{\max}$, then $\overrightarrow{\mu}_k = \overline{\mu}_k$, otherwise $\overrightarrow{\mu}_k = \underline{\mu}_k$,
5. find the closest $y_{\text{next}} = \min\limits_{k=1,\ldots,K} y_k : y_k > y_{\max}$,
6. calculate $y_{\max}$ as an average of y_k weighted by new grades $\overrightarrow{\mu}_k$,
7. if $y_{\max} \leq y_{\text{next}}$, continue, else go to step 4,
8. for each $k = 1, 2, \ldots, K$, if $y_k < y_{\min}$, then $\overleftarrow{\mu}_k = \overline{\mu}_k$, otherwise $\overleftarrow{\mu}_k = \underline{\mu}_k$,
9. find the closest $y_{\text{next}} = \max\limits_{k=1,\ldots,K} y_k : y_k < y_{\min}$,
10. calculate $y_{\min}$ as an average of y_k weighted by new grades $\overleftarrow{\mu}_k$,
11. if $y_{\min} \geq y_{\text{next}}$, finish, else go to step 8.

Algorithm 1.1: The KM type reduction

structural difference between type-1 and type-2 logic fuzzy sets. One of the most common type reduction methods is the centroid type-reducer. The centroid of a type-1 fuzzy set when the domain X is discretised into k points is:

$$C_A = \frac{\sum_{i=1}^{k} x_i \mu_A(x_i)}{\sum_{i=1}^{k} \mu_A(x_i)} \tag{5}$$

Referring to the literature [6, 23] the centroid of a type-2 fuzzy set $\tilde{A}$ with domain X discretised into k points $x_1, \ldots x_k$ with $x_1 < \ldots < x_k$ as

$$C_{\tilde{A}} = \int_{u_1 \in J_{x_1}} \cdots \int_{u_k \in J_{x_k}} [\mu_{\tilde{A}}(x_1)(u_1) \cdot \ldots \cdot \mu_{\tilde{A}}(x_k)(u_k)] / \frac{\sum_{i=1}^{k} x_i u_i}{\sum_{i=1}^{k} u_i} \tag{6}$$

In case $\tilde{A}$ is interval type-2 logic fuzzy set, then the centroid is the crisp set:

$$C_{\tilde{A}} = \int_{u_1 \in J_{x_1}} \cdots \int_{u_k \in J_{x_k}} / \frac{\sum_{i=1}^{k} x_i u_i}{\sum_{i=1}^{k} u_i} \tag{7}$$

It has been shown that this iterative procedure can converge in at most K iterations [8]. Once y_l and y_r are available, they can be used to compute the approximate output. Since the reduced type set is an interval fuzzy set of type 1, the fuzzy output value is [17]:

$$y(x) = \frac{y_{max} + y_{min}}{2} \tag{8}$$

The KM type reduction in its simplest form can be summarized as follows in Algorithm 1.1.

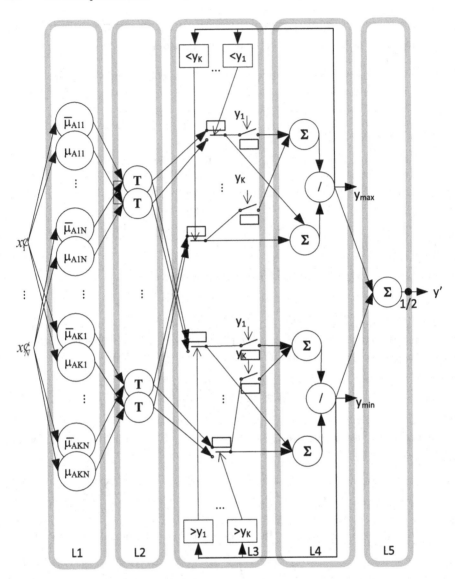

Fig. 2. Adaptive interval type-2 fuzzy logic system using KM type-reduction

In Nowicki's work [10] on defuzzification for binary class membership of objects, it can be seen that the result does not require any ordering of $y_{j,k}$ as is done in the KM method (Fig. 2).

According to a theorem stated in the literature [15] with a proof, it turns out that for given rough approximations, $\underline{\mu}_{j,k}$ and $\overline{\mu}_{j,k}$ of the binary set $y_{j,k} = 0, 1$ representing by a single rule class membership, where k is the index for rules $k = 1, ..., K$ and j is the index for classes $j = 1, ...J$, the lower and upper approximations of the object's class membership C_j are given by

$$y_{\min}(j) = \frac{\sum_{k=1}^{K} \underline{\mu}_{j,k} y_{j,k}}{\sum_{k=1}^{K} \underline{\mu}_{j,k} y_{j,k} + \sum_{k=1}^{K} \overline{\mu}_{j,k} \neg y_{j,k}}, \tag{9}$$

$$y_{\max}(j) = \frac{\sum_{k=1}^{K} \overline{\mu}_{j,k} y_{j,k}}{\sum_{k=1}^{K} \underline{\mu}_{j,k} \neg y_{j,k} + \sum_{k=1}^{K} \overline{\mu}_{j,k} y_{j,k}}. \tag{10}$$

3   Smooth Type Reduction

To characterize the smooth type reduction, assume that the described system has an output value V_p, where $p = 1...P$. Then its smooth maximum of $v_1, ..., v_p$ would be a differentiable approximation of maximum of a function with continuous derivatives. In addition, the universal smooth maximum/minimum function is defined as

$$y_\alpha(v_1, \ldots, v_P) = \frac{\sum_{p=1}^{P} v_p e^{\alpha v_p}}{\sum_{p=1}^{P} e^{\alpha v_p}} \tag{11}$$

which y_α has the following properties:

1. $y_\alpha \to max$ as $\alpha \to \infty$,
2. $y_\alpha \to min$ as $\alpha \to -\infty$,
3. $y_0 = \frac{\sum_{p=1}^{P} v_p}{P}$

Notably, this means that the values of $y_{-\infty}$ and y_∞ are the endpoints of the reduced set, respectively $y_{\min}$ and $y_{\max}$. In the search for end points, e.g., y_{max}, only those tuples should be considered that have lower memberships for consequents being no larger than y_0, which is the output of the type 0 fuzzy system in an interval fuzzy system of type 2. For values of v_k arranged in ascending order, we run the algorithm. Perform a right-shift operation to compute the output values v_p that maximize the result. An example shift is demonstrated in the Table 1 and the proposed algorithm using the smooth extremum function is presented in Algorithm 1.2 (Fig. 3).

Another approach to smooth maximum is to use LogSumExp, which is as follows: $LSE(v_1, \ldots, v_P) = \frac{1}{\alpha} \log \sum_P \exp(\alpha v_p)$, which can be normalized for all non-negative V_P, yielding a function with domain $[0, \infty)^n$ and range $[0, \infty)$: $g(v_1, \ldots, v_P) = \log(\sum_P \exp(v_p) - (P-1))$. There is also another approach that uses the p-norm, $\|(v_1, \ldots, v_R)\|_p = (\sum_r |v_r|^p)^{\frac{1}{p}}$. The LogSumExp approach as well as the p-Norm approach generate similar results.

4   Experimental Results

The source Wisconsin Breast Cancer data are reports of clinical cases [Mangasarian and Wolberg 1990] [18]. The original data set contained 699 cases divided into two categories: benign breast cancer (65.5% of instances) and malignant cancer (34.5%). Each case was described by nine attributes: clump thickness,

Table 1. Right-shifted mask to calculate y_{max}

r\k	1	2	R	R_{+1}	K_{-1}	K
R	0	0	1	1	1	1
R_{+1}	0	0	0	1	1	1
K_{-1}	0	0	0	0	1	1
K	0	0	0	0	0	1

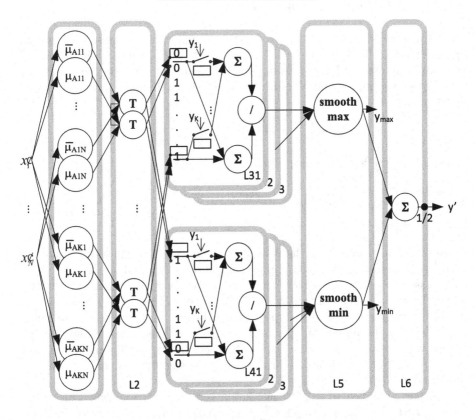

Fig. 3. Adaptive interval type-2 fuzzy logic system using smooth type-reduction

uniformity of cell size, uniformity of cell shape, marginal adhesion, single epithelial cell size, bare nuclei, bland chromatin, normal nucleoli, and mitosis, note that 16 individuals are missing the attribute.

The specificity of interval-valued fuzzy logic systems allows us for an analysis on a lower level of classification if only we make use of the interval outputs of the system: ymin and ymax. Using this information, instead of strict classification, we get three groups of objects classified with the following labels:

1 Let the consequent values be arranged in the ascending order $y_1 < y_2 < \ldots < y_K$ and the values in vector forms, i.e.,

$$\mathbf{y} = [y_1, \ldots, y_K]$$
$$\overline{\mu} = [\overline{\mu}_1, \ldots, \overline{\mu}_K]$$
$$\underline{\mu} = \left[\underline{\mu}_1, \ldots, \underline{\mu}_K\right]$$

To compute the right and the left endpoints of the type-reduced set, perform the following steps:

1. calculate type-1 system output y_0 as an average of elements of $\mathbf{y}$ weighted by mean membership grades, i.e., $\left(\underline{\mu} + \overline{\mu}\right)/2$,
2. find index R of the closest $y_R = \min_{k=1,\ldots,K} y_k : y_k > y_0$,
3. for $r = R, \ldots, K-1$:
   (a) set a mask $M_r = \underset{1 \quad \ldots \quad R \quad \ldots \quad K}{0 \ldots 0 1 \ldots 1}$,
   (b) apply the mask to upper and lower memberships
   $\overrightarrow{\mu} = (1 - M_r) \odot \underline{\mu} + M_r \odot \overline{\mu}$ (where $\odot$ is the Hadamard product),
   (c) calculate $y_{\max,r}$ as an average of elements $\mathbf{y}$ weighted by $\overrightarrow{\mu}$,
4. return $y_{\max}$ as an aggregation of all $y_{\max,r}$ with the use of smooth maximum, $r = R, \ldots, K-1$,
5. find index L of the closest $y_L = \min_{k=1,\ldots,K} y_k : y_k < y_0$,
6. for $l = 2, \ldots, L$:
   (a) set a mask $M_l = \underset{1 \quad \ldots \quad L \quad \ldots \quad K}{1 \ldots 1 0 \ldots 0}$,
   (b) apply the mask to upper and lower memberships
   $\overleftarrow{\mu} = (1 - M_l) \odot \underline{\mu} + M_l \odot \overline{\mu}$,
   (c) calculate $y_{\min,l}$ as an average of elements $\mathbf{y}$ weighted by $\overleftarrow{\mu}$,
7. return $y_{\max}$ as an aggregation of all $y_{\max,r}$ with the use of smooth maximum, $r = R, \ldots, K-1$.

Algorithm 1.2: Smooth type reduction

- certain classification if $y_{min} > 0.5$,
- uncertain classification if $y_{max} \geq 0.5 \geq y_{min}$,
- certain rejection if $y_{max} < 0.5$.

As a result, we get three rate groups: classified, misclassified, and unclassified ("NoClass.") when classification cannot be performed certainly. This can help in practical classification systems such as the medical diagnosis when uncertain classification cases can be again directed to a thorough examination. The classification results in the imputation of input values by means of rough-fuzzy sets are presented in Table 2

Table 2. Wisconsin Breast Cancer classification with optional uniform noise applied to single input $X_1, \ldots, X_9$ as well as to all inputs X_{all}

Original data	Singleton	Interrval KM-T2FLC	Interval T2FLC based Smooth Type-Reduction
	Class./Misclass.	Class./NoClass./Misclass.	Class./NoClass./Misclass.
	0.988/0.012	0.975/0.10/0.015	**0.986/0.011/0.003**
σ_1			
1.0	0.978/0.022	0.964/0.019/0.017	**0.974/0.020/0.006**
5.0	0.931/0.069	0.673/0.315/0.012	**0.675/0.317/0.008**
σ_2			
1.0	0.977/0.023	0.963/0.016/0.020	**0.973/0.018/0.009**
5.0	0.960/0.040	0.647/0.336/0.017	**0.660/0.333/0.007**
σ_3			
1.0	0.970/0.030	0.912/0.074/0.014	**0.933/0.062/0.005**
5.0	0.911/0.089	0.589/0.406/0.005	**0.694/0.302/0.004**
σ_4			
1.0	0.977/0.023	0.975/0.009/0.016	**0.985/0.009/0.006**
5.0	0.967/0.033	0.838/0.152/0.010	**0.844/0.150/0.006**
σ_5			
1.0	0.977/0.023	0.970/0.011/0.019	**0.981/0.010/0.009**
5.0	0.962/0.038	0.795/0.195/0.010	**0.799/0.191/0.010**
σ_6			
1.0	0.978/0.022	0.948/0.034/0.019	**0.961/0.032/0.008**
5.0	0.938/0.062	0.824/0.166/0.010	**0.825/0.167/0.008**
σ_7			
1.0	0.980/0.020	0.961/0.024/0.015	**0.970/0.025/0.005**
5.0	0.965/0.035	0.634/0.360/0.006	**0.642/0.352/0.006**
σ_8			
1.0	0.978/0.022	0.968/0.012/0.020	**0.978/0.011/0.009**
5.0	0.970/0.030	0.854/0.137/0.009	**0.853/0.139/0.008**
σ_9			
1.0	0.980/0.020	0.969/0.014/0.018	**0.978/0.015/0.008**
5.0	0.944/0.056	0.717/0.272/0.011	**0.722/0.271/0.007**
σ_{all}			
1.0	0.973/0.027	0.518/0.480/0.002	**0.538/0.462/0.000**
5.0	0.749/0.251	0.001/0.999/0.000	**0.021/0.979/0.000**

5   Conclusion

In this paper, a smooth type-reduction method that is competitive with the KM type-reduction system is presented. It shows good results as it achieves low training error values. It is worth noting that both type-2 fuzzy systems significantly exceed the learning ability of the type-1 fuzzy system. The proposed system is worth considering for solving problems with increased model uncertainty or when there is uncertain input data.

The initial learning of type 2 systems treated as type 1 fuzzy systems, followed by the application of generating type-2 fuzzy rules methods for uncertain data using the fuzzy-rough approximation [13,16] or possibilistic fuzzification [15], shows that fuzzy systems are important in the process of extracting explanatory fuzzy rules.

References

1. Bilski, J., Kowalczyk, B., Marchlewska, A., Zurada, J.: Local Levenberg-Marquardt algorithm for learning feedforward neural networks. J. Artif. Intell. Soft Comput. Res. **10**(4), 299–316 (2020)
2. Bilski, J., Rutkowski, L., Smoląg, J., Tao, D.: A novel method for speed training acceleration of recurrent neural networks. Inf. Sci. **553**, 266–279 (2020)
3. Chen, Y., Wang, D.: Study on centroid type-reduction of general type-2 fuzzy logic systems with weighted enhanced Karnik-Mendel algorithms. Soft. Comput. **22**(4), 1361–1380 (2018)
4. Greenfield, S., Chiclana, F.: Type-reduction of the discretised interval type-2 fuzzy set: approaching the continuous case through progressively finer discretisation. J. Artif. Intell. Soft Comput. Res. **1**(3), 183–193 (2011)
5. Greenfield, S., Chiclana, F., Coupland, S., John, R.: The collapsing method of defuzzification for discretised interval type-2 fuzzy sets. Inf. Sci. **179**, 2055–2069 (2009)
6. Karnik, N., Mendel, J.: Centroid of a type-2 fuzzy set. Inf. Sci. **132**, 195–220 (2001)
7. Liang, Q., Mendel, J.: Interval type-2 fuzzy logic systems: theory and design. IEEE Trans. Fuzzy Syst. **8**, 535–550 (2000)
8. Mendel, J.: Uncertain rule-based fuzzy logic systems. Introduction and New Directions (2001)
9. Mendel, J., John, R.: Type-2 fuzzy sets made simple. IEEE Trans. Fuzzy Syst. **10**(2), 117–127 (2002)
10. Nowicki, R.: On combining neuro-fuzzy architectures with the rough set theory to solve classification problems with incomplete data. IEEE Trans. Knowl. Data Eng. **20**, 1239–1253 (2008)
11. Nowicki, R., Seliga, R., Żelasko, D., Hayashi, Y.: Performance analysis of rough set-based hybrid classification systems in the case of missing values. J. Artif. Intell. Soft Comput. Res. **11**(4), 307–318 (2021)
12. Nowicki, R.K., Starczewski, J.T.: On non-singleton fuzzification with DCOG defuzzification. In: Rutkowski, L., Scherer, R., Tadeusiewicz, R., Zadeh, L.A., Zurada, J.M. (eds.) ICAISC 2010. LNCS (LNAI), vol. 6113, pp. 168–174. Springer, Heidelberg (2010). https://doi.org/10.1007/978-3-642-13208-7_22
13. Nowicki, R., Starczewski, J.: A new method for classification of imprecise data using fuzzy rough fuzzification. Inf. Sci. **414**, 33–52 (2017)
14. Runkler, T., Coupland, S., John, R.: Properties of interval type-2 defuzzification operators (2015)
15. Starczewski, J.: Advanced Concepts in Fuzzy Logic and Systems with Membership Uncertainty. Studies in Fuzziness and Soft Computing, vol. 284. Springer, Heidelberg (2013). https://doi.org/10.1007/978-3-642-29520-1

16. Starczewski, J., Nowicki, R., Nieszporek, K.: Fuzzy-rough fuzzification in general FL classifiers. In Guervós, J.J.M., Garibaldi, J., Linares-Barranco, A., Madani, K., Warwick, K. (eds.) Proceedings of the 11th International Joint Conference on Computational Intelligence, IJCCI 2019, Vienna, Austria, 17–19 September 2019, pp. 335–342. ScitePress (2019)

17. Tan, W.W., Wu, D.: Design of type-reduction strategies for type-2 fuzzy logic systems using genetic algorithms. In: Jain, L.C., Palade, V., Srinivasan, D. (eds.) Advances in Evolutionary Computing for System Design, pp. 169–187. Springer, Heidelberg (2007). https://doi.org/10.1007/978-3-540-72377-6_7

18. Wolberg, W.H.: Breast Cancer Wisconsin. University of Wisconsin Hospitals Madison, Wisconsin, USA (1990)

19. Wu, D., Mendel, J.: Recommendations on designing practical interval type-2 fuzzy systems. Eng. Appl. Artif. Intell. **85**, 182–193 (2019)

20. Wu, L., Qian, F., Wang, L., Ma, X.: An improved type-reduction algorithm for general type-2 fuzzy sets. Inf. Sci. **593**, 99–120 (2022)

21. Zadeh, L.: The concept of a linguistic variable and its application to approximate reasoning. Inf. Sci. **8**, 199–249 (1975)

22. Zadeh, L.: The concept of a linguistic variable and its application to approximate reasoning-II. Inf. Sci. **8**, 301–357 (1975)

23. Zadeh, L.: The concept of a linguistic variable and its application to approximate reasoning-III. Inf. Sci. **9**(1), 43–80 (1975)

Evolutionary Algorithms and Their Applications

A Multi-population-Based Algorithm with Different Ways of Subpopulations Cooperation

Krzysztof Cpałka[ID], Krystian Łapa[(✉)][ID], and Leszek Rutkowski[ID]

Department of Computational Intelligence, Częstochowa University of Technology,
Częstochowa, Poland
{krzysztof.cpalka,krystian.lapa,leszek.rutkowski}@pcz.pl

Abstract. Metaheuristic methods are designed to solve continuous and discrete problems. Such methods include population based algorithms (PBAs). They are distinguished by the flexibility of defining the fitness function, therefore they are a good alternative to gradient methods. However, creating new variants of PBAs that work similarly and differ in detail might be problematic. Therefore, it is interesting to combine existing PBAs in order to increase their effectiveness. One of the hybrid methods is the Multi-population Nature-Inspired Algorithm (MNIA), which uses search operators from different PBAs. The formula of MNIA's operation is based on the appropriate cooperation of its subpopulations. That is why in this paper we focus on expanding MNIAs with various schemes of such cooperation. In particular, we analyze various combinations of migration models, intervals, and topologies. The proposed solutions were tested and compared using generally known benchmark functions. The obtained results showed an advantage of certain patterns of cooperation of the - subpopulations, which confirmed the validity of the adopted assumptions.

Keywords: Metaheuristic method · Population-based algorithm · Multi-population algorithm · Cooperation of subpopulations

1 Introduction

Metaheuristic methods are designed to solve continuous and discrete problems. Such methods include population based algorithms (PBAs - see e.g. [1,10,11, 17,18,29,35,37,39,61,68,71]). They are distinguished by the lack of necessity to determine derivative functions, flexibility of defining the fitness function and the ease of its modification. PBAs are therefore a good alternative to gradient methods. Alternatively, PBA can be used to initially investigate a given problem and find an initial optimum. Such optima can be, for example, a starting point for gradient methods. Various combinations of these algorithms can be found in the literature (see e.g. [1,26,41,58]). PBAs can be combined with other methods in a number of other ways, which is the reason why many hybrid solutions are

L. Rutkowski et al. (Eds.): ICAISC 2022, LNAI 13588, pp. 205–218, 2023.
https://doi.org/10.1007/978-3-031-23492-7_18

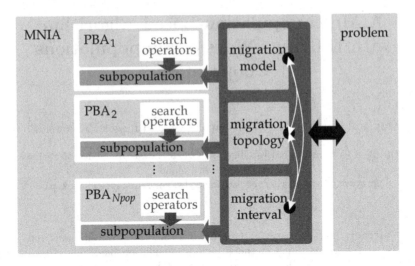

Fig. 1. The idea of various ways of cooperation in MNIA.

created [52]. In such approaches, various methods of computational intelligence are often used [4,57,64,65,67,69].

The problem with PBAs is creating their new variations (see e.g. [8,13,27, 33,40]). They often work in a similar way and differ only in details [15,54]. Therefore, combining capabilities offered by existing PBAs in order to increase their effectiveness is an interesting research topic. This approach was used in the MNIA (Multi-population Nature-Inspired Algorithm [53]), which uses operators from different PBAs to process a given population. The formula of its operation is based on the appropriate cooperation of its subpopulations (islands). In this paper, we focus on expanding a MNIA by using various schemes of such cooperation. In particular, we investigate various combinations of migration models, intervals and topologies (see e.g. [2,9,28,30]). Migration topologies define how subpopulations are interconnected, and migration models define how individuals are exchanged within subpopulations. In contrast, migration intervals determine the frequency of interactions between subpopulations. They are therefore the number of steps between two consecutive cooperation of subpopulations.

1.1 Motivation

In our previous work, we proposed an algorithm that used search operators from different PBAs. In that algorithm, cooperation between subpopulations was of a fundamental nature. Since the issue of interactions between subpopulations seems to determine the effectiveness of the multi-population algorithm, we decided to focus on it in this paper.

1.2   Contribution of the Paper

For the purpose of this paper, we collected information available in the literature on the migration topology and the strategy of exchanging individuals between subpopulations collected. Next, we combined that information with the capabilities of a multi-population algorithm using search operators derived from different PBAs (MNIA). In addition, we proposed combinations of topology and strategy, and then the proposed combinations were tested with different migration intervals.

1.3   Structure of the Paper

Section 2 presents the characteristics of the MNIA method which was used to test the proposed combinations of the cooperation between subpopulations. Section 3 discusses the considered combinations in more detail. Section 4 presents the obtained results and in Sect. 5 the conclusions are drawn and future research ideas are presented.

2   Multi-population Nature-Inspired Algorithm (MNIA)

The MNIA has many subpopulations. Each subpopulation is processed by a different PBA equipped with individual search operators. This is an approach that follows the Island Model with Migrations [60]. Therefore, the operation of MNIA can be interpreted as follows [53]:

- Groups of $Npop$ experts (subpopulations or islands) in various research centres are looking for a solution to a certain problem.
- Each group of experts looks for a solution using methods typical of a research center ($PBA_1, PBA_2, \ldots PBA_{Npop}$, see Fig. 1).
- Research centres interact with each other (according to the migration topology) - groups of experts regularly exchange obtained solutions. Such exchange is represented by an interaction between subpopulations (according to the migration model). The intensity of such exchange is determined by a certain interval.
- If any group of experts finds a solution that meets the requirements defined in the adopted fitness function, the search for the solution is completed - MNIA ends its operation. The algorithm's stop condition may also take into account the number of iterations, the number of calls to the fitness function, runtime, etc.

More details about MNIA can be found in [53]. In the further part of the paper, this algorithm will be the base method for testing solutions for cooperation between subpopulations. In particular, various combinations of models, intervals, and migration topology are tested. They are described in the detail in the next section.

Table 1. Migration models for MNIA adopted in the simulations.

Mark	Description
BS	During migration, the best individual of the source subpopulation replaces the worst of the target subpopulation
RW	During migration, the individual selected by the roulette wheel from the source subpopulation replaces the worst of the target subpopulation
TS	During migration, an individual selected by a tournament from the source subpopulation replaces the worst of the target subpopulation
RN	During migration, an individual randomly selected from the source subpopulation replaces the worst of the target subpopulation

Table 2. Migration intervals for MNIA adopted in the simulations.

Mark	Description
M1	Migrations are performed every 1 step of the algorithm
M4	Migrations are performed every 4 steps of the algorithm
M10	Migrations are performed every 10 steps of the algorithm
M20	Migrations are performed every 20 steps of the algorithm

3   Methods of Subpopulation Cooperation

The idea of the subpopulation cooperation is shown in Fig. 1. The considered migration topologies for MNIA are shown in Fig. 2. Most of them have been presented in the literature [14, 16, 32, 36, 42]. We do not comment on the legitimacy of these topologies, but it should be noted that some of them are similar to each other and that the presented set can be easily extended. In Fig. 2 the following markings are used:

- The arrowhead indicates the subpopulation that has been modified.
- A dark circle (e.g. for 'starw') means an additional population replacement after the adopted interval for the population that is the worst of all others in terms of the mean fitness function value.
- A light-colored circle (e.g. for 'starb') means an additional replacement of the population with the best subpopulation.
- Light lines indicate connections from subpopulations selected at random (e.g. for 'swrl02').

Migration topologies determine how subpopulations are connected, but do not specify the rules for how their individuals are traded between them. These rules result from the adopted migration model. The considered migration models are shown in Table 1. The migration interval complements the topology and the migration model. As already said, it determines the intensity of subpopulations interaction. The considered intervals are shown in Table 2.

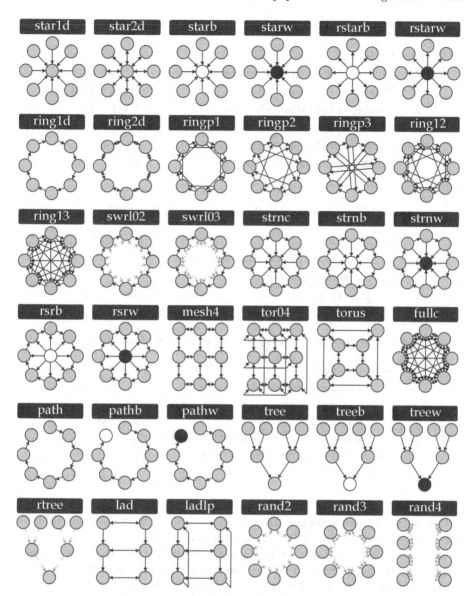

Fig. 2. Migration topologies for MNIA adopted in the simulations.

4   Simulations

The considered solutions were tested using the MNIA algorithm and the known CEC05 benchmark functions [56], hereinafter referred to as F01-F25. The parameters of the simulations performed are presented in Table 3. The migration variants shown in Fig. 2, the migration models shown in Table 1 and migration inter-

Table 3. Parameters related to the performed simulations.

No.	Parameter	Value
1	Number of dimensions of benchmark functions	$D = 50$
2	Number of iterations	$Niter = 1000$
3	Number of individuals when $Npop = 1$	$Nind = 256$
4	Number of individuals when $Npop = 8$	$Nind = 32$
5	Number of individuals when $Npop = 16$	$Nind = 16$
6	Number of simulations repetitions	100

Table 4. PBAs used in the construction of the MNIA.

No.	Algorithm	Acronym	Literature
1	Whale Optimization Algorithm	WO	[34]
2	Fireworks Algorithm	FA	[59]
3	Cuckoo Search	CS	[62]

Table 5. Parameters of the PBAs used in the MNIA construction.

No.	Parameter	Description	Value	Algorithm
1	b	Logarithmic spiral shape constant	1.0	WO
2	$\hat{A}$	Maximum amplitude of explosion	0.2	FA
3	n	Number of fireworks	$0.2 \cdot Npop$	FA
4	m	Parameter for controlling number of sparks	$Npop - n$	FA
5	p_a	Fraction of worst abandoned nests	0.1	CS

vals shown in Table 2 were tested. Moreover, each simulation was performed in two variants: (a) 8 subpopulations of 32 individuals each ('8x32', see Table 6) and (b) 16 subpopulations of 16 individuals each ('16x16').

The PBAs shown in Table 4 with the parameters shown in Table 5 were used in the construction of the MNIA. The same simulation variants were tested for: (a) MNIA with different PBAs and (b) MNIA with the same PBAs. Variant (a) is typical of MNIA, while value (b) corresponds to the scheme used in other multi-population PBAs. This approach made it possible to compare the results and assess the impact of differentiation of the MNIA component methods on the final result of the optimization.

The obtained results are presented in Table 6. The comparison of the considered simulation variants is shown in Table 7. A comparison with the results of PBAs designed by other authors is presented in Table 8. In order to make the comparison, all tested algorithms were implemented in a test environment of our own design, which unified the test procedure.

Table 6. Summary of the percentage improvement of the results for MNIA in relation to the results obtained without migration. The results are averaged for the considered problems. Best results are shown in bold.

Npop x Nind	Topology	M1 BS	M4 BS	M10 BS	M20 BS	M1 RW	M4 RW	M10 RW	M20 RW	M1 TS	M4 TS	M10 TS	M20 TS	M1 RN	M4 RN	M10 RN	M20 RN	Avg
8x32	star1d	-5	13	35	35	4	34	45	14	57	11	21	33	68	18	26	58	29
8x32	star2d	9	121	88	97	20	128	224	294	-6	48	84	82	-28	113	135	108	95
8x32	starb	-25	-4	14	-1	-15	-12	20	2	-20	-3	18	-4	-13	-19	6	20	-2
8x32	starw	141	156	193	106	228	179	195	143	174	228	90	67	125	153	116	66	147
8x32	rstarb	28	175	159	79	62	404	**760**	264	17	61	163	153	42	107	110	88	**167**
8x32	rstarw	-81	-2	19	24	-78	-35	20	36	-79	-29	0	2	-77	-5	-22	-31	-21
8x32	ring1d	-2	145	101	149	35	124	159	157	-10	60	137	70	-35	97	71	50	82
8x32	ring2d	-46	85	157	142	-30	182	188	161	-33	106	111	87	-53	66	81	64	79
8x32	ringp1	-31	80	115	84	-32	136	240	118	-28	129	73	76	-51	27	120	23	67
8x32	ringp2	-36	114	88	117	-1	146	272	394	-24	90	103	89	-46	49	39	76	92
8x32	ringp3	-34	120	107	65	-3	88	271	169	-42	60	79	72	-43	52	49	24	65
8x32	ring12	-42	93	130	113	-16	128	345	157	-44	82	61	112	-53	71	115	47	81
8x32	ring13	-59	23	100	54	-54	150	300	204	-50	74	25	136	-64	58	98	90	68
8x32	swrl02	-48	74	111	169	-29	246	322	223	-40	127	167	221	-56	70	82	122	109
8x32	rwrl03	-63	66	195	168	-60	153	186	317	-60	98	245	196	-54	144	217	211	122
8x32	strnc	-3	184	95	129	13	101	153	122	-5	124	306	195	6	106	69	64	104
8x32	strnb	-25	64	126	103	-31	81	61	80	-27	39	63	44	-34	51	69	43	44
8x32	strnw	8	113	248	95	3	126	226	258	13	149	355	161	10	108	100	73	128
8x32	rsrb	-48	93	153	62	-7	271	519	235	-2	268	158	73	-20	42	98	148	128
8x32	rsrw	-64	38	64	28	-54	83	75	110	-53	58	85	64	-65	62	6	102	34
8x32	mesh4	-53	38	165	83	-49	138	290	114	-38	147	125	75	-38	25	173	132	83
8x32	tor04	-55	126	147	188	-32	96	264	128	-49	92	148	351	-31	231	102	144	116
8x32	torus	-56	142	228	254	-40	114	266	209	-52	133	431	163	-53	112	90	268	138
8x32	fullc	-81	22	156	90	-80	110	226	227	-82	71	244	265	-80	63	144	257	97
8x32	path	-39	69	77	110	-7	102	100	79	-5	93	164	169	1	148	107	55	76
8x32	pathb	-32	9	103	56	-30	77	80	109	-23	46	110	82	-27	93	94	78	52
8x32	pathw	18	95	159	151	44	137	163	121	121	164	178	117	34	196	177	48	120
8x32	tree	-19	65	51	48	-5	81	42	61	-8	35	16	75	3	43	33	-3	32
8x32	treeb	-11	29	80	51	-11	53	58	93	-24	37	69	54	-10	27	24	59	36
8x32	treew	-43	6	49	85	-42	13	39	14	-49	22	47	61	-37	25	43	0	15
8x32	rtree	107	84	152	106	112	234	88	55	196	195	183	149	161	156	177	89	140
8x32	lad	-35	181	92	85	-35	69	178	121	-46	90	162	226	-40	62	93	122	83
8x32	ladlp	8	110	120	100	1	272	300	248	-13	118	319	325	-11	147	347	150	**159**
8x32	rand2	-28	81	435	300	-30	206	130	175	-6	107	438	189	-4	158	174	178	156
8x32	rand3	-42	143	245	227	-38	162	326	250	-38	138	534	158	-23	60	195	193	156
8x32	rand4	-50	111	207	274	-57	167	201	216	-54	119	131	235	-51	43	143	280	120
16x16	star1d	-12	-29	-42	-45	-14	-23	-36	-40	-22	-32	-33	-13	-47	-20	-33	-38	-30
16x16	star2d	-30	143	213	78	-26	86	158	81	-33	85	81	135	-53	48	64	69	69
16x16	starb	-30	-46	-47	-48	-45	-42	-41	-36	-31	-58	-42	-36	-42	-42	-55	-30	-42
16x16	starw	123	114	108	76	123	183	92	35	166	77	154	73	144	121	22	46	104
16x16	rstarb	-30	86	159	181	79	168	99	183	-17	119	121	135	1	95	129	197	107
16x16	rstarw	-95	-72	-46	-42	-94	-66	-41	-46	-95	-62	-60	-60	-95	-61	-60	-49	-65
16x16	ring1d	-41	57	83	45	-44	23	57	30	-47	32	61	26	-60	11	37	7	17
16x16	ring2d	-66	29	115	90	-72	33	46	84	-66	54	87	67	-67	33	83	60	32
16x16	ringp1	-63	57	123	257	-64	29	42	63	-67	65	97	64	-70	5	85	46	42
16x16	ringp2	-64	44	80	102	-53	47	80	121	-64	9	118	98	-58	20	113	46	40
16x16	ringp3	-58	55	170	58	-61	56	66	41	-60	83	128	86	-66	27	73	37	40
16x16	ring12	-78	90	135	162	-70	47	134	100	-74	25	135	89	-73	21	40	115	50
16x16	ring13	-80	50	83	178	-76	6	46	103	-79	14	114	125	-76	6	88	167	42
16x16	swrl02	-71	56	183	183	-71	31	180	194	-71	70	106	147	-62	42	63	67	65
16x16	rwrl03	-78	66	207	158	-69	-3	101	72	-78	39	131	110	-72	15	71	65	46
16x16	strnc	-48	60	63	56	-34	27	30	39	-51	61	61	69	-54	20	32	16	22
16x16	strnb	-51	21	31	29	-53	5	36	-11	-61	9	-1	29	-63	16	-11	14	-4
16x16	strnw	-29	139	79	99	-42	122	119	100	-44	101	117	64	-35	63	108	53	63
16x16	rsrb	-62	103	189	156	-49	68	191	250	-60	70	165	190	-37	49	145	144	95
16x16	rsrw	-83	-30	-10	37	-85	-30	32	14	-85	-26	10	37	-87	-54	-11	-7	-24
16x16	mesh4	-74	26	9	16	-70	-6	46	24	-77	-7	64	30	-76	-17	18	22	-5
16x16	tor04	-68	51	118	96	-65	23	131	74	-71	63	49	156	-72	37	46	92	41
16x16	torus	-78	55	189	74	-78	18	159	159	-76	3	96	129	-74	19	80	97	48
16x16	fullc	-90	-43	132	185	-90	-46	95	166	-95	-58	68	149	-89	-42	53	131	27
16x16	path	-61	0	10	45	-60	7	11	0	-63	35	41	-1	-62	4	-1	-6	-6
16x16	pathb	-70	11	17	56	-62	-3	11	2	-69	-18	13	21	-73	-21	1	-15	-12
16x16	pathw	5	69	58	47	-8	70	33	11	-7	71	116	57	-57	26	30	66	38
16x16	tree	-54	-25	-14	1	-36	-40	-12	-8	-52	-25	-10	-10	-44	-34	-24	-7	-25
16x16	treeb	-53	-12	-4	-23	-37	-26	-6	-17	-45	-42	5	1	-44	-27	0	8	-20
16x16	treew	-67	-36	-13	-29	-62	-24	-31	-35	-69	-33	-14	-22	-68	-43	-28	-29	-38
16x16	rtree	49	121	94	96	104	68	51	16	111	155	88	62	85	62	77	45	80
16x16	lad	-80	54	58	127	-73	14	70	75	-76	-7	75	78	-77	-19	25	47	18
16x16	ladlp	-66	35	91	131	-63	32	54	88	-70	67	46	91	-65	59	55	40	33
16x16	rand2	-29	85	49	66	-46	47	70	112	-46	-14	64	79	-55	48	55	67	34
16x16	rand3	-59	71	131	82	-59	26	177	216	-62	32	99	141	-55	26	99	105	61
16x16	rand4	-73	62	211	55	-69	44	146	142	-64	15	127	186	-68	15	62	146	58
x	**Avg**	-37	64	108	95	-27	79	**135**	112	-31	61	113	100	-35	48	73	73	x

Table 7. Comparison of the considered simulation variants in the context of the percentage improvement of the MNIA results in relation to the results obtained without migration. The results were averaged for the considered problems and specified variants. Best results are shown in bold.

Migration interval				Migration model				$Npop$ x $Nind$	
M1	M4	M10	M20	BS	RW	TS	RN	8 x 32	16 x 16
−32	63	**107**	95	57	**75**	61	40	**89**	28

Table 8. Average ranking position of the considered PBAs (including MNIA) for the best model of RW migration (see Table 7). The best results including MNIA are marked in bold, the best results without MNIA are underlined (to show the problem of selecting a single population-based algorithm).

F	CS				WO				FW				MNIA			
	M1	M4	M10	M20	M1	M4	M10	M20	M1	M4	M10	M20	M1	M4	M10	M20
F01	11.7	10.8	9.9	9.2	5.1	2.5	**2.1**	2.4	15.4	14.5	13.8	14.1	7.9	5.6	5.3	5.5
F02	5.8	8.4	10.2	11.2	8.8	4.1	2.4	**2.1**	15.5	14.7	13.7	14.2	7.9	6.8	5.3	5.0
F03	<u>4.7</u>	5.8	7.0	7.9	11.5	10.6	9.8	9.8	14.0	13.7	15.6	14.7	4.6	2.7	1.8	**1.7**
F04	15.7	14.6	14.0	13.7	<u>5.6</u>	6.2	6.4	7.1	11.0	9.4	9.4	9.6	4.1	3.3	**3.0**	3.0
F05	13.0	13.4	14.1	14.4	5.4	2.7	**2.1**	2.1	14.5	10.3	9.9	10.4	7.7	5.8	5.1	5.1
F06	15.5	14.6	14.0	13.9	4.1	2.6	2.2	**1.8**	11.4	9.9	9.7	9.9	8.3	6.2	6.2	5.7
F07	10.8	9.9	10.4	11.0	5.5	2.2	2.2	**2.0**	15.1	14.3	13.9	14.6	7.5	6.1	5.0	5.4
F08	14.5	14.5	14.5	14.5	3.9	2.5	2.5	**2.0**	11.3	9.6	9.3	9.7	8.4	6.5	6.2	6.1
F09	12.4	11.1	10.4	9.7	5.9	2.6	**1.8**	2.5	14.8	14.0	13.4	13.8	7.9	5.5	4.9	5.2
F10	**2.6**	3.6	4.9	6.6	10.2	8.6	7.7	7.9	15.5	14.6	13.9	13.9	9.8	6.1	5.1	5.1
F11	9.3	10.2	11.2	11.5	5.8	3.0	2.0	**1.9**	15.5	14.6	13.9	13.9	7.7	6.1	4.9	4.7
F12	<u>5.1</u>	6.9	8.3	9.2	9.7	7.1	5.9	5.8	14.1	13.6	14.0	13.9	6.0	4.7	3.5	**3.3**
F13	14.5	14.1	14.0	14.1	4.7	2.6	2.2	**1.9**	12.6	10.1	9.8	10.1	8.0	6.0	5.8	5.5
F14	11.7	10.6	10.4	10.2	5.7	2.5	**2.0**	2.3	14.9	14.1	13.6	14.1	7.8	5.8	4.9	5.3
F15	12.3	12.5	12.9	13.0	5.6	3.9	**3.3**	3.5	13.8	11.7	11.4	11.5	6.6	5.1	4.4	4.3
F16	10.6	11.2	11.7	12.1	6.8	4.4	3.7	**3.5**	13.4	11.5	11.5	11.6	7.2	5.5	4.8	4.6
F17	13.4	12.6	12.6	12.8	4.8	2.5	2.2	**1.9**	13.1	11.7	11.4	11.8	7.9	6.1	5.6	5.6
F18	12.4	11.9	12.0	12.2	4.9	2.4	2.3	**2.1**	13.6	12.4	12.0	12.6	7.9	6.2	5.5	5.7
F19	13.2	12.7	12.5	12.3	5.1	2.8	**2.4**	2.5	13.2	11.8	11.4	11.7	7.9	5.8	5.3	5.4
F20	8.1	8.1	8.5	8.9	7.8	5.3	**4.6**	4.9	14.8	13.8	13.2	13.4	8.5	5.7	5.0	5.0
F21	<u>4.6</u>	6.3	7.7	8.8	10.0	7.5	6.3	6.3	14.7	14.0	14.3	14.1	6.9	5.0	3.9	**3.7**
F22	11.7	11.8	12.1	12.2	6.4	5.0	<u>4.5</u>	4.5	13.1	11.6	11.7	11.7	6.5	4.8	4.3	**4.1**
F23	11.4	11.2	11.4	11.5	6.4	4.2	<u>3.7</u>	3.9	13.9	12.3	12.1	12.4	6.7	5.1	4.3	4.4
F24	13.5	13.4	13.6	13.7	5.1	3.1	<u>2.6</u>	<u>2.6</u>	13.2	10.7	10.4	10.7	7.5	5.7	5.2	5.0
F25	11.8	11.5	11.6	11.9	5.8	3.2	2.7	<u>2.6</u>	13.7	12.3	12.1	12.4	7.6	5.8	5.2	5.1
AVG	10.8	10.9	11.2	11.5	6.4	4.2	**3.6**	**3.6**	13.9	12.5	12.2	12.4	7.4	5.5	4.8	4.8

The simulation conclusions can be summarized as follows:

- The formula of subpopulation cooperation has a large impact on the effectiveness of the multi-population algorithm used in the simulations (see Table 6).

For most of the considered test functions, satisfactory results were obtained using the following formula: 'rstarb'+M10+RW+8x32 (see Table 6).

- The results obtained for the MNIA using different PBAs are comparable to the results obtained for the best population algorithm for a given function. Moreover, for several functions, MNIA gave the best results (see Table 8). There were also large differences in the results, depending on the PBA used. Therefore, the use of MNIA eliminates the need to test different PBAs, which makes it easier to find a satisfactory solution.

- The best migration topology is 'rstarb' for variant $Npop = 8$, $Nind = 32$ (+167% - see Table 6). This is the topology proposed in this paper and it was developed by analogy with the topologies considered in the literature. By contrast, the best standard topology was 'ladp' the variant $Npop = 8$, $Nind = 32$ (+159% - see Table 6).

- The adopted migration interval affects the obtained results. The best interval is M10, especially for the BS model (see Table 6). It can be seen that applying an interval that is too large or too small does not have a positive effect on the obtained results.

- Replacing the best or random individuals does not bring much improvement, causing premature algorithm convergence (BS) or lack of it (RN) (see Table 6).

5   Conclusions

In this paper, various ways of affecting subpopulations in the multi-population nature-inspired algorithm (MNIA) were considered. More specifically, the following selected combinations of 36 migration topologies, 4 migration models, 4 migration intervals, and 2 variants of the number of subpopulations were tested. The conducted simulations show that the use of the island model with different PBAs and different ways of influencing the subpopulation may significantly increase the effectiveness of the optimization. This confirmed the validity of the assumptions made in this paper and has encouraged further research on interactions between subpopulations.

Our plans for further research include automatic selection of how subpopulations interact and using the resulting solutions to increase the effectiveness of biometric methods [63,66,70], population-based algorithms [47–51], fuzzy systems [43–46,55], neural networks and other deep learning methods [5–7,12,21,31,38] and their various implementations. For example, in a molecular implementation [24,25] intelligent systems can be realized by the individual molecules [3,19,22] or even precisely designed atoms layouts [20,23] constructed by multi-population approach.

Acknowledgment. This paper was financed under the program of the Minister of Science and Higher Education under the name 'Regional Initiative of Excellence' in the years 2019-2022, project number 020/RID/2018/19 with the amount of financing PLN 12 000 000.

References

1. Ahmadianfar, I., Bozorg-Haddad, O., Chu, X.: Gradient-based optimizer: a new metaheuristic optimization algorithm. Inf. Sci. **540**, 131–159 (2020)
2. Asadzadeh, L.: A parallel artificial bee colony algorithm for the job shop scheduling problem with a dynamic migration strategy. Comput. Ind. Eng. **102**, 359–367 (2016)
3. Bałanda, M., Pełka, R., Fitta, M., Laskowski, Ł, Laskowska, M.: Relaxation and magnetocaloric effect in the Mn 12 molecular nanomagnet incorporated into mesoporous silica: a comparative study. RSC Adv. **6**(54), 49179–49186 (2016)
4. Bartczuk, Ł, Przybył, A., Cpałka, K.: A new approach to linear modelling of dynamic systems based on fuzzy rules. Int. J. Appl. Math. Comput. Sci. (AMCS) **26**(3), 603–621 (2016)
5. Bilski, J., Wilamowski, B.M.: Parallel learning of feedforward neural networks without error backpropagation. In: Rutkowski, L., Korytkowski, M., Scherer, R., Tadeusiewicz, R., Zadeh, L.A., Zurada, J.M. (eds.) ICAISC 2016. LNCS (LNAI), vol. 9692, pp. 57–69. Springer, Cham (2016). https://doi.org/10.1007/978-3-319-39378-0_6
6. Bilski, J., Kowalczyk, B., Marchlewska, A., Żurada, J.M.: Local Levenberg-Marquardt algorithm for learning feedforward neural networks. J. Artif. Intell. Soft Comput. Res. **10**(4), 299–316 (2020)
7. Bilski, J., Rutkowski, L., Smoląg, J., Tao, D.: A el method for speed training acceleration of recurrent neural networks. Inf. Sci. **553**, 266–279 (2021). https://doi.org/10.1016/j.ins.2020.10.025
8. Braik, M.S.: Chameleon swarm algorithm: a bio-inspired optimizer for solving engineering design problems. Expert Syst. Appl. **174**, 114685 (2021)
9. Chen, Y., He, F., Li, H., Zhang, D., Wu, Y.: A full migration BBO algorithm with enhanced population quality bounds for multimodal biomedical image registration. Appl. Soft Comput. **93**, 106335 (2020)
10. Dziwiński, P., Bartczuk, Ł., Paszkowski, J.: A new auto adaptive fuzzy hybrid particle swarm optimization and genetic algorithm. J. Artif. Intell. Soft Comput. Res. **10** (2020)
11. Dziwiński, P., Przybył, P., Trier, P., Paszkowski, J., Hayashi, Y.: Hardware implementation of a Takagi-Sugeno neuro-fuzzy system optimized by a population algorithm. J. Artif. Intell. Soft Comput. Res. **11**(3), 243–266 (2021). https://doi.org/10.2478/jaiscr-2021-0015
12. Duda, P., Jaworski, M., Cader, A., Wang, L.: On training deep neural networks using a streaming approach. J. Artif. Intell. Soft Comput. Res. **10**(1), 15–26 (2020). https://doi.org/10.2478/jaiscr-2020-0002
13. Hasanzadeh, M.R., Keynia, F.: A new population initialisation method based on the Pareto 80/20 rule for meta-heuristic optimisation algorithms. IET Software **15**(5), 323–347 (2021)
14. Holly, S., Nieße, A.: Dynamic communication topologies for distributed heuristics in energy system optimization algorithms, pp. 191–200 (2021)
15. Hussain, K., Mohd Salleh, M.N., Cheng, S., Shi, Y.: Metaheuristic research: a comprehensive survey. Artif. Intell. Rev. **52**(4), 2191–2233 (2019)
16. Karaboga, D., Aslan, S.: A new emigrant creation strategy for parallel artificial bee colony algorithm. In: 9th International Conference on Electrical and Electronics Engineering (ELECO), pp. 689–694 (2015). https://doi.org/10.1109/eleco.2015.7394477

17. Korytkowski, M., Senkerik, R., Scherer, M.M., Angryk, R.A., Kordos, M., Siwocha, A.: Efficient image retrieval by fuzzy rules from boosting and metaheuristic. J. Artif. Intell. Soft Comput. Res. **10**(1), 57–69 (2020). https://doi.org/10.2478/jaiscr-2020-0005

18. Krell, E., Sheta, A., Balasubramanian, A.P.R., King, S.A.: Collision-free autonomous robot navigation in unknown environments utilizing PSO for path planning. J. Artif. Intell. Soft Comput. Res. **9** (2019)

19. Laskowska, M., et al.: Magnetic behaviour of Mn12-stearate single-molecule magnets immobilized inside SBA-15 mesoporous silica matrix. J. Magn. Magn. Mater. **478**, 20–27 (2019)

20. Laskowska, M., Oyama, M., Kityk, I., Marszalek, M., Dulski, M., Laskowski, L.: Surface functionalization by silver-containing molecules with controlled distribution of functionalities. Appl. Surf. Sci. **481**, 433–436 (2019)

21. Laskowski, L: Hybrid-maximum neural network for depth analysis from stereoimage. In: Rutkowski, L., Scherer, R., Tadeusiewicz, R., Zadeh, L.A., Zurada, J.M. (eds.) ICAISC 2010. LNCS (LNAI), vol. 6114, pp. 47–55. Springer, Heidelberg (2010). https://doi.org/10.1007/978-3-642-13232-2_7

22. Laskowski, L., Kityk, I., Konieczny, P., Pastukh, O., Schabikowski, M., Laskowska, M.: The separation of the Mn12 single-molecule magnets onto spherical silica nanoparticles. Nanomaterials **9**(5), 764 (2019)

23. Laskowski, L, et al.: Multi-step functionalization procedure for fabrication of vertically aligned mesoporous silica thin films with metal-containing molecules localized at the pores bottom. Microporous Mesoporous Mater. **274**, 356–362 (2019)

24. Laskowski, L, Laskowska, M., Jelonkiewicz, J., Boullanger, A.: Molecular approach to hopfield neural network. In: Rutkowski, L., Korytkowski, M., Scherer, R., Tadeusiewicz, R., Zadeh, L.A., Zurada, J.M. (eds.) ICAISC 2015. LNCS (LNAI), vol. 9119, pp. 72–78. Springer, Cham (2015). https://doi.org/10.1007/978-3-319-19324-3_7

25. Laskowski, L, Laskowska, M., Jelonkiewicz, J., Boullanger, A.: Spin-glass implementation of a hopfield neural structure. In: Rutkowski, L., Korytkowski, M., Scherer, R., Tadeusiewicz, R., Zadeh, L.A., Zurada, J.M. (eds.) ICAISC 2014. LNCS (LNAI), vol. 8467, pp. 89–96. Springer, Cham (2014). https://doi.org/10.1007/978-3-319-07173-2_9

26. Liu, T., Gao, X., Yuan, Q.: An improved gradient-based NSGA-II algorithm by a new chaotic map model. Soft. Comput. **21**(23), 7235–7249 (2017)

27. Liu, Y., et al.: Chaos-assisted multi-population salp swarm algorithms: framework and case studies. Expert Syst. Appl. **168**, 114369 (2021)

28. Lynn, N., Ali, M.Z., Suganthan, P.N.: Population topologies for particle swarm optimization and differential evolution. Swarm Evol. Comput. **39**, 24–35 (2018)

29. Lapa, K., Cpałka, K., Laskowski, L., Cader, A., Zeng, Z.: Evolutionary algorithm with a configurable search mechanism. J. Artif. Intell. Soft Comput. Res. **10** (2020)

30. Ma, H., Shen, S., Yu, M., Yang, Z., Fei, M., Zhou, H.: Multi-population techniques in nature inspired optimization algorithms: a comprehensive survey. Swarm Evol. Comput. **44**, 365–387 (2019)

31. Mańdziuk, J., Żychowski, A.: Dimensionality reduction in multilabel classification with neural networks. In: International Joint Conference on Neural Networks (IJCNN 2019), pp. 1–8 (2019). https://doi.org/10.1109/IJCNN.2019.8852156

32. Medina, A., Tosca P.G., Ramírez-Torres, J.: A Comparative Study of Neighborhood Topologies for Particle Swarm Optimizers, pp. 152–159 (2009)

33. Migallón, H., Jimeno-Morenilla, A., Rico, H., Sánchez-Romero, J.L., Belazi, A.: Multi-level parallel chaotic Jaya optimization algorithms for solving constrained engineering design problems. J. Supercomput. **77**(11), 12280–12319 (2021). https://doi.org/10.1007/s11227-021-03737-0

34. Mirjalili, S., Lewis, A.: The whale optimization algorithm. Adv. Eng. Softw. **95**, 51–67 (2016)

35. Mizera, M., Nowotarski, P., Byrski, A., Kisiel-Dorohinicki, M.: Fine tuning of agent-based evolutionary computing. J. Artif. Intell. Soft Comput. Res. **9** (2019)

36. Najmeh, S.J., Salwani, A., Abdul, R.H.: Multi-population cooperative bat algorithm-based optimization of artificial neural network model. Inf. Sci. **294**, 628–644 (2015)

37. Nasim, A., Burattini, L., Fateh, M.F., Zameer, A.: Solution of linear and-linear boundary value problems using population-distributed parallel differential evolution. J. Artif. Intell. Soft Comput. Res. **9** (2019)

38. Niksa-Rynkiewicz, T., Szewczuk-Krypa, N., Witkowska, A., Cpałka, K., Zalasiński, M., Cader, A.: Monitoring regenerative heat exchanger in steam power plant by making use of the recurrent neural network. J. Artif. Intell. Soft Comput. Res. **11**(2), 143–155 (2021). https://doi.org/10.2478/jaiscr-2021-0009

39. Okulewicz, M., Mańdziuk, J.: The impact of particular components of the PSO-based algorithm solving the dynamic vehicle routing problem. Appl. Soft Comput. **58**, 586–604 (2017). https://doi.org/10.1016/j.asoc.2017.04.070

40. Ono, K., Hanada, Y., Kuma M., Kimura, M.: Enhancing island model genetic programming by controlling frequent trees. J. Artif. Intell. Soft Comput. Res. **9** (2019)

41. Pourchot, A., Sigaud, O.: CEM-RL: combining evolutionary and gradient-based methods for policy search. arXiv preprint arXiv:1810.01222 (2018)

42. Sanu, M., Jeyakumar, G.: Empirical performance analysis of distributed differential evolution for varying migration topologies. Int. J. Appl. Eng. Res. **10**, 11919–11932 (2015)

43. Scherer, R., Rutkowski, L.: Neuro-fuzzy relational classifiers. In: Rutkowski, L., Siekmann, J.H., Tadeusiewicz, R., Zadeh, L.A. (eds.) ICAISC 2004. LNCS (LNAI), vol. 3070, pp. 376–380. Springer, Heidelberg (2004). https://doi.org/10.1007/978-3-540-24844-6_54

44. Scherer, R., Rutkowski, L.: Neuro-fuzzy relational systems. In: Proceedings of FSKD 2002, pp. 44–48 (2002)

45. Scherer, R., Rutkowski, L.: Relational equations initializing neuro-fuzzy system. In: 10th Zittau Fuzzy Colloquium, Zittau, Germany, pp. 18–22 (2002)

46. Scherer, R.: Neuro-fuzzy systems with relation matrix. In: Rutkowski, L., Scherer, R., Tadeusiewicz, R., Zadeh, L.A., Zurada, J.M. (eds.) ICAISC 2010. LNCS (LNAI), vol. 6113, pp. 210–215. Springer, Heidelberg (2010). https://doi.org/10.1007/978-3-642-13208-7_27

47. Słowik, A.: Steering of balance between exploration and exploitation properties of evolutionary algorithms - mix selection. In: Rutkowski, L., Scherer, R., Tadeusiewicz, R., Zadeh, L.A., Zurada, J.M. (eds.) ICAISC 2010. LNCS (LNAI), vol. 6114, pp. 213–220. Springer, Heidelberg (2010). https://doi.org/10.1007/978-3-642-13232-2_26

48. Słowik, A., Białko, M.: Design and optimization of combinational digital circuits using modified evolutionary algorithm. In: Rutkowski, L., Siekmann, J.H., Tadeusiewicz, R., Zadeh, L.A. (eds.) ICAISC 2004. LNCS (LNAI), vol. 3070, pp. 468–473. Springer, Heidelberg (2004). https://doi.org/10.1007/978-3-540-24844-6_69

49. Słowik, A., Białko, M.: Modified version of roulette selection for evolution algorithms – the fan selection. In: Rutkowski, L., Siekmann, J.H., Tadeusiewicz, R., Zadeh, L.A. (eds.) ICAISC 2004. LNCS (LNAI), vol. 3070, pp. 474–479. Springer, Heidelberg (2004). https://doi.org/10.1007/978-3-540-24844-6_70

50. Słowik, A., Białko, M.: Design and optimization of IIR digital filters with non-standard characteristics using continuous ant colony optimization algorithm. In: Darzentas, J., Vouros, G.A., Vosinakis, S., Arnellos, A. (eds.) SETN 2008. LNCS (LNAI), vol. 5138, pp. 395–400. Springer, Heidelberg (2008). https://doi.org/10.1007/978-3-540-87881-0_39

51. Słowik, A., Białko, M.: Design of IIR digital filters with-standard characteristics using differential evolution algorithm. Bull. Pol. Acad. Sci.-Tech. Sci. 55(4), 359–363 (2007)

52. Słowik, A., Cpałka, K.: Hybrid approaches to nature-inspired population-based intelligent optimization for industrial applications. IEEE Trans. Ind. Inf. 18(1), 546–558 (2022)

53. Słowik, A., Cpałka, K., Łapa, K.: Multi-population nature-inspired algorithm (MNIA) for the designing of interpretable fuzzy systems. IEEE Trans. Fuzzy Syst. 28(6), 1125–1139 (2020)

54. Sörensen, K.: Metaheuristics-the metaphor exposed. Int. Trans. Oper. Res. 22(1), 3–18 (2015)

55. Starczewski, J., Scherer, R., Korytkowski, M., Nowicki, R.: Modular type-2 neuro-fuzzy systems. In: Wyrzykowski, R., Dongarra, J., Karczewski, K., Wasniewski, J. (eds.) PPAM 2007. LNCS, vol. 4967, pp. 570–578. Springer, Heidelberg (2008). https://doi.org/10.1007/978-3-540-68111-3_59

56. Suganthan, P.N., et al.: Problem definitions and evaluation criteria for the CEC 2005 special session on real-parameter optimization. KanGAL report (2005)

57. Szczypta, J., Przybył, A., Cpałka, K.: Some aspects of evolutionary designing optimal controllers. In: Rutkowski, L., Korytkowski, M., Scherer, R., Tadeusiewicz, R., Zadeh, L.A., Zurada, J.M. (eds.) ICAISC 2013. LNCS (LNAI), vol. 7895, pp. 91–100. Springer, Heidelberg (2013). https://doi.org/10.1007/978-3-642-38610-7_9

58. Tabassum, M.F., Saeed, M., Akgül, A., Farman, M., Akram, S.: Solution of chemical dynamic optimization systems using el differential gradient evolution algorithm. Phys. Scr. 96(3), 035212 (2021)

59. Tan, Y.: Fireworks Algorithm. Springer, Heidelberg (2015). https://doi.org/10.1007/978-3-662-46353-6

60. Wang, L., Maciejewski, A., Siegel H., Roychowdhury, V.: A comparative study of five parallel genetic algorithms using the traveling salesman problem. In: Proceedings of the 11th International Parallel Processing Symposium. IEEE Computer Society Press (1998)

61. Wei, Y., et al.: Vehicle emission computation through microscopic traffic simulation calibrated using genetic algorithm. J. Artif. Intell. Soft Comput. Res. 9 (2019)

62. Yang, X.S., Deb, S.: Cuckoo search: recent advances and applications. Neural Comput. Appl. 24(1), 169–174 (2014)

63. Zalasiński, M., Cpałka, K.: A new method for signature verification based on selection of the most important partitions of the dynamic signature. Neurocomputing 10, 13–22 (2018)

64. Zalasiński, M., Cpałka, K.: A new method of on-line signature verification using a flexible fuzzy one-class classifier. Academic Publishing House EXIT, pp. 38–53 (2011)

65. Zalasiński, M., Cpałka, K.: Novel algorithm for the on-line signature verification using selected discretization points groups. In: Rutkowski, L., Korytkowski, M., Scherer, R., Tadeusiewicz, R., Zadeh, L.A., Zurada, J.M. (eds.) ICAISC 2013. LNCS (LNAI), vol. 7894, pp. 493–502. Springer, Heidelberg (2013). https://doi.org/10.1007/978-3-642-38658-9_44

66. Zalasiński, M., Cpałka, K., Hayashi, Y.: A new approach to the dynamic signature verification aimed at minimizing the number of global features. In: Rutkowski, L., Korytkowski, M., Scherer, R., Tadeusiewicz, R., Zadeh, L.A., Zurada, J.M. (eds.) ICAISC 2016. LNCS (LNAI), vol. 9693, pp. 218–231. Springer, Cham (2016). https://doi.org/10.1007/978-3-319-39384-1_20

67. Zalasiński, M., Cpałka, K., Hayashi, Y.: New method for dynamic signature verification based on global features. In: Rutkowski, L., Korytkowski, M., Scherer, R., Tadeusiewicz, R., Zadeh, L.A., Zurada, J.M. (eds.) ICAISC 2014. LNCS (LNAI), vol. 8468, pp. 231–245. Springer, Cham (2014). https://doi.org/10.1007/978-3-319-07176-3_21

68. Zalasiński, M., Cpałka, K., Laskowski, Ł., Wunsch, D.C., Przybyszewski, K.: An algorithm for the evolutionary-fuzzy generation of on-line signature hybrid descriptors. J. Artif. Intell. Soft Comput. Res. **10**(3), 173–187 (2020). https://doi.org/10.2478/jaiscr-2020-0012

69. Zalasiński, M., Łapa, K., Cpałka, K.: New algorithm for evolutionary selection of the dynamic signature global features. In: Rutkowski, L., Korytkowski, M., Scherer, R., Tadeusiewicz, R., Zadeh, L.A., Zurada, J.M. (eds.) ICAISC 2013. LNCS (LNAI), vol. 7895, pp. 113–121. Springer, Heidelberg (2013). https://doi.org/10.1007/978-3-642-38610-7_11

70. Zalasiński, M., Łapa, K., Cpałka, K.: Prediction of values of the dynamic signature features. Expert Syst. Appl. **104**, 86–96 (2018)

71. Zalasiński, M., Łapa, K., Cpałka, K., Przybyszewski, K., Yen, G.G.: On-line signature partitioning using a population based algorithm. J. Artif. Intell. Soft Comput. Res. **10**(1), 5–13 (2020). https://doi.org/10.2478/jaiscr-2020-0001

Automated Design of Dynamic Heuristic Set Selection for Cross-Domain Selection Hyper-Heuristics

Ahmed Hassan[(⊠)] and Nelishia Pillay

University of Pretoria, Pretoria, South Africa
ahmedhassan@aims.ac.za, npillay@cs.up.ac.za

Abstract. Selection hyper-heuristics have been used successfully to solve hard optimization problems. These techniques choose a heuristic or a group of heuristics to create a solution and/or improve it. In a prior study, we proposed an approach that changes the heuristic set from which the hyper-heuristic is allowed to choose dynamically and that led to improving the performance of the hyper-heuristic. Previously, we manually designed the proposed approach which involved challenging design decisions and parameter tuning. In this study, we automate the design of the previously proposed approach using grammatical evolution to reduce human involvement in the design process. The proposed automated approach is evaluated on the domains of the CHeSC challenge. It is found that the automated design reduces the design time remarkably and performs as good as the manual design.

Keywords: Selection hyper-heuristics · Dynamic heuristic sets · Grammatical evolution

1 Introduction

Selection hyper-heuristics have been used successfully to solve hard optimization problems such as timetabling [19,22], bin packing [7,20], and scheduling [12]. These techniques choose a *low-level heuristic* or a group of low-level heuristics to create a solution and/or improve it [3,5]. The performance of selection hyper-heuristics is influenced by the quality of the low-level heuristics utilized by the hyper-heuristic [14,21,25].

Most often, the entire set of all available low-level heuristics (referred to as the *universal set*) is used by the hyper-heuristic. The universal set may include poor low-level heuristics or some low-level heuristics that are effective only at particular phases during the search. For instance, Remde et al. [24] noted that some low-level heuristics that are ineffective at the beginning of the search can be valuable at the end of the search. Therefore, the universal set may harm the performance of the hyper-heuristic if used as is. Recent investigations have demonstrated that eliminating poor heuristics from the universal set before running the hyper-heuristic enhances the performance [8,25]. These studies generate a *static set*, which is a subset of the universal set, that remains unchanged for the whole lifespan of the hyper-heuristic.

Recently, we developed a dynamic heuristic set selection approach (DHSS) which changes the heuristic set, called the *active set*, from which the hyper-heuristic can choose dynamically such that different heuristics can be part of the active set at different phases during the lifespan of the hyper-heuristic [9].

The design of DHSS was challenging as we were faced with difficult design decisions such as how to measure the quality of low-level heuristics, when to update the active set, which criterion to use to update the active set, and whether the poor heuristics should be removed at some point during the search. Furthermore, these design decisions are parameterized and require parameter tuning.

In this paper, we propose grammatical evolution to automate the design of DHSS to overcome the challenges of the manual design. The automated design includes finding appropriate design choices and setting the parameters of those design choices. The choice of GE is motivated by its prior success in the automated design of search techniques [6,16]. In addition, the expressiveness of grammars eases the representation of complex systems [23].

The automated design is evaluated on the domains of the CHeSC challenge. The automated approach performs as good as the manual approach (outperforms the manual approach slightly) and reduces the design time significantly. The main contribution of this paper is an automated approach for DHSS that reduces the design time substantially without incurring a performance loss when compared to the best manually designed approach.

The rest of this paper is organized as follows. In Sect. 2, we provide background information. Section 3 describes the proposed GE for automating the active set approach. The experimental setup is outlined in Sect. 4. Section 5 discusses the results. The paper is concluded in Sect. 6.

2   Background

In this section, we present background information required for subsequent discussions.

2.1   Dynamic Heuristic Set Selection

In a prior study [9], we proposed DHSS to decide which heuristic to include in the active set at different points during the lifespan of the hyper-heuristic. The design of DHSS involves challenging design decisions listed below.

1. *Update strategy* to decide when to update the active set and which low-level heuristics to include in the active set.
2. *Removal strategy* to decide which low-level heuristics to permanently remove and when to do that.
3. *Reset strategy* to decide when to *reset* the active set such that it will include all low-level heuristics that are not permanently removed.
4. *Performance measure* to evaluate the quality of low-level heuristics.

To create a fully functional DHSS, each *design decision* is assigned a *design choice*. The design choices for the update strategy are described in Table 1. The design choices for the removal strategy are described in Table 2. The design choices for the reset strategy are listed in Table 3. There are 24 design choices for the performance measures that are based on three performance indicators: the percentage or the frequency of improvements, the percentage or the frequency of disimprovements, and the execution time of the heuristics.

Table 1. The design choices for the update strategy.

Choice	Description
PhDom	Removes all dominated heuristics (those that use more time but yield poorer results) from the active set and updates the active set every pl iterations
PhQi	Converts the heuristic performance into quality indexes and excludes all heuristics that perform below the average index. Updates the active set every pl iterations
PhGrd	Selects greedily the top tp heuristics every pl iterations
PtDom	Same as PhDom but includes a patience factor pt that forces the updates if the best solution is not improved for $pt \times wait_{max}$ where $wait_{max}$ is the maximum number of iterations we waited so far before the best solution is improved
PtQi	Same as PhQi but includes a patience factor pt (as explained for PtDom)
PtGrd	Same as PhGrd but includes a patience factor pt (as explained for PtDom)

Table 2. Design choices for removal strategy.

Choice	Description
NoRem	No removal strategy
PtRem	Wait for $pt \times wait_{max}$ before removing the worst heuristic permanently where pt and $wait_{max}$ are as explained for PtDom
FqRem	Divides the search into phases of equal length and removes the worst heuristic at the end of each phase where the number of phases is determined by a frequency parameter fq
IndRem	Removes all heuristics that perform worse than the average performance calculated from the ratio between the percentage/frequency of improvements and the percentage/frequency of disimprovements
GrpRem	Removes all heuristics that perform worse than the average performance calculated from the ratio between the percentage/frequency of improvements/disimprovement and the percentage/frequency of improvement/disimprovements for all other heuristics
ConsRem	Removes all heuristics that perform worse than the average performances calculated from IndRem and GrpRem

Table 3. The design choices for the reset strategy.

Choice	Description
NoRes	No reset strategy
PtRes	Waits for $pt \times wait_{max}$ before resetting the active set where pt and $wait_{max}$ are as explained for PtDom
FqRes	Divides the search into phases of equal length and resets the active set at the end of each phase where the number of phases is determined by a frequency parameter fq

2.2   Cross-Domain Hyper-Heuristics

The purpose of cross-domain hyper-heuristics is to improve the generality of hyper-heuristics by designing hyper-heuristics that perform well across a wide spectrum of problem domains [4]. The CHeSC challenge was proposed to promote cross-domain hyper-heuristics. The challenge used six problem domains which are the boolean satisfiability problem (SAT), one-dimensional bin packing problem (BP), personnel scheduling problem (PS), permutation flow shop problem (PFS), traveling salesman problem (TSP), and vehicle routing problem (VRP).

The cross-domain performance of the competing hyper-heuristics is evaluated by ranking the hyper-heuristics in each problem domain and adding up all per-domain scores to obtain a single overall score which is used to determine the winner. The ranking is done using Formula 1 which assigns scores of 10, 8, 6, 5, 4, 3, 2, 1 points to the top 8 hyper-heuristics and the rest of the hyper-heuristics receive no points. Ties are broken by adding the scores in the respective positions and sharing them equally among all hyper-heuristics that tie.

3   Grammatical Evolution for Automated Design of DHSS

This section outlines Auto-GE proposed for the automated design of DHSS. The task of Auto-GE is to find appropriate design choices (including parameter values) for the design decisions involved in DHSS. The generational control model is used to replace the current population with a new population in every generation.

3.1   Grammar for DHSS

The grammar used to specify DHSS is presented in Listing 1.1. The terminals of the grammar represent the design choices explained in Tables 1, and 2, 3.

Listing 1.1. Grammar for DHSS

```
<dynset>   ::= <update> <remove> <reset> <measure>
<update>   ::= PhDom(<pl>) | PhQi(<pl>,<asp>) | PhGrd(<pl>,<tp
              >) | PtDom(<pf1>,<pl>) | PtQi(<pf1>,<pl>,<asp>) | PtGrd(<
              pf1>,<pl>,<asp>)
<pl>   ::= 1 | 2 | 4 | 8 | 16 | ... | 512 | 1024
<asp>  ::= 0.1 | 0.2 | ... | 0.9 | 1.0
<tp>   ::= 2 | 3 | 4 | 5 | 6
<pf1>  ::= 1 | 2 | 3 | ... | 10
<remove>   ::= none | PtRem(<pf2>) | FqRem(<fq1>) | IndRem(<r
              >,<asp>,<a1>,<a2>,<a3>,<a4>) | GrpRem(<r>,<asp>,<b1>,<b2
              >,<b3>,<b4>) | ConsRem(<r>,<asp>,<a1>,<a2>,<a3>,<a4>,<b1
              >,<b2>,<b3>,<b4>)
<pf2>  ::= 1 | 2 | 3 | ... | 10
<fq1>  ::= 0.1 | 0.15 | 0.2 | 0.25 | 0.3 | 0.5
<r>    ::= 0.1 | 0.2 | ... | 0.8 | 0.9
<asp>  ::= 0.1 | 0.2 | ... | 0.9 | 1.0
<a1>   ::= 0.1 | 0.2 | ... | 0.8 | 0.9
<a2>   ::= 0.1 | 0.2 | ... | 0.8 | 0.9
<a3>   ::= 0.1 | 0.2 | ... | 0.8 | 0.9
<a4>   ::= 0.1 | 0.2 | ... | 0.8 | 0.9
<b1>   ::= 0.1 | 0.2 | ... | 0.8 | 0.9
<b2>   ::= 0.1 | 0.2 | ... | 0.8 | 0.9
<b3>   ::= 0.1 | 0.2 | ... | 0.8 | 0.9
<b4>   ::= 0.1 | 0.2 | ... | 0.8 | 0.9
<reset>    ::= none | PtRes(<pf3>) | FqRes<fq2>
<pf3>  ::= 1 | 2 | 3 | ... | 10
<fq2>  ::= 0.05 | 0.1 | 0.15 | 0.2 | 0.25 | 0.3 | 0.5
<measure>  ::= Imp | ImpDu | Dimp | DimpDu | FqImp | FqImpDu
              | FqDimp | FqDimpDu | Prf | PrfDu | WtdPrf(<w1>,<w2>) |
              WtdPrfDu(<w1>,<w2>) | FqPrf | FqPrfDu | WtdFqPrf(<w1>,<w2
              >) | WtdFqPrfDu(<w1>,<w2>) | BestPrf(<w1>,<w2>,<w3>) |
              BestPrfDu(<w1>,<w2>,<w3>) | RbestPrf(<w1>,<w2>,<w3>,<w4>)
              | RbestPrfDu(<w1>,<w2>,<w3>,<w4>) | BestFqPrf(<w1>,<w2
              >,<w3>) | BestFqPrfDu(<w1>,<w2>,<w3>) | RbestFqPrf(<w1>,<
              w2>,<w3>,<w4>) | RbestFqPrfDu(<w1>,<w2>,<w3>,<w4>)
<w1>   ::= 0.1 | 0.2 | ... | 1.0
<w2>   ::= 0.1 | 0.2 | ... | 1.0
<w3>   ::= 0.1 | 0.2 | ... | 1.0
<w4>   ::= 0.1 | 0.2 | ... | 1.0
```

3.2   Initial Population

Each element in the initial population is a variable-length chromosome consisting of several codons where a codon is an integer in the range $[0, 255]$. The integer representation is widely adopted in GE [23] and reported to perform better than the 8-bit string representation [11]. The initial population is created at random where the length of each chromosome is chosen at random and each codon is chosen at random from the range $[0, 255]$. Duplicates are not allowed in the initial population to enhance diversity.

3.3 Mapping

The integer genotype is mapped into a phenotype representing DHSS using the grammar presented in Sect. 3.1. The integer chromosomes are processed from left to right using one codon at a time to expand the current derivation rule of the current nonterminal. A nonterminal is fully converted into terminals before considering the next nonterminal in the same production choice. If all codons in the chromosome are used without converting all nonterminals into terminals, i.e. without generating a complete DHSS, the chromosome is wrapped from the beginning and this process is repeated until a complete DHSS is generated or the number of wraps hits a predefined limit and in this case, the individual is declared *invalid* and assigned the worst fitness. The rule used to convert genotypes into phenotypes is *codon MOD the number of choices of the current derivation rule*.

3.4 Fitness Evaluation

The individuals are evaluated by using the DHSS encoded in the individual with FS-ILS [1] to solve 20 instances chosen at random from the public instances of the CHeSC challenge (5 instances are chosen at random from the 4 public domains). The fitness of the individual is measured by the average normalized objective value across the 20 instances.

We normalize the objective values in the range $[0, 1]$ since the objective functions differ remarkably on a per-domain basis. In particular, the objective functions for SAT and BP are orders of magnitude lower than the objective functions for PS or PFS. This will lead to a bias toward individuals that perform well on PS and PFS if the objective values are not normalized.

3.5 Tournament Selection

Tournament selection is used. A fixed number of individuals are chosen at random from the population and the best individual is the winner of the tournament which will be used as a parent to produce offspring for the next generation.

3.6 Elitism Selection

Elitism selection is used to preserve the fittest individuals across generations. A fixed percentage of the best individuals are cloned to the next generation.

3.7 Crossover

After choosing two parents using tournament selection, the crossover operator is applied with a probability rate. We use the standard one-point crossover operator which determines a crossover point at random and swaps the tail segments of the two parents such that the tail segment of the first parent becomes the tail segment of the second parent and vice versa. If the crossover operator is not applied, the two parents are cloned to the next generation as in [6,17]. It is interesting to note that despite its simplicity, the one-point crossover operator was used successfully in GE for automated design [6,17].

3.8 Mutation

The individuals generated by the crossover operator undergo mutation using the standard integer mutation operator [11]. With a very small mutation probability, each codon is replaced with an integer chosen at random from the range $[0, 225]$.

3.9 Replacement

At each generation, a new population is created via the elitism selection, crossover, and mutation operators. The new population replaces the current population.

4 Experimental Setup

4.1 Parameter Tuning

The proposed Auto-GE is configured manually via trial and error where several values for each parameter are tried and the best values are chosen. Table 4 reports the best values which are close to the parameter values found for similar GE for automated design [6,17].

Table 4. Auto-GE parameter setting.

Parameter	Value	Parameter	Value
Population size	50	Tournament size	3
Crossover rate	0.85	Mutation probability	0.05
Elitism	10%	Individual length	10–30
Wrapping	Yes	Generations	20

4.2 Technical Specifications

The experiments are executed in Java 8 and ran partly on the Lengau Cluster of the Center for High-Performance Computing, and partly on the MITC cluster. The Lengau cluster has Intel Xeon CPUs (2.6 GHz) and runs CentOS operating system. The MITC cluster has Intel Xeon CPUs (2.4 GHz) and runs Ubuntu 18.04 operating system.

5   Results and Discussion

5.1   Automated Design Performance

We evaluate the automatically designed DHSS (Auto-DHSS) and the best manually designed DHSS (Man-DHSS) [9] on the domains of the CHeSC challenge. The performance of Auto-DHSS and Man-DHSS is compared with that of the best methods for the CHeSC challenge. The best method for CHeSC is the method with the highest cross-domain score computed as explained in Sect. 2.2. The results are reported in Table 5.

From the results, Auto-DHSS outperforms Man-DHSS. The cross-domain performance of FS-ILS without using either Auto-DHSS or Man-DHSS is worse than when using any one of them. This is consistent with previous findings [9] that DHSS is beneficial for hyper-heuristics.

The per-domain scores consider each problem domain separately. Even though the individual per-domain scores are of less importance when dealing with cross-domain hyper-heuristics, they are still useful to see which problems present a challenge for which hyper-heuristic. The per-domain scores are presented in Fig. 1.

Auto-DHSS does not outperform Man-DHSS in all problem domains despite having better cross-domain performance. In SAT, PS, TSP, Man-DHSS performs better than Auto-DHSS, whereas in BP, PFS, and VRP, Auto-DHSS performs better than Man-DHSS. No hyper-heuristic dominates all other hyper-heuristics in all domains. For each hyper-heuristic, there is at least one problem domain that is too challenging to solve. For instance, the winner of CHeSC (adapHH)

Table 5. The performance of Auto-DHSS and Man-DHSS compared to the best hyper-heuristics for CHeSC.

Method	Score	Method	Score
Auto-DHSS	167.66	Man-DHSS [9]	158.58
adapHH [15]	126.0	FS-ILS [1]	120.58
VNS [10]	81.83	ML [18]	79.92
PHunter [2]	62.33	EPH [13]	48.50

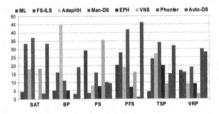

Fig. 1. The per-domain performance of Auto-DHSS and Man-DHSS compared to the best hyper-heuristics for CHeSC.

performs poorly in PS. Some hyper-heuristics receive no points at all in some domains such as EPH in SAT and VNS in BP.

5.2 Design Time

The automated design can be contrasted to the manual design not only by considering the results but also by considering other factors including the design time, reusability, and the reduction of the level of expertise required to complete the task. The manual design was an iterative process that lasted for two weeks, whereas the automated approach took less than 24 h to run on a cluster.

6 Conclusion and Future Work

In this paper, we employed grammatical evolution to automate the design of the DHSS that was previously proposed in [9]. The automatically designed DHSS (Auto-DHSS) and the manually designed DHSS (Man-DHSS) were used within FS-ILS [1], which is the best hyper-heuristic for the CHeSC challenge, to manage the heuristic set dynamically. It was found that both Auto-DHSS set and Man-DHSS improved the performance of FS-ILS and Auto-DHSS set outperformed Man-DHSS. Furthermore, the automated design reduced the design time substantially.

Although the automated approach improves the performance of the manual approach, the improvement is not as remarkable as we hoped. This could possibly be attributed to the fact that we automate one aspect of the hyper-heuristics without considering other aspects that can influence the overall performance. In the future, we will widen the scope of our automated approach by considering other aspects of hyper-heuristics.

Acknowledgment. This work is funded as part of the Multichoice Research Chair in Machine Learning at the University of Pretoria, South Africa. This work is based on the research supported wholly/in part by the National Research Foundation of South Africa (Grant Numbers 46712). Opinions expressed and conclusions arrived at, are those of the author and are not necessarily to be attributed to the NRF. This work is run on the Lengau Cluster of the Center for High Performance Computing, South Africa.

References

1. Adriaensen, S., Brys, T., Nowé, A.: Fair-share ILS: a simple state-of-the-art iterated local search hyperheuristic. In: Proceedings of the 2014 Annual Conference on Genetic and Evolutionary Computation, pp. 1303–1310 (2014)
2. Chan, C.Y., Xue, F., Ip, W.H., Cheung, C.F.: A hyper-heuristic inspired by pearl hunting. In: Hamadi, Y., Schoenauer, M. (eds.) LION 2012. LNCS, pp. 349–353. Springer, Heidelberg (2012). https://doi.org/10.1007/978-3-642-34413-8_26
3. Drake, J.H., Kheiri, A., Özcan, E., Burke, E.K.: Recent advances in selection hyper-heuristics. Eur. J. Oper. Res. **285**(2), 405–428 (2020)

4. Burke, E.K., et al.: The cross-domain heuristic search challenge – an international research competition. In: Coello, C.A.C. (ed.) LION 2011. LNCS, vol. 6683, pp. 631–634. Springer, Heidelberg (2011). https://doi.org/10.1007/978-3-642-25566-3_49

5. Burke, E.K., et al.: Hyper-heuristics: a survey of the state of the art. J. Oper. Res. Soc. **64**(12), 1695–1724 (2013)

6. Burke, E.K., Hyde, M.R., Kendall, G.: Grammatical evolution of local search heuristics. IEEE Trans. Evol. Comput. **16**(3), 406–417 (2012)

7. López-Camacho, E., Terashima-Marin, H., Ross, P., Ochoa, G.: A unified hyper-heuristic framework for solving bin packing problems. Expert Syst. Appl. **41**(15), 6876–6889 (2014)

8. Gutierrez-Rodríguez, A.E., et al.: Applying automatic heuristic-filtering to improve hyper-heuristic performance. In: 2017 IEEE Congress on Evolutionary Computation (CEC), pp. 2638–2644. IEEE (2017)

9. Hassan, A., Pillay, N.: Dynamic heuristic set selection for cross-domain selection hyper-heuristics. In: Aranha, C., Martín-Vide, C., Vega-Rodríguez, M.A. (eds.) TPNC 2021. LNCS, vol. 13082, pp. 33–44. Springer, Cham (2021). https://doi.org/10.1007/978-3-030-90425-8_3

10. Hsiao, P.C., Chiang, T.C., Fu, L.C.: A VNS-based hyper-heuristic with adaptive computational budget of local search. In: 2012 IEEE Congress on Evolutionary Computation, pp. 1–8. IEEE (2012)

11. Hugosson, J., Hemberg, E., Brabazon, A., O'Neill, M.: Genotype representations in grammatical evolution. Appl. Soft Comput. **10**(1), 36–43 (2010)

12. Koulinas, G., Kotsikas, L., Anagnostopoulos, K.: A particle swarm optimization based hyper-heuristic algorithm for the classic resource constrained project scheduling problem. Inf. Sci. **277**, 680–693 (2014)

13. Meignan, D.: An evolutionary programming hyper-heuristic with co-evolution for CHeSC11. In: The 53rd Annual Conference of the UK Operational Research Society (OR53), vol. 3 (2011)

14. Mısır, M., Verbeeck, K., De Causmaecker, P., Vanden Berghe, G.: The effect of the set of low-level heuristics on the performance of selection hyper-heuristics. In: Coello, C.A.C., Cutello, V., Deb, K., Forrest, S., Nicosia, G., Pavone, M. (eds.) PPSN 2012. LNCS, vol. 7492, pp. 408–417. Springer, Heidelberg (2012). https://doi.org/10.1007/978-3-642-32964-7_41

15. Mısır, M., Verbeeck, K., De Causmaecker, P., Vanden Berghe, G.: An intelligent hyper-heuristic framework for CHeSC 2011. In: Hamadi, Y., Schoenauer, M. (eds.) LION 2012. LNCS, pp. 461–466. Springer, Heidelberg (2012). https://doi.org/10.1007/978-3-642-34413-8_45

16. Sabar, N.R., Ayob, M., Kendall, G., Qu, R.: Grammatical evolution hyper-heuristic for combinatorial optimization problems. IEEE Trans. Evol. Comput. **17**(6), 840–861 (2013)

17. Nyathi, T., Pillay, N.: Comparison of a genetic algorithm to grammatical evolution for automated design of genetic programming classification algorithms. Expert Syst. Appl. **104**, 213–234 (2018)

18. Ochoa, G., et al.: HyFlex: a benchmark framework for cross-domain heuristic search. In: Hao, J.-K., Middendorf, M. (eds.) EvoCOP 2012. LNCS, vol. 7245, pp. 136–147. Springer, Heidelberg (2012). https://doi.org/10.1007/978-3-642-29124-1_12

19. Pillay, N.: Evolving hyper-heuristics for a highly constrained examination timetabling problem. In: Proceedings of the 8th International Conference on the Practice and Theory of Automated Timetabling (PATAT 2010), pp. 336–346 (2010)

20. Pillay, N.: A study of evolutionary algorithm selection hyper-heuristics for the one-dimensional bin-packing problem. S. Afr. Comput. J. **48**(1), 31–40 (2012)
21. Pillay, N.: A review of hyper-heuristics for educational timetabling. Ann. Oper. Res. **239**(1), 3–38 (2016)
22. Pillay, N., Özcan, E.: Automated generation of constructive ordering heuristics for educational timetabling. Ann. Oper. Res. **275**(1), 181–208 (2019)
23. McKay, R.I., Hoai, N.X., Whigham, P.A., Shan, Y., O'neill, M.: Grammar-based genetic programming: a survey. Genet. Program. Evolvable Mach. **11**(3–4), 365–396 (2010)
24. Remde, S., Cowling, P., Dahal, K., Colledge, N.: Exact/heuristic hybrids using rVNS and hyperheuristics for workforce scheduling. In: Cotta, C., van Hemert, J. (eds.) EvoCOP 2007. LNCS, vol. 4446, pp. 188–197. Springer, Heidelberg (2007). https://doi.org/10.1007/978-3-540-71615-0_17
25. Soria-Alcaraz, J.A., Ochoa, G., Sotelo-Figeroa, M.A., Burke, E.K.: A methodology for determining an effective subset of heuristics in selection hyper-heuristics. Eur. J. Oper. Res. **260**(3), 972–983 (2017)

Evolutionary Approach to Melodic Line Harmonization

Jan Mycka[1] , Adam Żychowski[1] , and Jacek Mańdziuk[1,2(✉)]

[1] Warsaw University of Technology, Warsaw, Poland
{a.zychowski,mandziuk}@mini.pw.edu.pl
[2] AGH University of Science and Technology, Krakow, Poland

Abstract. The paper presents a novel evolutionary algorithm (EA) for melodic line harmonization (MLH) - one of the fundamental tasks in music composition. The proposed method solves MLH by means of a carefully constructed fitness function (FF) that reflects theoretical music laws, and dedicated evolutionary operators. A modular design of the FF makes the method flexible and easily extensible. The paper provides a detailed analysis of technical EA implementation, its parameterization, and experimental evaluation. A comprehensive study proves the algorithm's efficacy and shows that constructed harmonizations are not only technically correct (in line with music theory) but also *nice to listen to*, i.e. they fulfill aesthetic requirements, as well. The latter aspect is verified and rated by a music expert - a harmony teacher.

Keywords: Evolutionary algorithm · Harmonization · Music composition

1 Introduction

The majority of real-life optimization problems are associated with engineering, however, certain aspects of creative activities, such as painting, music composition, poetry, or film making, can be modeled as optimization problems [4,19], as well. In this paper, one such task – the melodic line harmonization is considered.

The melodic line harmonization is a part of the process of composing music and is about determining the musically appropriate chord accompaniment for a given melody. It is a creative process that requires intuition and experience of the musician, although, the music theory defines certain strict constraints and rules which the composed music should follow in order to sound well [21]. In this perspective, melodic line harmonization can be treated as an optimization problem with maximizing the number of fulfilled constraints.

Evolutionary Algorithms (EAs), thanks to their effectiveness, are widely applied to various practical problems [5,11,17,27]. This paper shows that EAs can also be successfully adapted to the field of art and create formally correct, well-structured, and musically aesthetic melody harmonizations.

© The Author(s), under exclusive license to Springer Nature Switzerland AG 2023
L. Rutkowski et al. (Eds.): ICAISC 2022, LNAI 13588, pp. 230–241, 2023.
https://doi.org/10.1007/978-3-031-23492-7_20

2   Related Work

Algorithmic music composition is a well-studied area of research with various computational intelligence methods proposed in the literature [13,25]. There are several research paths in this domain and researchers focus on various aspects of music generation, for instance, style transfer [9], imitating a particular composer (e.g. F. Chopin [15,16]), real-time music accompaniment [12], timbre, pitch, rhythm, chord [14]. In this paper, we consider the problem of melodic line harmonization which is an essential part of the music composition process. The definition and details of the examined problem are presented in Sect. 2.1.

The most common approach to solving this task is learning harmonizations based on existing melody lines using neural networks [6,8,10], which requires a set of training data and is usually limited to a particular genre or music style, e.g. Bach chorales [8]. Music composition can also be approached with Markov chains [3,18,26] or evolutionary algorithms [7,20,22]. Evolutionary approaches propose various representations of melodic line and fitness function definitions to assess evolving solutions. Moreover, in [7] a multiobjective genetic algorithm is constructed which, for a given melody, generates a set of harmonic functions without adding new melodic lines.

Due to slightly different problem definitions and the lack of well-established benchmarks, making a direct comparison between methods is usually difficult. Thus, the evaluation process is often performed by human experts who rate the obtained results (music pieces). This approach is also taken in this paper.

2.1   Melodic Line Harmonization

Harmonization of a melodic line is one of the fundamental tasks in music. The input data in a harmonization problem is one melodic line, and the product of harmonization is usually four melodic lines (voices): soprano (the highest), alto, tenor, bass (the lowest). A given (input) melodic line could be also accompanied by harmonic functions which are added to every or almost every note in that line. These functions determine which notes can be included in the chord formed across all four lines (vertically).

Harmonization of a melodic line depends not only on the composer's creativity but also on various theoretical rules derived from music theory. These rules regulate (1) the form of individual melodic lines, (2) chord's construction, and (3) how successive chords should be connected to each other.

The problem considered in this paper is a harmonization of a soprano line, with harmonic functions added to each note. The solution is created based on a selected set of theoretical rules for melodic line harmonization.

2.2   Contribution

The main contribution of the paper includes: (1) a novel evolutionary algorithm capable of designing correct melodic line harmonizations; (2) a specially designed fitness function that reflects theoretical music rules and can be easily tuned toward certain aspects of the output harmonization; (3) an extensive evaluation of the proposed method which shows its quality and robustness; (4) a detailed analysis of the algorithm's performance and parameterization.

3 Evolutionary Harmonization

3.1 The Search Space and the Initial Population

Not every note can be used in a created chord. Harmonic functions define which notes fit into a chord and which do not. Harmonizations containing notes in chords that do not correspond to the required functions are incorrect.

After receiving the input (soprano line with harmonic functions), for each unique function, a set of all possible chords that fulfill that function is created. Created harmonizations are, therefore, not generated from individual notes but from the whole chords. The above rules significantly narrow down the search space, however, due to still many possible arrangements of notes in each chord, the number of potential solutions is still too large (between 3^l - 7^l, where l is a harmonization length) to evaluate all of them. Examples of created chords for one of the functions are shown in Fig. 1.

Fig. 1. Various chords for function S_{II} with a fixed (green) note in soprano. (Color figure online)

Individuals are represented as 4 sequences of notes, one per each harmonized voice. The chord is formed by the notes located across all four voices (vertically). Individuals in the initial population are created randomly. Soprano notes are completed to a randomly selected chord satisfying the following two conditions:

(*) the chord corresponds to the function assigned to the completed note,
(**) the note given in the input voice is located in the chord in the same voice.

3.2 Next Generation Population

After the generation of the initial population, the EA is run for a predefined number of n generations. In each generation, first s_e *elite* (i.e. currently best) individuals are promoted from the previous generation without any adjustments, so as to ensure that the best individuals found in the entire run of the algorithm will not be lost. The rest of the population is generated by means of selection procedure and genetic operators (mutation and crossover), following Algorithm 1.

```
1  GenerateNewPopulation (P)
2  |   CalculateFitnessValues(P)  // calculates fitness of each individual
3  |   P_new ← GetElite(P, s_e)     // population of new individuals
4  |   while |P_new| < |P| do
5  |   |   c_1 ← Selection(P)
6  |   |   if rand([0,1]) < p_c then // crossover
7  |   |   |   c_2 ← Selection(P)
8  |   |   |   c_new = Crossover(c_1, c_2)
9  |   |   else
10 |   |   |   c_new ← c_1
11 |   |   end
12 |   |   c_new ← Mutation(c_new)
13 |   |   P_new = P_new ∪ {c_new}
14 |   end
15 |   return P_new
```

Algorithm 1: Next generation population procedure.

3.3 Selection Method

Selection of individuals from the population is performed in a t_s-tournament with a roulette, i.e. first t_s individuals are uniformly sampled with replacement to participate in the tournament. The drawn individuals are sorted from best to worst according to their score. Let's denote by c_i, $i = 1, \ldots, t_s$ the i-th ranked individual. The chance of winning the tournament by c_i is calculated as follows:

$$p(c_i) = \begin{cases} p_s & \text{if } i = 1 \\ (1 - \sum_{j=1}^{i-1} p(c_j)) \cdot p_s & \text{if } 1 < i < t_s \\ (1 - \sum_{j=1}^{i-1} p(c_j)) & \text{if } i = t_s \end{cases} \quad (1)$$

where $p_s \geq 0.5$ is the so-called *selection pressure*.

3.4 Mutation

Generated harmonizations are built using the whole chords, rather than individual notes. For this reason, mutations are also performed on the entire chords and each chord in the harmonization is mutated with the same probability equal to $\frac{p_m}{l}$, where l is the length of the harmonization (the number of notes in the input melodic line) and p_m is mutation coefficient. Mutation of a chord consists in replacing it with another randomly selected chord that satisfies conditions (*)-(**).

3.5 Crossover

Crossover is performed with probability p_c. Two crossover methods are proposed and tested: the classic operator and the one-point operator. Analogously to mutation, the crossover is performed using whole chords rather than individual notes.

Both crossover operators are presented in Algorithm 2, where $c[i], i = 1, \dots, l$ is the chord located at position i in a harmonization of length l.

```
1 Crossover₁ (c₁, c₂)
2     for i ∈ [1, ..., l] do
3         if rand([0,1]) < 0.5 then
4             c[i] = c₁[i]
5         else
6             c[i] = c₂[i]
7         end
8     end
9     return c
```

```
 1 Crossover₂ (c₁, c₂)
 2     k ← rand(1, ..., l)
 3     for i ∈ [1, ..., l] do
 4         if i < k then
 5             c[i] = c₁[i]
 6         else
 7             c[i] = c₂[i]
 8         end
 9     end
10     return c
```

Algorithm 2: Crossover: left - classic method, right - one-point method.

3.6 Fitness Function

The fitness function is based on music theory and is composed of 22 rules of harmonization, taken from a harmony textbook [24]. Similar rules can be found in [2,23]. Each rule is assigned a weight (positive or negative) that affects the final score of the generated harmonization. Examples of violations of three of these rules are shown in Fig. 2. A detailed description and implementation of all rules can be found in a project repository [1].

The fitness function can be divided into 3 main modules:

1. Strong constraints C_s (strong penalty terms) - stemming from the rules that must be strictly met in the created harmonization to be considered correct.
2. Weak constraints C_w (weak penalty terms) - derived from rules that do not have to be strictly satisfied in the created harmonization, but their non-fulfillment lowers the harmonization assessment.
3. Aesthetic value V_a (reward terms) - the rules specifying chord arrangements or connections between chords that improve the harmonization sound.

The fitness function f_t for individual c has the following form:

$$f_t(c) = V_a + C_w + (p \cdot t)C_s,$$

$$C_s = \sum_{i=1}^{m_s} \phi_i(c), \quad C_w = \sum_{j=1}^{m_w} \chi_j(c), \quad V_a = \sum_{k=1}^{m_a} \psi_k(c), \quad (2)$$

where $\phi_i(c) \leq 0$ is the penalty for not fulfilling strong constraint $i, i = 1, \dots, m_s$, $\chi_j(c) \leq 0$ is the penalty for not fulfilling weak constraint $j, j = 1, \dots, m_w$, $\psi_k(c) \geq 0$ is the reward associated with the rule $k, k = 1, \dots, m_a, (m_s = 9, m_w = 9, m_a = 4)$, $t \leq n$ is the generation number, and p is a constant parameter. Please note that during the evolution, the fitness function value is calculated for each individual regardless of the fulfillment of the strong constraints. These constraints, however, define the correctness of each individual.

(a) Strong constraint: At least one voice has to move in different direction than other voices. (b) Weak constraint: A maximal interval that can take in two consecutive moves is tenth. (c) Weak constraint: There should not be septim interval between two consecutive notes in one voice.

Fig. 2. Examples of rules violations.

4 Experimental Results

Since there are no standard benchmarks for the considered problem we decided to use a set of exercises from the harmony textbook [24] as a test set (similar exercises can be found in other harmony textbooks, e.g. [2,23]). The selected problems were divided into 3 groups based on their complexity and length:

1. long examples (about 20 chords), using only basic functions,
2. short examples (about 10 chords), with more complicated functions,
3. long examples (about 20 chords), with more complicated functions.

4.1 Algorithm Parametrization

The choice of the evolutionary parameters is crucial for the algorithm performance. The values of the following parameters were selected based on preliminary tests: population size (s_p), tournament size (t_s), elite size (s_e), selection pressure (p_s), mutation coefficient (p_m), crossover method and crossover probability (p_c), number of generations (n).

The following baseline values were selected: $s_p = 1000$, $t_s = 4$, $s_e = 3$, $p_s = 0.7$, $p_m = 1$, classic crossover with $p_c = 0.8$, $n = 5000$. Individual parameters were then optimized (with the remaining parameters frozen) to select the best values for each of them. The tests were run on 3 different examples, one from each group. These examples were different from the ones used as the test set. Each test was repeated 5 times with different seed values for the random number generator.

Population Size (s_p). The following population sizes were tested: 10, 100, 500, 1000, 1750, 2500, 3500, 5000. As expected, for smaller population sizes, the algorithm performed noticeably worse because the solution space was not searched sufficiently. For larger values (above 1000), the results were not substantially different from each other. Results for an example from the third group are presented in Fig. 3a. The resulting size of the population was chosen as 1000.

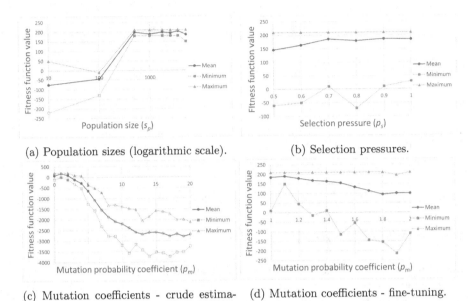

(a) Population sizes (logarithmic scale).   (b) Selection pressures.

(c) Mutation coefficients - crude estima-   (d) Mutation coefficients - fine-tuning.
tion.

Fig. 3. Parameter tuning averaged over 5 runs for an example from the third group. The minimum and maximum are the worst and best fitness function values, resp., for the individuals returned in 5 runs. Empty shape (e.g. ○) denotes that the algorithm did not return any correct solution over 5 runs and filled shape that at least one solution was correct.

Tournament Size (t_s). Four values of tournament size, equal to 2, 4, 8, and 10 were tested (see Table 1). The results for $t_s = 4$ and $t_s = 8$ were similar to each other. At the same time, $t_s = 4$ led to higher standard deviation of the population (larger diversity of individuals) and was therefore selected for the final experiments.

Elite Size (s_e). Four values of elite size, equal to 0, 3, 5 and 10 were tested. The results are presented in Table 1. The algorithm with the elite mechanism is more stable and achieves better results. The value of $s_e = 3$ was finally selected.

Selection Pressure (p_s). This parameter describes the probability of the best individual winning the tournament. Values between 0.5 and 1 with a step of 0.1 were tested. Results of the algorithm are presented in Fig. 3b. The higher the value of p_s, the lower the standard deviation in the population. Too low standard deviation can have a negative impact on the results due to the lack of diversity in the population. At the same time, an increase of p_s results in an increase of the percentage of correct individuals in the population, as shown in Table 2. Finally, to balance the value of standard deviation and the percentage of correct individuals, $p_s = 0.8$ was chosen.

Table 1. Fitness function with respect to the tournament size (top part) and the elite size (bottom part).

Example	$t_s = 2$			$t_s = 4$			$t_s = 8$			$t_s = 10$		
	Mean	Min	Max	Mean	Min	Max	Mean	Min	Max	Mean	Min	Max
1	11	−150	165	355	355	355	343	305	355	335	305	355
2	188	110	210	210	210	210	210	210	210	210	210	210
3	−171	−370	155	188	180	210	197	175	210	174	95	210
Example	$s_e = 0$			$s_e = 3$			$s_e = 5$			$s_e = 10$		
	Mean	Min	Max	Mean	Min	Max	Mean	Min	Max	Mean	Min	Max
1	152	75	280	355	355	355	324	250	355	325	305	355
2	110	30	210	210	210	210	210	210	210	210	210	210
3	43	−30	125	188	180	210	187	175	210	206	190	210

Table 2. Percentage of correct individuals in the population, in relation to p_s.

Example	p_s					
	0.5	0.6	0.7	0.8	0.9	1
1	0.03	0.14	0.28	0.37	0.37	0.46
2	0.02	0.8	0.18	0.27	0.32	0.37
3	0.01	0.06	0.16	0.25	0.32	0.38

Mutation Coefficient (p_m). Values between 0 and l were tested, where l is the harmonization length (number of chords), with a step equal to 1. The best results were achieved with $p_m = 0, 1, 2$. For higher values, the results were significantly weaker, and for the highest ones, the returned results were incorrect.

As a further refinement of p_m, the values from 1 to 2 with step 0.1 were tested, which led to the final selection of $p_m = 1.1$. The results for an example from the third group are presented in Fig. 3c (initial tests with larger values) and Fig. 3d (fine-tuning tests).

Crossover Method and Probability (p_c). To select the crossover method and its probability, various probability values, between 0 and 1 with a step of 0.1, for the two crossover versions were tested. For each value, tests were run thirty times and the values for all three tuning examples were normalized using min-max normalization. The average results are shown in Table 3.

The algorithm achieved similar results for values between $0.4 - 0.8$. For this reason, t-Student tests were performed to select the best values for each model with a significance level of 0.05. A value of 0.8 was selected for both models. In the last step, the t-Student test was conducted between two crossover variants (both with the chosen probability of 0.8) with hypothesis H_0: "the results obtained are not significantly different" and the resulting p-value=0.113. Finally, one-point model with $p_c = 0.8$ was selected.

Table 3. Normalized mean values of crossover tuning procedure.

p_c	0	0.1	0.2	0.3	0.4	0.5	0.6	0.7	0.8	0.9	1
Mean, classic	0.86	0.84	0.9	0.88	0.91	0.89	0.91	0.92	0.93	0.88	0.9
Mean, one-point	0.86	0.9	0.9	0.89	0.91	0.93	0.93	0.95	0.96	0.95	0.95

Generation Number (n). This parameter was chosen as a compromise between the quality of results and the running time. The value of $n = 5000$ was selected from the set $\{1000, 3000, 5000, 10000\}$.

The final selection of the steering parameters was as follows: $s_p = 1000$, $s_e = 3$, $t_s = 4$, $p_s = 0.8$, $p_m = 1.1$, $p_c = 0.8$ (one-point crossover), $n = 5000$.

4.2 Algorithm Efficacy

The efficacy of the algorithm was checked on 9 samples taken from the harmony textbook [24]. For each sample, the algorithm was able to find the correct solution in a relatively short time. The generation numbers in which the first correct solution and the best solution were found, resp. are shown in Table 4. In each case, the first correct solution was found in less than 90 generations.

The number of generations required to find the correct solution varies between groups and depends mainly on the length of an example (cf. groups 1 and 2) and, to a lesser extent, the example's complexity (cf. groups 1 and 3). At the same time, for more complex problems (group 3) the solution is likely to improve even after 3500 iterations, which does not happen for easier samples (groups 1 and 2).

4.3 Evaluation by the Human Expert

The algorithm evaluates harmonizations based merely on their numerical fitness. Hence, we asked a harmony teacher to assess their aesthetic value, as well. The evaluation was performed according to a school scale from 1 (lowest score) to 5 (highest score). Out of 9 solutions, 4, 4 and 1 were rated 5, 4.5 and 4, resp., with the average grade of 4.67. This means that the solutions are theoretically and sonically correct. An example solution rated 5 is presented in Fig. 4.

4.4 Algorithm Running Time

The average running times of the algorithm in three groups are presented in Table 5. One can observe a quasi-linear relationship between the example length and the execution time. Harmonizations in the group 2 are obtained in about half of the time required for harmonizing samples from groups 1 and 3. On the other hand, it seems that the degree of the example's complexity does not affect the running time - the average times in groups 1 and 3 are similar.

Table 4. The number of generations required to find a solution.

Group no.	Example no.	Generation number in which the result was found:					
		First correct harmonization			Finally returned harmonization		
		Mean	Min	Max	Mean	Min	Max
1	1	16.6	14	22	208.2	86	343
	2	16.8	13	22	268.2	129	744
	3	14.8	12	18	218.4	109	397
2	4	8.2	6	10	177.6	27	593
	5	3.2	1	5	24	13	36
	6	6.8	5	8	826.2	75	3200
3	7	33.6	19	86	1933.6	96	3576
	8	19.2	17	22	699.8	249	1224
	9	21	18	25	3098.6	2179	3838

T SD T T   S S T T   D   T D D   T   S S   T T S D   T

Fig. 4. Harmonization created by the algorithm for an example from the third group. Given line (soprano) is marked in green. (Color figure online)

Table 5. The average algorithm's running time (harmonization time) in seconds.

Group 1			Group 2			Group 3		
Mean	Min	Max	Mean	Min	Max	Mean	Min	Max
531.91	485.47	567.02	225.76	180.66	252.05	536.18	508.68	582.39

4.5 Parameters' Relevance and Robustness

The experiments showed that changing some parameters has a greater effect on the results than changing other parameters. The crossover method, crossover probability p_c, and the selection pressure (p_s) have a relatively small impact on the results. For selection pressure, any value above 0.5 yields satisfactory results.

In contrast, changes of mutation probability value ($\frac{p_m}{l}$) have significant impact. The results achieved for mutation coefficient between 1 and 2 are stable, but increasing p_m above 2 results in a gradual results deterioration. The elite mechanism has been shown to be crucial for the algorithm's performance. Its lack causes significant performance degradation and lower repeatability of results.

5 Conclusions

Creating melodic line harmonization is a non-trivial task. In this paper, we employ EAs to approach this problem. There are two key components of the proposed algorithm: (a) restriction of the search space (*)-(**) to feasible solutions, and (b) specially-designed fitness function, based on theoretical music rules, that defines proper harmonizations. The fitness function consists of three modules: one responsible for the correctness of harmonization and the other two for its quality. The harmonization process is performed for the whole chords and likewise the mutation and crossover operators are applied to the whole chords, not to individual notes.

Harmonizations constructed by the algorithm were evaluated by the harmony teacher so as to additionally assess their aesthetic properties (sound). All but one harmonization were rated at least 4.5 on a scale from 1 to 5, with a good number of them rated 5. This means that in terms of musical quality generated harmonizations meet all expectations. The algorithm finds the solution quickly in terms of both the number of generations and the overall computational time.

The modular design of the fitness function allows it to be easily expanded and modified in the future. Adding more theoretical rules should allow harmonizations to be generated for more advanced and complex harmonic functions. Moreover, the task definition can be extended to the generation of harmonizations for melodic lines without the presence of harmonic functions.

References

1. https://github.com/MelodicLineHarmonization/melodicLineHarmonization.git
2. Benham, H.: A Student's Guide to Harmony and Counterpoint. Rhinegold Publishing Limited, London (2006)
3. Buys, J., van der Merwe, B.: Chorale harmonization with weighted finite-state transducers. In: Twenty-Third Annual Symposium of the Pattern Recognition Association of South Africa, pp. 95–101. PRASA South Africa (2012)
4. Carnovalini, F., Rodà, A.: Computational creativity and music generation systems: an introduction to the state of the art. Front. AI **3**, 14 (2020)
5. Coello, C.A.C., Lamont, G.B.: Applications of Multi-objective Evolutionary Algorithms, vol. 1. World Scientific, Singapore (2004)
6. De Prisco, R., Eletto, A., Torre, A., Zaccagnino, R.: A neural network for bass functional harmonization. In: Di Chio, C., et al. (eds.) EvoApplications 2010. LNCS, vol. 6025, pp. 351–360. Springer, Heidelberg (2010). https://doi.org/10.1007/978-3-642-12242-2_36
7. Freitas, A., Guimaraes, F.: Melody harmonization in evolutionary music using multiobjective genetic algorithms. In: Proceedings of the Sound and Music Computing Conference (2011)
8. Gang, D., Lehmann, D., Wagner, N.: Tuning a neural network for harmonizing melodies in real-time. In: ICMC (1998)
9. Grinstein, E., Duong, N.Q., Ozerov, A., Pérez, P.: Audio style transfer. In: 2018 IEEE International Conference on Acoustics, Speech and Signal Processing (ICASSP), pp. 586–590. IEEE (2018)

10. Hild, H., Feulner, J., Menzel, W.: HARMONET: a neural net for harmonizing chorales in the style of J.S. Bach. In: NIPS 1991: Proceedings of the 4th International Conference on Neural Information Processing Systems, pp. 267–274 (1991)
11. Hu, Y., Liu, K., Zhang, X., Su, L., Ngai, E., Liu, M.: Application of evolutionary computation for rule discovery in stock algorithmic trading: a literature review. Appl. Soft Comput. **36**, 534–551 (2015)
12. Jiang, N., Jin, S., Duan, Z., Zhang, C.: RL-Duet: online music accompaniment generation using deep reinforcement learning. In: Proceedings of the AAAI Conference on Artificial Intelligence, vol. 34, pp. 710–718 (2020)
13. Liu, C.H., Ting, C.K.: Computational intelligence in music composition: a survey. IEEE Trans. Emerg. Top. Comput. Intell. **1**(1), 2–15 (2016)
14. Lopez-Rincon, O., Starostenko, O., Ayala-San Martín, G.: Algoritmic music composition based on artificial intelligence: a survey. In: 2018 International Conference on Electronics, Communications and Computers, pp. 187–193. IEEE (2018)
15. Mańdziuk, J., Goss, M., Woźniczko, A.: Chopin or not? A memetic approach to music composition. In: 2013 IEEE Congress on Evolutionary Computation, pp. 546–553 (2013)
16. Mańdziuk, J., Woźniczko, A., Goss, M.: A neuro-memetic system for music composing. In: Iliadis, L., Maglogiannis, I., Papadopoulos, H. (eds.) AIAI 2014. IAICT, vol. 436, pp. 130–139. Springer, Heidelberg (2014). https://doi.org/10.1007/978-3-662-44654-6_13
17. Mańdziuk, J., Żychowski, A.: A memetic approach to vehicle routing problem with dynamic requests. Appl. Soft Comput. **48**, 522–534 (2016)
18. Moray, A., Williams, C.K.I.: Harmonising chorales by probabilistic inference. Adv. Neural Inf. Process. Syst. **17**, 25–32 (2005)
19. Oliveira, H.G.: A survey on intelligent poetry generation: Languages, features, techniques, reutilisation and evaluation. In: Proceedings of the 10th International Conference on Natural Language Generation, pp. 11–20 (2017)
20. Olseng, O., Gambäck, B.: Co-evolving melodies and harmonization in evolutionary music composition. In: International Conference on Computational Intelligence in Music, Sound, Art and Design (2018)
21. Pachet, F., Roy, P.: Musical harmonization with constraints: a survey. Constraints **6**(1), 7–19 (2001)
22. Prisco, R.D., Zaccagnino, G., Zaccagnino, R.: Evocomposer: an evolutionary algorithm for 4-voice music compositions. Evol. Comput. **28**(3), 489–530 (2020)
23. Rimsky-Korsakov, N.: Practical Manual of Harmony. C. Fischer, New York (2005)
24. Sikorski, K.: Harmony part 1. PWM (2020)
25. Siphocly, N.N., Salem, A.B.M., El-Horabty, E.S.M.: Applications of computational intelligence in computer music composition. Int. J. Intell. Comput. Inf. Sci. **21**(1), 59–67 (2021)
26. Wassermann, G., Glickman, M.: Automated harmonization of bass lines from Bach chorales: a hybrid approach. Comput. Music J. **43**(2–3), 142–157 (2020)
27. Żychowski, A., Gupta, A., Mańdziuk, J., Ong, Y.S.: Addressing expensive multi-objective games with postponed preference articulation via memetic co-evolution. Knowl.-Based Syst. **154**, 17–31 (2018)

Using Automatic Programming to Improve Gradient Boosting for Classification

Roland Olsson[✉️][iD] and Shubodha Acharya[iD]

Østfold University College, Halden, Østfold, Norway
Roland.Olsson@hiof.no

Abstract. In this paper, we present our new and automatically tuned gradient boosting software, Classifium GB, which beats its closest competitor, H2O, for all datasets that we ran. The primary reason that we found it easy to develop Classifium GB is that we employed meta machine learning, based on evolution, to automatically program its most important parts.

Gradient boosting is often the most accurate classification algorithm for tabular data and quite popular in machine learning competitions. However, its practical use has been hampered by the need to skilfully tune many hyperparameters in order to achieve the best accuracy.

Classifium GB contains novel regularization methods and has automatic tuning of all regularization parameters. We show that Classifium GB gives better accuracy than another automatically tuned algorithm, H2O, and often also outperforms manually tuned algorithms such as XGBoost, LightGBM and CatBoost even if the tuning of these is done with exceptional care and uses huge computational resources.

Thus, our new Classifium GB algorithm should rapidly become the preferred choice for practically any tabular dataset. It is quite easy to use and even say Random Forest or C5.0 require more skilled users. The primary disadvantage is longer run time.

Keywords: Machine learning · Gradient boosting · XGBoost · LightGBM · CatBoost · AutoML · Hyperparameters · Automatic programming · Automatic design of algorithms through evolution · Meta machine learning

1 Introduction

This paper introduces our novel gradient boosting algorithm, Classifium GB, that gives better accuracy than the commercial H2O implementation for every dataset that we have tested.

The key ingredients of Classifium GB were produced using meta machine learning, that is through running another machine learning algorithm to produce general code suitable for gradient boosting and apparently superior to anything that human beings have been able to come up with during the decades that gradient boosting and ensemble algorithms have been hot research topics. Thus,

L. Rutkowski et al. (Eds.): ICAISC 2022, LNAI 13588, pp. 242–253, 2023.
https://doi.org/10.1007/978-3-031-23492-7_21

automatic programming based on evolution is now so advanced that even some of the sharpest minds in computer science and machine learning cannot match its capability and creativity.

In practice, the two most accurate machine learning methods for classification of tabular datasets are gradient boosting and neural nets, often used together through stacking to create ensembles that win machine learning competitions. In this paper, we present Classifium GB and compare it with all known state-of-the-art gradient boosting implementations, namely XGBoost, LightGBM, CatBoost and H2O.

A problem with algorithm comparisons in the literature is that the hyperparameter tuning often only is carefully done for the novel method that is presented. We have gone to great lengths to avoid this pitfall and to ensure a fair comparison by quite thorough tuning for all of the above implementation. Our experimental results show that Classifium GB generally is the most accurate even if one spends more than 100 000 CPU hours on careful manual tuning of XGBoost, LightGBM and CatBoost. H2O and Classifium GB are automatically tuned.

Our paper makes the following key contributions.

1. Novel regularization.
2. A node candidate evaluation function generated through our automatic programming system, ADATE [7].
3. A new and automated hyperparameter tuning pipeline.

Our thorough comparison between all the leading gradient boosting implementations may be of additional interest.

The paper is organized as follows. First, Sect. 2 presents the various related gradient boosting implementations and also gives a brief introduction to automatic programming with ADATE.

Section 3 contains a mathematical description of our special version of gradient boosting along with regularization techniques. It also presents our new regularization methods and the key contribution of the paper, which is a novel node selection function generated by automatic programming. Section 3 finishes with our new but somewhat simplistic hyperparameter tuning pipeline.

In Sect. 4, we present our datasets and explain our experimental methodology and experimental results. Finally, Sect. 5 contains some conclusions and outlook for the future.

2 Related Work

2.1 Gradient Boosting

The original Gradient Boosting Machine (GBM) [3] was invented by Friedman at about the same time as Breiman created the first Random Forest version [1], which then generally was viewed as the most accurate tree ensemble method for more than a decade. Thus, even if GBM was an outstanding algorithm, it still did not become as popular as Random Forest.

However, this changed dramatically in 2014 when Tianqi Chen made the first extreme gradient boosting (XGBoost) [2] implementation, which turned out to be superior to Random Forest for every single dataset as long as it was well tuned. This lead to quick adoption of XGBoost among leading hyperscalers such as AWS and Azure and also to newer alternatives such as LightGBM [5] and CatBoost [8].

The primary difference between XGBoost and GBM is that the former contains many more regularization methods to prevent overfitting. For example, XGBoost adopted random sampling of predictors from Random Forest, where this is controlled by the well-known "mtry" hyperparameter. The corresponding parameter in XGBoost is "colsample_bynode". However, the number of regularization hyperparameters has grown rapidly and there are now about 20 of them, which makes it difficult to tune optimally. There have been many attempts to automatically tune XGBoost using for example Bayesian optimization or differential evolution, but skilled manual tuning has so far remained superior as we show in our experimental results, at least for the case of H2O.

LightGBM was made by Microsoft with an overall focus on efficiency, but also builds trees in a slightly different way, where a node candidate is chosen based on its reduction of the global loss instead of just the loss along a particular branch. It is also able to combine features that are mutually exclusive such as the ones that result from one-hot encoding. However, CatBoost and Classifium never need one-hot encoding since they have built-in handling of categorical predictors. Another speed improving technique, among several others in LightGBM, is to avoid exact sorting of numerical features and use histograms instead. This method has now also been incorporated into XGBoost, which appears able to rapidly adapt to any advances made by its competition. Our experiments show that Microsoft has been highly successful in improving time complexity since LightGBM is around 5 times faster than the best of the others for big datasets.

The main difference between CatBoost and XGBoost is that the former directly handles categorical predictors. CatBoost has several creative methods that are not widely known, such as a form of automatic feature extraction for categorical predictors that can be merged into one single predictor on-the-fly as is found to reduce the loss. It also contains special methods to avoid that gradients become too biased towards the data used to build a tree. CatBoost contains fewer hyperparameters and appears somewhat easier to tune and is efficient both on CPUs and GPUs.

The overall goal with H2O is that it should be very easy to use with excellent graphical presentations, while also giving exceptionally high accuracy. It contains a version of XGBoost along with sophisticated and automatic hyperparameter tuning based on evolutionary computation. Another special technology that is available in H2O is automatic construction of ensembles of different machine learning models, such as deep neural nets and gradient boosting trees stacked together. The H2O [6] software is free but H2O.ai gets revenue from consulting services they provide for users of their software. Overall, it represents the current state-of-the-art among automatically tuned boosters.

2.2   Automatic Programming

ADATE is a system for machine learning of purely functional, recursive and symbolic programs that operate on algebraic data types, for example lists or trees. However, the ADATE runs that produced the Classifium GB evaluation function only used ADATE to generate arithmetical expressions.

ADATE is able to automatically construct help functions as needed and can for instance invent a long division algorithm for binary natural numbers from tabula rasa without knowing say addition, subtraction, multiplication or even the positional system for representing numbers. It has been used to synthesize several hundred different recursive algorithms of a similar complexity since 1991.

ADATE basically consists of program transformations, rules for combining them and systematic search algorithms to drive the evolution of gradually better and better formulas or algorithms.

Let S be a newly synthesized expression and E and H(E) be subexpressions of the program to be transformed, where H is a unary lambda expression and S also is in cases two and three below. ADATE then tries the following kinds of so-called replacements, also known as R transformations, in a systematic and exhaustive way.

1. $E \longrightarrow S$
2. $E \longrightarrow S(E)$
3. $H(E) \longrightarrow S(E)$

For example, the third case above is implemented by first selecting a node in the syntax tree of the program as the root of H, another node below it to be the root of E and then synthesizing expressions that contain one or more copies of E. If the tree contains n nodes, the number of possible choices of H and E is $O(n^2)$, whereas the number of possible expressions S almost always grows exponentially with the number of nodes in S.

However, evolution as well as local search is often able to make great progress even if only a very small neighbourhood of a given solution, in our case a program, is explored. Of course, there are thousands of examples of this in the literature on combinatorial optimization, including say the satisfiability (SAT) and the traveling salesman (TSP) problems.

Given that a number of R transformations, typically a few hundred, have been performed and evaluated, ADATE keeps the ones that did not make the program worse and labels these as so-called REQ transformations. The REQs are sorted according to the size of their synthesized expressions. Then, ADATE systematically generates combinations of k REQ transformations in order of bigger total size for k equal to one, two, three and four. ADATE also has heuristics for keeping all the REQs local, that is possibly restricted to a small subtree of the syntax tree.

A so-called compound transformation consists of such a sequence of REQs followed by newly generated R transformations restricted to occur in the same subtree as the REQs. R transformations and the mostly neutral short walks provided by these compound transformations are the fundamental mechanisms that ADATE uses to generate the neighbourhood of a program chosen for expansion.

Additionally, there are transformations for inventing new auxiliary functions and for lifting and distribution of case-expressions, but these are mostly useful for inventing more general functional programs and not needed when considering only formulas as in this paper.

ADATE has always used Pareto fronts that balance syntactic complexities and evaluation values. Obviously, size matters in order to alleviate overfitting by sharpening Occam's razor. Each program in a Pareto front is iteratively expanded with exponentially increasing neighbourhood cardinality until it is knocked out from the front by a new program that is smaller and at least as good. Thus, the programs in a Pareto front are always gradually bigger and better.

3   The Classifium Gradient Boosting Algorithm

We will first review standard gradient boosting as it is used in Classifium GB and then present the novelties in Classifium GB, that is our own regularization method, a new node candidate evaluation function and the automatic tuning pipeline.

3.1   Standard Gradient Boosting

Classifium GB and related boosters incrementally build ensembles of trees where each new tree takes steps that reduce the error that remains after the previously generated trees. If there are c different classes for the output variable (response) in a classification problem, each leaf in Classifium GB contains c weights, one for each class. When a new example is to be classified after training is finished, it is fed to all trees and a leaf is reached for each one. The weights in all reached leaves are summed and the class with the overall max weight is chosen as the classification of the new example.

In order to simplify the presentation, we will in the following assume that $c = 3$ and the reader will afterwards find it trivial to generalize the algorithm to any value of c.

Assume that training example number e has class k as its correct response and that this example during training has weights $w_{e,1}$, $w_{e,2}$ and $w_{e,3}$ accumulated from the trees generated so far. The contribution to the overall current error (loss) from this example for weight $w_{e,i}$ is given by the usual log loss defined as follows for $i = 1, 2, 3$.

$$f_e(w_{e,i}) = -\ln \frac{e^{w_{e,k}}}{e^{w_{e,1}} + e^{w_{e,2}} + e^{w_{e,3}}}.$$

Note that f_e goes to zero for $i = k$ if and only if $w_{e,k}$ goes to positive infinity or the other weigths both go to negative infinity, which is exactly what we want.

Assume that the next tree has been built except for the leaves and that example e reached a leaf with weights d_1, d_2 and d_3 which are to be chosen. We wish to minimize $f_e(w_{e,i} + d_i)$. Let f'_e and f''_e denote the first and second order derivatives with respect to $w_{e,i}$. Using a second order Taylor series, we get the following approximation.

$$f_e(w_{e,i} + d_1) \approx f_e(w_{e,i}) + f'_e(w_{e,i})d_i + \frac{1}{2}f''(w_{e,i})d_i{}^2.$$

To minimize the right hand side, we set its derivative with respect to d_i to zero and obtain

$$d_i = -\frac{f'_e(w_{e,i})}{f''_e(w_{e,i})}$$

However, the above was for only one training example e. In general, we get the following for t training examples.

$$d_i = -\frac{\sum_{e=0}^{t} f'_e(w_{e,i})}{\sum_{e=0}^{t} f''_e(w_{e,i})}$$

Following standard practice, we use G to denote the sum of the first order derivatives and H for the sum of the second order derivatives. By substituting the expression for d_i into the approximation of the loss and removing constants, it turns out that a good heuristic is to choose the node split candidate that maximizes G^2/H.

Classifium GB borrows learning rate, L1 and L2 regularization, mtry, subsampling, max number of leaves and min samples in leaf from XGBoost and Random Forest.

In contrast to XGBoost, it has built-in handling of nominal predictors and uses basically the same approach as in the latest version of Random Forest.

3.2 Novelties in Classifium GB

Max Expected Number of Split Candidates. A minor novelty in Classifium GB is a new regularization parameter that we call maxExpectedNumCands and which is a complement to mtry, especially when there are very many split candidates per predictor.

If there are n split candidates for a node, we allow each one to be considered with a probability that is maxExpectedNumCands divided by n, which on average will mean that the subset of candidates considered has cardinality maxExpectedNumCands. Typical optimal values for this parameter lie between 2 and 36.

Weight Update. In order to automatically generate better functions for weight updates as well as better functions to choose the best node candidate, we employed the ADATE automatic programming system and used 8 UCI datasets, each with about 10k lines, to evaluate its automatically generated programs on-the-fly. The total run time for automatic improvement of gradient boosting was several months on a cluster with 1000 CPU cores.

In addition to the basic arithmetical operations, ADATE was allowed to use the following help function, which is an approximation of the tanh function.

$$t(x) = x(27 + x^2)/(27 + 9x^2)$$

For $|x| > 3$, the asymptotic value is returned.

ADATE then came up with the following new weight update, which uses t to softly clip the G/H ratio after adding the learning rate lr to it.

$$d_i = -t(t(t(t(0.25\texttt{lr})) \cdot (G/H + \texttt{lr})))$$

Node Split Candidate Selection. In order to explain the new split evaluation function, it helps to first look at the implementation of the standard one.

We partition the examples that reach a child of the current node according to class as usual when building decision trees. Let $G_{l,i}$ denote the sum of first order derivatives for the examples of class i that reach the left child and use analogous notation for second order derivatives and the right child.

For a given split, let p_l denote the proportion of examples that go left and p_r the proportion that goes right. The standard split evaluation can then be written as follows.

$$p_l \cdot \sum_{i=1}^{c} -G_{l,i}^2/H_{l,i} + p_r \cdot \sum_{i=1}^{c} -G_{r,i}^2/H_{r,i}$$

However, the new one generated by ADATE is formulated as a recursive functional program where the function g corresponds to one of the two sums above. ADATE generated the following novel definition of g that takes a list of (G, H) values as its argument. We present it in the usual functional programming notation where :: is the infix list constructor. ExtraPar1 is a new hyperparameter.

```
g [] = 0.0
g( ( G, H ) :: Xs ) =
  t( g Xs / ExtraPar1 ) - G * G / H
```

Each tree uses its own random permutation of the classes. Thus, which class that is the first one varies from tree to tree. Apparently, the new g function chooses to prioritize the classes according to the current permutation and the degree of prioritization is tuned by ExtraPar1.

The overall definition of split candidate evaluation found by ADATE is as follows, where XsL and XsR are the (G, H) values for the left and right children respectively.

$$t(p_l \cdot t(t(t(t(t(g(\texttt{XsL})))))) + p_r \cdot g(\texttt{XsR})$$

Note that ADATE has generated code that squashes the g value for the left child. The effect of this is that the evaluation focuses on the right child and that it does not care so much about left branches in a tree. Thus, a tree becomes more like a rule in C5.0.

Hyperparameter Tuning Pipeline. Each stage in the pipeline is a single, double or triple loop that simultaneously tunes one, two or three hyperparameters respectively. The tuning is done using two-fold cross validation in order to save time. Of course, this may lead to less accurate tuning. Also to save time, almost all of the pipeline runs with just 100 trees but it has a few final stages with 500 trees followed by a final stage that tunes the number of trees.

The stages are as follows, where "to" means "up to and including".

1. Tune `ExtraPar1` from 0 to 2 in steps of 0.2.
2. Let n be the number of trees. Tune the learning rate from $0.2/n$ to $5/n$ in steps of $0.4/n$.
3. Tune the following combinations. For each `maxNumLeaves` in 1, 2, 4, 8, 16, 32, 64, ∞ let `minSamplesInLeaf` also grow exponentially with a factor 2 from 1 to the max possible number of samples in a leaf.
4. Tune `ExtraPar1` again in the same way as before.
5. Tune the learning rate again in the same way as before.
6. Tune the following for `maxExpectedNumCands` equal to 2 and 36. For each `maxNumLeaves` in 1, 2, 4, 8, 16, 32, 64, ∞, let `mtry` be αm, where m is the number of predictors and α goes from 0.4 to 1 in steps of 0.1.
7. Tune `subsampling` from 0.4 to 1 in steps of 0.1.
8. Tune `ExtraPar1` again in the same way as before.
9. Tune the learning rate again in the same way as before.
10. Tune `subsampling` again in the same way as before.
11. Try `regL1` equal to 0 and then from 10^{-6} to $2^{16} \cdot 10^{-6}$, growing exponentially by a factor 2.
12. Tune `regL2` in the same way as `regL1`.
13. Set the number of trees to 500 and tune the learning rate as above.
14. Tune `ExtraPar1` again in the same way as before.
15. Tune `subsampling` again in the same way as before.
16. Finally, tune the number of trees from 500 to 1500 in steps of 100.

Of course, manual tuning may give better results than this automated pipeline, but commercial machine learning practitioners may not have the time or the patience that is needed to outperform it.

4   Experimental Results

We chose 7 UCI datasets that are well known to be useful for comparing classification algorithms and then split them with one half for training and the other half for testing. Our motivation for choosing a rather large test set was to make the comparison more statistically reliable.

The methodology was to use 5-fold stratified cross-validation for hyperparameter tuning on the training set and then run with the best found parameters

Table 1. A Table presenting the information about the number of data used for training and testing from each data set.

Data sets	Training rows	Training columns	Testing rows	Testing columns
Adult	24420	15	24420	15
Bank	20593	21	20593	21
Dota	51471	117	51471	117
Flavours of Physics	33775	51	33775	51
Forest Cover Type	290505	55	290505	55
MiniB-ooNE	65031	51	65031	51
Porto Seguro	21695	59	21695	59

Table 2. Test accuracies for the 7 data sets.

Algorithms with dataset	XGBoost	Light GBM	Cat boost	H2O (Default)	H2O (XGB)	Classifium
Adult	0.8752	0.8754	0.8748	0.8766	0.8746	0.8752
Bank	0.9165	0.9159	0.9174	0.9159	0.9159	0.9160
Dota	0.5949	0.5938	0.5928	0.5973	0.5835	0.5975
Flavours of Physics	0.8927	0.8907	0.8928	0.8893	0.8872	0.8916
Forest Cover	0.9682	0.919	0.9601	0.9543	0.9553	0.9719
MiniBoo NE	0.9468	0.9476	0.9470	0.9456	0.9442	0.9471
Porto Seguro	0.5953	0.5935	0.5923	0.5970	0.5954	0.5985

on the entire training set. The resulting forest was then run on the test set and all results reported in Table 2 are for test data. Table 1 briefly describes each data set.

Adult, Bank, Dota, Forest Cover Type, and Porto Seguro have both categorical and numerical predictors whereas Flavours of Physics and MiniBooNE only have numerical ones. Forest Cover Type has seven different response classes but the other ones are all binary. Only Adult, Bank, and Porto Seguro have missing values.

The most relevant comparison in Table 2 is between XGBoost in H2O and Classifium GB since they are the only gradient boosting algorithms that are automatically tuned. As can be seen in the table, Classifium GB is more accurate every single time albeit with a small margin sometimes.

If H2O is run in its default mode, it builds an ensemble with deep learning and gradient boosting. However, Classifium GB still emerges as the overall winner in the table, which means that it often also can beat automatically tuned deep neural nets, at least on tabular data sets such as these.

The comparison between Classifium GB and the manually tuned algorithms is less clear but Classifium GB was the overall winner for 4 out of the 7 datasets and quite close to the winner for the other 3.

When compared with XGBoost, run after optimal hyperparameter tuning, Classifium GB beats XGBoost for the Dota 2, Forest Cover Type, MiniBooNE, and Porto seguro data sets. When compared with LightGBM, also run after optimal hyperparameter tuning, Classifium GB performed better for Dota 2, Flavours of Physics, Forest Cover Type, and Porto Seguro. Finally, when compared with CatBoost, also after optimal hyperparameter tuning, Classifium GB outperformed it for the Adult, Dota 2, Forest Cover Type, and Porto Seguro data sets.

4.1 Tuning of XGBoost, LightGBM and CatBoost

Since XGBoost, LightGBM, and CatBoost, in contrast to H2O and Classifium GB, are dependent on good manual tuning, we will now provide the details of how this tuning was done, which will help the reader to assess how reliable the above experimental results are. Of course, the most important comparison, that is the one between H2O and Classifium GB is not dependent on manual tuning.

We tuned XGBoost, LightGBM, and CatBoost using the following different hyperparameter tuning pipelines, which are based on the official documentation of the algorithms, [2,5], and [8], and some best practice presented online by Analytics Vidyha [4], and our own working experience with the algorithms.

Our general tuning approach was to use 5-fold cross validation on the training set and use grids covering one or two parameters. Each two dimensional grid typically had size 5 × 5 and was moved until the center point had the lowest cross validation error rate for the training set. Then, we halved the resolution of the grid and repeated the translation to once again get the minimum in the center. This was done repeatedly until the experimenter was satisfied that no more improvement was possible using the one or two parameters under consideration and tuning then proceeded to the next stage in the pipeline.

Pipeline for Tuning XGBoost. We started with all hyperparameters having their default values and then followed the pipeline below.

Step 1: Tune the number of trees and the learning rate together using a grid as described above. The hyperparameter set was updated with the new values of the number of trees and the learning rate.
Step 2: Tune max_depth and min_child_weight together.
Step 3: Tune colsample_bynode and subsampling.
Step 4: Tune gamma.
Step 5: Re-calibrate the number of trees
Step 6: Tune regularization parameters L1 and L2.
Step 7: Go back to step 1 and repeat until no improvement is found.

Pipeline for Tuning LightGBM. The following is rather similar to the pipeline for XGBoost but given here anyway to provide a more exact description of our tuning.

Step 1: Tune the number of trees and the learning rate.
Step 2: Tune maximum depth and minimum gain to split.

Step 3: Tune minimum data in leaves and the number of leaves.
Step 4: Re-calibrate the number of trees.
Step 5: Tune bagging fraction and feature fraction.
Step 6: Tune regularization parameters lambda_l1 and lambda_l2.
Step 7: Go back to step 1 and repeat until no improvement is found.

Pipeline for Tuning CatBoost. Since CatBoost has somewhat different hyperparameters, we used the following custom designed pipeline for it.

Step 1: Tune the number of trees and the learning rate.
Step 2: Tune the maximum depth of the trees.
Step 3: Tune the L1 and L2 leaf regularization hyperparameters.
Step 4: Tune the random strength.
Step 5: Tune the border count. Generally, the default value of this parameter gives the best result.
Step 6: Tune the bagging temperature.
Step 7: Go back to step 1 and repeat until no improvement is found.

5   Conclusions and Future Work

We have used automatic programming to develop a novel and automatically tuned gradient boosting algorithm that in general seems to be more accurate for tabular datasets than the commercial H2O software even if the latter is allowed to use both deep learning and gradient boosting. Our new algorithm is highly competitive also with other leading boosters such as XGBoost, LightGBM and CatBoost.

However, a unique aspect of our Classifium GB booster is that its most essential code was not written by human beings. Instead, it was designed by our old automatic programming system, ADATE. Thus, this is an example of machines learning how to learn or in other words, meta machine learning.

The primary contribution of the paper is the automatically synthesized weight update and split candidate selection, that is the two most important parts of gradient boosting. Most likely, these new heuristics can be incorporated also into the other gradient boosting packages and make them more accurate.

The paper also proposes two minor additions to gradient boosting, namely new regularization and a simple but still effective hyperparameter tuning pipeline.

Future work includes using an analogous meta machine learning approach for other algorithms, for example LSTM, GNN or Transformers. We have already made preliminary experiments on LSTM with amazing results.

References

1. Breiman, L.: Random forests. Mach. Learn. **45**(1), 5–32 (2001)
2. Chen, T., He, T., Benesty, M., Khotilovich, V., Tang, Y., Cho, H., et al.: Xgboost: extreme gradient boosting. R package version 0.4-2 **1**(4), 1–4 (2015)
3. Friedman, J.H.: Greedy function approximation: a gradient boosting machine. Ann. Stat. 1189–1232 (2001)
4. Jain, A., et al.: Complete guide to parameter tuning in xgboost with codes in python (2016). https://www.analyticsvidhya.com/blog/2016/03/completeguide-parameter-tuning-xgboost-with-codes-python. Accessed 27 May 2020
5. Ke, G., et al.: LightGBM: a highly efficient gradient boosting decision tree. Adv. Neural. Inf. Process. Syst. **30**, 3146–3154 (2017)
6. LeDell, E., Poirier, S.: H2O AutoML: scalable automatic machine learning. In: 7th ICML Workshop on Automated Machine Learning (AutoML) (2020)
7. Olsson, R.: Inductive functional programming using incremental program transformation. Artif. Intell. **74**(1), 55–81 (1995)
8. Prokhorenkova, L., Gusev, G., Vorobev, A., Dorogush, A.V., Gulin, A.: Catboost: unbiased boosting with categorical features. In: Advances in Neural Information Processing Systems, vol. 31 (2018)

Forced Movement Extensions of the Particle Swarm Optimizers with Inertia Weight

Krzysztof Wójcik[(✉)] [ID]

Cardinal Stefan Wyszyński University in Warsaw, Warsaw, Poland
krzysztof.wojcik@uksw.edu.pl

Abstract. This paper investigates some ideas for extending the IPSO method. We design an IPSO-based procedure that combines the adaptive coefficients and forced particle movements. We also propose a new criterion for the evaluation of the forced movement. Then, using a set of benchmark functions, we test the performance of algorithm variants based on the designed procedure. Our experiments show that optimizers with forced movement mechanisms consistently find better values of the global best solutions. Moreover, the addition of time adaptive coefficients might benefit the algorithm's performance.

Keywords: Particle swarm optimization · Forced particle movements · Adaptive coefficients · Runtime stasis

1 Introduction

Particle Swarm Optimization (PSO) [10] is a population-based stochastic optimization technique. It has been applied to many optimization problems, and the method itself was subject to modifications, such as introducing an inertia weight (PSO with Inertia Weight, IPSO) [15]. This method has been a subject of analysis and improvement itself, spawning several methods since. One of them is a forced movement PSO method (f-PSO) [13].

Recently, new ideas for possible improvements have been proposed. First is a concept of *stasis* [11], designed to indicate possible stagnation of the particle, that could serve as a criterion for applying a forced movement. Moreover, a new guideline for designing a particle swarm optimizer with time adaptive acceleration and inertia weight coefficients has been published [3]. According to the author of the guideline, an adaptive PSO method should outperform the non-adaptive IPSO.

In this paper, we adapt the recent concept of stasis and utilize it as a criterion for applying forced movement to particles. Furthermore, following a recently published guideline for designing a time-adaptive IPSO based on the movement patterns analysis (MAPSO, [3]), we design an algorithm with adaptive values of inertia weight and acceleration coefficients. Ultimately, we combine the two ideas into a novel type of PSO optimizer. In order to test the performance of the proposed algorithms, we use a set of standard benchmark functions.

The text consists of seven sections. Section 2 presents the IPSO-based optimizer with forced movements, adaptive coefficients, and a novel method combining these

© The Author(s), under exclusive license to Springer Nature Switzerland AG 2023
L. Rutkowski et al. (Eds.): ICAISC 2022, LNAI 13588, pp. 254–264, 2023.
https://doi.org/10.1007/978-3-031-23492-7_22

components. In Sect. 3, we propose a new criterion for applying forced movement to particles. Section 4 presents a procedure for generating the algorithms tested in experiments. Section 5 describes the setups of the experiments, all algorithms are tested against the benchmark functions, and the results are presented and analyzed. Section 6 compares the algorithms in the light of the results and discusses the advantages and disadvantages of all approaches. Section 7 summarizes the presented research.

2   IPSO-Based Algorithms

Particles are the primary elements of a swarm, each representing a certain proposition of a solution. Their movement is described by the equation

$$X^{t+1} = X^t + V^t, \tag{1}$$

where X^s is a vector representing locations of particles in step s, and V^s is a vector of velocities of particles in step s.

2.1   Particle Swarm Optimization with Inertia Weight (IPSO)

Let's consider a swarm of N particles positioned in the D-dimensional space $\mathbb{R}^D$. In the IPSO method, the position X_i and velocity V_i of particle i is updated according to the set of equations [15]

$$\begin{cases} V_i^{t+1} = wV_i^t + c_1 r^t \otimes (L_i - X_i^t) + c_2 s^t \otimes (G - X_i^t), \\ X_i^{t+1} = X_i^t + V_i^{t+1}, \end{cases} \tag{2}$$

where L_i is the best location that the particle i has found so far (*neighborhood best*), G—the best location found by particles overall (*global best*), w is an inertia weight coefficient, c_1, c_2 are acceleration coefficients, r^t and s^t are vectors of numbers generated independently from a uniform distribution over an interval of $[0, 1]$, and $\otimes$ is a Hadamard product (a component wise multiplication).

2.2   Forced Movement PSO

One of the modifications of the IPSO method is the **forced movement particle swarm optimization (f-PSO)** [13, 14]. The authors extend the regular IPSO with forced movements (so-called "kicks") designed to repulse the particles from their local attractors.

A vital element is defining a set of rules deciding when to apply a forced movement. Hence, the authors define **potential of the swarm**.

Definition 1 (Potential). *For $d \in D$, the **current potential** ϕ_d^t of the swarm in dimension d is*

$$\phi_d^t = \sum_{i=1}^N \underbrace{(|V_{i,d}^t| + |G_d - X_{i,d}^t|)}_{\varphi_{i,d}^t} = \sum_{i=1}^N \varphi_{i,d}^t. \tag{3}$$

$\phi^t = (\phi_1^t, \ldots, \phi_D^t)$ *is the **total potential** of the swarm, $\varphi_{i,d}^t$ (partial potential) is the contribution of particle i to the potential of the swarm in dimension d.*

Using the current potential as a criterion for applying forced movement, the formula for movement of particles in the f-PSO method is given by the set of equations [1]

$$V_{i,d}^{t+1} = \begin{cases} (2p_d^t - 1) \cdot \gamma, & \text{if } \forall_{i \in N} \ \varphi_{i,d}^t < \delta \\ wV_{i,d}^t + c_1 r_d^t \otimes (L_{i,d} - X_{i,d}^t) + c_2 s_d^t \otimes (G_d - X_{i,d}^t), & \text{otherwise} \end{cases} \quad (4)$$
$$X_{i,d}^{t+1} = X_{i,d}^t + V_{i,d}^{t+1},$$

where γ—strength coefficient of the applied "kick" and δ—"kick" threshold are small positive constants and p^t is a vector of numbers generated independently from a uniform distribution over an interval of $[0, 1]$.

2.3 Movement-Pattern Adaptation PSO

For some time, PSO optimizers with adaptive coefficient values have been analyzed [4, 5,7,9,12,16,18]. Intuitively, the adaptive behavior of the particles should be beneficial to the overall performance of the swarm. However, the advantage has not yet been proven. A regular IPSO optimizer initiated with parameters that are known to give good results ($c_1 = c_2 = 1.494$, $w = 0.729$, [6]) proved to perform better than the adaptive propositions [8].

In the recent literature, a new IPSO-based time-adaptive swarm optimizer, namely the **movement pattern adaptation PSO** (**MAPSO**) has been proposed [3]. The movement of particles in the method is given by the set of equations

$$\begin{cases} V_i^{t+1} = w^t V_i^t + c_{1_i}^t r^t \otimes (L_i - X_i^t) + c_{2_i}^t s^t \otimes (G - X_i^t), \\ X_i^{t+1} = X_i^t + V_i^{t+1}. \end{cases} \quad (5)$$

In order to establish a formula for the evolution of the movement coefficients, the authors investigate the relationship between consecutive particle locations (autocorrelation) ρ, expected movement distance V_c and focus of the search F (i.e., the ratio of how much should particles concentrate their search around their global best solutions against the local best solution). Then, the authors divide the desired particle behavior into three phases:

1. *Global exploration*—the particles should explore the search space dynamically and not follow any particular direction. To achieve this kind of behavior, the particles should take large steps (large V_c), search in all directions (low ρ), and balance the search between local and global best solutions ($F = 1$).
2. *Improvement*—each particle is encouraged in the direction of the best improvement. Thus, autocorrelation ρ should increase to maintain the search direction, and V_c should decrease to prevent the movement distance from being too large.
3. *Local exploration*—the particles should now explore the vicinity of the best solutions. The author suggests the possibility of setting large F values to concentrate the search around the global best solution. ρ and V_c should be set to some low values.

2.4 Forced Movement MAPSO

The MAPSO method, similarly to the regular IPSO, can also be expanded with a forced movement mechanism. Combining the two ideas results in a novel method—the **forced movement MAPSO (f-MAPSO)**. The movement of particles in the method is given by the set of equations

$$
V_{i,d}^{t+1} = \begin{cases} (2p_d^t - 1) \cdot \gamma_i^t, & \text{if } \forall_{i \in N} \; \varphi_{i,d}^t < \delta_i^t \\ w^t V_{i,d}^t + c_{1i}^t r_d^t \otimes (L_{i,d} - X_{i,d}^t) + c_{2i}^t s_d^t \otimes (G_d - X_{i,d}^t), & \text{otherwise} \end{cases}
$$
$$
X_{i,d}^{t+1} = X_{i,d}^t + V_{i,d}^{t+1}.
$$

(6)

Observe that the acceleration coefficients and the inertia weight are not the only values that can change throughout the lifetime of the swarm. The strength of the "kicks" γ^t, as well as the threshold δ^t, can also be adjusted accordingly.

3 Utilizing Stasis in Forced Movement IPSO Variants

Recently, a new concept of **stasis** [11] describing particle behavior has been introduced. Detecting when a particle reaches a state of stasis is based on observing the evolution of the value of its location's variance over time. Suppose the variance stabilizes on some constant level. In that case, it is a clear signal that the particle may have stopped exploring the solution space, which often results in stagnation.

3.1 Relevant Definitions

Definition 2 (The particle location variance stasis time). *Let ζ be a positive real number. The particle location variance stasis time $pvst(\zeta)$ is a minimal number of steps necessary for all subsequent differences between variances of particle locations to be lower than ζ, that is*

$$
pvst(\zeta) = \min\{t \mid |d^{s+1} - d^s| < \zeta \text{ for all } s \geq t\},
$$

(7)

where $d^s = Var[X^s]$.

For empirical analysis, a measure including a time frame was introduced:

Definition 3 (The particle location weak variance stasis time). *Let l_w be a given positive integer and ζ be a positive real number. The particle location weak variance stasis time $pwvst(\zeta)$ is the minimal number of steps necessary to get l_w subsequent differences between variances of particle locations lower than ζ, that is*

$$
pwvst(l_w, \zeta) = \min\{t \mid |d^{t+k+1} - d^{t+k}| < \zeta \\ \text{for all } k \in \{0, 1, \ldots, l_w - 1\}\}.
$$

(8)

3.2   Adapting Stasis to Empirical Evaluation

The purpose of stasis is to indicate possible stagnation. Thus, it could be utilized as a condition for applying forced movement to a particle. However, it needs to be adapted for runtime evaluation before it can be put to practical use. Hence, we propose **runtime stasis** criterion.

Definition 4 (Runtime Stasis). *Let l_w be a given positive integer and ζ be a positive real number. We say that the* **runtime stasis** *criterion is satisfied for particle i if:*

$$\forall_{d \in D} \; \eta_{i,d}^t < \zeta, \tag{9}$$

where

$$\eta_{i,d}^t = \max_{k \in \{1,2,\dots,l_w\}} |D_{i,d}^{t-k}(k_w) - D_{i,d}^{t-k-1}(k_w)| \tag{10}$$

and $D_{i,d}^t(k_w)$ denotes the value of the empirical variance of the last k_w locations of the particle i in the dimension d:

$$D_{i,d}^t(k_w) = \frac{1}{k_w} \sum_{j=1}^{k_w} \left(X_{i,d}^{t-j} - \frac{1}{k_w} \sum_{j=1}^{k_w} X_{i,d}^{t-j} \right)^2 \tag{11}$$

Now, we can incorporate runtime stasis into the f-PSO and f-MAPSO algorithms.

3.3   Stasis f-PSO

The runtime stasis can be utilized as a mean to determine whether to apply forced movement to particles. The formula for the movement of particles in the **stasis based forced movement PSO** method (**Stasis f-PSO**) is given by the set of equations

$$V_i^{t+1} = \begin{cases} (2p^t - 1) \cdot \gamma, & \text{if } \forall_{d \in D} \; \eta_{i,d}^t < \zeta \\ wV_i^t + c_1 r^t \otimes (L_i^t - X_i^t) + c_2 s^t \otimes (G - X_i^t), & \text{otherwise} \end{cases} \tag{12}$$
$$X_i^{t+1} = X_i^t + V_i^{t+1}.$$

3.4   Stasis f-MAPSO

In the **stasis based forced movement MAPSO** method (**Stasis f-MAPSO**), we also use runtime stasis to control the forced movement of particles:

$$V_i^{t+1} = \begin{cases} (2p^t - 1) \cdot \gamma_i^t, & \text{if } \forall_{d \in D} \; \eta_{i,d}^t < \zeta_i^t \\ w^t V_i^t + c_{1i}^t r^t \otimes (L_i^t - X_i^t) + c_{2i}^t s^t \otimes (G - X_i^t), & \text{otherwise} \end{cases} \tag{13}$$
$$X_i^{t+1} = X_i^t + V_i^{t+1}$$

4   Algorithms

All f-PSO, MAPSO, f-MAPSO, Stasis f-PSO, and Stasis f-MAPSO methods are based on the regular IPSO. The evaluation of the forced movement condition and modification of coefficient values are two independent actions. Thus, a comprehensive procedure can be designed.

4.1   IPSO-based Algorithm

The first addition to the standard IPSO algorithm is the forced movement mechanism, which is relevant when calculating a new velocity vector. Should a condition for applying a "kick" be satisfied, then, instead of the regular IPSO update, a new velocity vector would be generated. The second addition is a decision process controlling the modification of the inertia weight, acceleration coefficients, the strength of the forced movements, and the threshold for forced movement criteria. An IPSO-based procedure containing these modifications is described by Algorithm 1.

Algorithm 1. IPSO-based procedure

1: **Input:** Objective function $f : \mathbb{R}^D \to \mathbb{R}$, number of particles N
2: **Output:** $G \in \mathbb{R}^D$
3: **for** $i = 1 \to N$ **do**
4:      Initialize X_i^0 randomly
5:      Initialize V_i^0 with $\overrightarrow{0}$
6:      Initialize $L_i := X_i^0$
7: **end for**
8: Initialize: $G := \arg\min_{\left\{ L_i | i \in \{1,\ldots,N\} \right\}} f.$
9: **repeat**
10:      **for** $i = 1 \to N$ **do**
11:          **for** $d = 1 \to D$ **do**
12:              **if** *kick condition satisfied* **then**   ▷ (f-PSO*)
13:                  $V_{i,d}^{t+1} := (2 \cdot t - 1) \cdot \delta$
14:              **else**
15:                  $V_{i,d}^{t+1} = w \cdot V_{i,d}^t + c_{1i} \cdot r \cdot (L_{i,d}^t - X_{i,d}^t) + c_{2i} \cdot s \cdot (G_d - X_{i,d}^t)$
16:              **end if**
17:              $X_{i,d}^{t+1} := X_{i,d}^t + V_{i,d}^{t+1}$
18:          **end for**
19:          **if** $f(X_i) \leq f(L_i)$ **then** $L_i := X_i$
20:          **end if**
21:          **if** $f(X_i) \leq f(G)$ **then** $G := X_i$
22:          **end if**
23:      **end for**
24:      **if** *MAPSO condition satisfied* **then**   ▷ (MAPSO*)
25:          Alter $c_1, c_2, w, \gamma, \delta, \zeta$
26:      **end if**
27: **until** termination criterion are not met (iterations, stagnation etc.)
28: **return** G

The evaluation of the forced movement mechanism takes place before the velocity update (line 12), and modification of coefficients happens after the particle locations are updated (line 24). Observe that the procedure is easily modifiable. In line 12, should the forced movement mechanism be turned off, then the condition would always be evaluated as *false*. The same can be done in line 24.

4.2   Controlling Movement Coefficient Values

Although the guideline [3] provides examples of coefficient values and their evolution in relation to the phase the swarm is currently in, a straightforward solution for detecting the phases is not given. In order to design the *MAPSO condition* in line 24, we take into account the *local best stagnation* and *improvement rate*. For each particle, the condition is evaluated independently.

Every particle starts in the *global exploration* phase. If it constantly improves the local best solution found, it enters the *improvement* phase. If local best stagnation occurs, then the particle is moved directly to the *local exploration* phase.

The particle remains in *improvement* phase as long as local stagnation does not occur. If so happens, it enters the *local exploration* phase. In the *local exploration* phase, the particle is given time to search locally for better solutions. If it cannot find one, the phase is reset to the *global exploration*.

5   Experiments

In order to compare the efficiency of the algorithms, a series of experiments on the standard benchmark functions (Ackley, Griewank, High Conditioned Elliptic, Schwefel, Rastrigin, Rosenbrock, Sphere [17]) was performed. The experimental swarms consisted of $N = 10$ particles, and the size of the search space was set to $D = 5$ dimensions.

For IPSO, f-PSO and Stasis f-PSO, the acceleration coefficients and inertia weight were set to $c_1 = c_2 = 1.711897$, $w = 0.711897$ [2]. The forced movement related parameters were $\gamma = 1e - 3$, $\delta = 1e - 12\zeta = 1e - 12$. For MAPSO, f-MAPSO and Stasis f-MAPSO, different acceleration and inertia weight coefficients were calculated according to the guideline [3], and γ, δ, ζ were chosen arbitrarily. For a particle in the given phase, the coefficient values were:

1. *Global exploration*—the authors suggest setting $V_c = 25$, $\rho = 0.1$ and $F = 1$, which results in $c_1 = c_2 = 2.07$, $w = 0.73$. Also we set $\gamma = 1e - 1$, $\delta = 1e - 6$, $\zeta = 1e - 6$.
2. *Improvement*—the mean values suggested are $V_c = 15$, $\rho = 0.8$ and $F = 1$, which results in coefficient values $c_1 = c_2 = 0.39$, $w = 0.96$. In this phase we set $\gamma = 0$, $\delta = 0$, $\zeta = 0$ to prevent forced movements from occurring.
3. *Local exploration*—we set $V_c = 5$, $\rho = 0.1$ and maintain $F = 1$. The derived coefficients are $c_1 = c_2 = 0.5$, $w = 0.67$. Also, we set $\gamma = 1e - 6$, $\delta = 1e - 36$, $\zeta = 1e - 36$.

For each algorithm, success and failure criteria were defined. A simulation "failed" if 100,000 iterations were completed or the global best solution did not change for 500 iterations, or "succeeded" if the function value evaluated for the global best particle location fell below a predefined threshold.

Each algorithm was run 300 times for every test function. The following were measured: "Iterations" (the mean number of steps until the success condition is met), "Time" (the mean time taken to reach the success condition), "Success rate" (percentage of successful runs), and "Iter/Sec" (short for iterations per second, measuring the mean number of time steps completed in one second). The results are summarized in Table 1.

Table 1. Experimental results

Function		IPSO	f-PSO	f-PSO(S)	MAPSO	f-MAPSO	f-MAPSO(S)
ACKLEY Threshold: 1e−4	Success rate	0.95	0.94	0.96	0.65	**1.0**	**1.0**
	Time	0.27	0.63	0.62	**0.26**	0.79	0.82
	Iterations	171	172	177	**146**	288	271
	Iter/Sec	**663**	278	292	593	371	340
GRIEWANK Threshold: 1e−4	Success rate	0.0	0.0	0.0	0.0	0.0	0.0
H.C.ELLIPTIC Threshold: 1e−10	Success rate	0.5	**1.0**	0.98	0.43	**1.0**	0.58
	Time	**0.34**	1.19	1.38	0.49	1.03	0.98
	Iterations	**292**	387	449	330	483	428
	Iter/Sec	**884**	332	335	706	482	461
RASTRIGIN Threshold: 1e−4	Success rate	0.07	0.05	0.07	0.02	**0.80**	0.20
	Time	**0.31**	0.64	0.81	0.35	2.38	3.15
	Iterations	281	**226**	310	372	1104	1351
	Iter/Sec	899	365	388	**1079**	481	446
ROSENBROCK Threshold: 1e−4	Success rate	0.68	0.81	0.83	0.65	**0.93**	0.78
	Time	3.08	16.95	17.2	**0.95**	3.73	3.2
	Iterations	4093	4225	4268	**863**	1433	1190
	Iter/Sec	**1360**	249	248	951	400	378
SCHWEFEL Threshold: 1e−4	Success rate	0.03	0.03	0.03	0.03	**0.04**	0.03
	Time	**0.18**	0.43	0.51	0.31	0.26	0.37
	Iterations	156	149	162	232	**109**	141
	Iter/Sec	**920**	350	335	690	436	398
SPHERE Threshold: 1e−10	Success rate	**1.0**	**1.0**	**1.0**	**1.0**	**1.0**	**1.0**
	Time	**0.20**	0.56	0.56	0.21	0.36	0.38
	Iterations	168	167	170	**140**	142	144
	Iter/Sec	**870**	300	306	709	403	385

5.1 Results Analysis

Analyzing the results, a few general observations can be made. The inclusion of forced movement mechanisms severely affected the number of iterations per second. f-MAPSO (potential-based) had the top success rates in 6 out of 7 test functions. IPSO and MAPSO did not achieve the highest success rate for any test function except the Sphere function. Furthermore, by analyzing the results for each function, one can observe that:

- For the Ackley function, almost all algorithms achieved almost 100% success rate. Only MAPSO struggled with this function but required the least time and iterations.
- For the Griewank function, none of the algorithms reached the success threshold.
- For the H.C.Elliptic function, the f-PSO, Stasis f-PSO, and f-MAPSO were the most successful. However, IPSO and MAPSO required the least time for success.
- For the Rastrigin function, the "adaptive" algorithms performed better than the "non-adaptive" ones. f-MAPSO achieved by far the highest success rate.
- For the Rosenbrock function, the algorithms with forced movement mechanism succeeded more often than IPSO and MAPSO.
- For the Schwefel function, no significant difference in success rate can be observed for the algorithms. MAPSO required the least amount of time and almost the least amount of iterations to reach success.
- For the Sphere function, all algorithms succeeded in every run. IPSO required the least time, while MAPSO needed the least iterations.

6 Discussion

In this part of the paper, we discuss the impact of forced movement extensions, adaptive coefficients, and stasis as a criterion for applying forced movements to the IPSO method.

6.1 Forced Movement Extensions

Firstly, we compare the effectiveness of PSO versions with and without forced movement extensions, that is, f-PSOs and f-MAPSOs vs. IPSO and MAPSO. For most of the test functions (Ackley, Elliptic, Rastrigin, Rosenbrock), optimizers with a forced particle movement mechanism achieved a higher success rate. On the other hand, the calculation of forced movement mechanisms requires additional computational resources. This resulted in a lower count of iterations per second and an extended time (Ackley, Elliptic, Rastrigin, Rosenbrock, Schwefel, Sphere) to reach success. A question arises about whether the additional resources should be spent on the forcing mechanism instead of adding more particles to the swarm.

6.2 Adaptive Coefficients

Secondly, we compare the effectiveness of PSO versions with and without adaptive coefficients, that is, IPSO vs. MAPSO, f-PSO vs. f-MAPSO, Stasis f-PSO vs. Stasis f-MAPSO. The introduction of adaptive movement coefficients did not significantly affect the results, apart from the Rastrigin function, for which the success rate was considerably higher. The challenge in designing an algorithm with adaptive coefficients is finding the correct recipe to control those values. A broader set of experiments must be performed and more ideas tested to confirm or deny the possible benefits of introducing the adaptive coefficients to IPSO and its forced movement variants.

6.3 Runtime Stasis Criterion

Thirdly, we compare the two forced movement criteria (potential and stasis), that is, f-PSO vs. Stasis f-PSO and f-MAPSO vs. Stasis f-MAPSO. The runtime stasis criterion proved to work as a mechanism triggering forced movements. However, it had a higher calculation complexity than the potential-based criterion and was at times inferior (Elliptic, Rastrigin) in terms of success rate. On the other hand, the potential-based mechanism requires a fully connected topology for communication between particles. Limited communication may affect the efficiency of this approach. Meanwhile, the runtime stasis criterion is calculated independently for each swarm member, which could be especially useful when the number of particles is limited, and they explore the search domain isolated from one another.

7 Summary and Conclusions

In this paper, we investigated extending the IPSO method with forced movement mechanisms and adaptive coefficients. Based on the recently published guideline [3] for designing a PSO optimizer with adaptive coefficient values and the theory of the f-PSO method [13, 14], we designed an IPSO-based **f-MAPSO** method which combines forced movements with adaptive coefficients. Furthermore, we introduced **runtime stasis** as a new criterion for determining when to force particle movement.

Next, we designed a procedure containing adaptive coefficients and forced movement mechanisms. We showed how to derive IPSO, MAPSO, f-PSO, and f-MAPSO algorithms from the procedure. We tested the performance of the algorithms with a series of experiments using a set of benchmark functions: Ackley, Griewank, High Conditioned Elliptic, Schwefel, Rastrigin, Rosenbrock, and Sphere [17]. The results were collected and summarized in Table 1, analyzed in Sect. 5 and discussed in Sect. 6.

The main novelties of this paper are (1) designing a movement pattern adaptation PSO optimizer according to the recently published guideline [3], (2) proposing an adaptation of the concept of stasis [11] and utilizing it as a criterion for applying forced movements in the f-PSO method and (3) presenting a way of combining adaptive parameters and forced particle movements into a new method (f-MAPSO).

More analysis will be done on forced movements and adaptive coefficients in future work. The presented methods of choosing the correct thresholds for coefficients for given swarm lifetime phases are not optimal. The optimizers were only tested on a narrow set of optimization problems. Their applicability and performance in real-world problem applications are still unexplored.

References

1. Bassimir, B., Schmitt, M., Wanka, R.: Self-adaptive potential-based stopping criteria for particle swarm optimization with forced moves. Swarm Intell. **14**(4), 285–311 (2020). https://doi.org/10.1007/s11721-020-00185-z
2. Bonyadi, M.R., Michalewicz, Z.: Impacts of coefficients on movement patterns in the particle swarm optimization algorithm. IEEE Trans. Evol. Comput. **21**(3), 378–390 (2017). https://doi.org/10.1109/TEVC.2016.2605668

3. Bonyadi, M.R.: A theoretical guideline for designing an effective adaptive particle swarm. IEEE Trans. Evol. Comput. **24**(1), 57–68 (2020). https://doi.org/10.1109/TEVC.2019.2906894

4. Chauhan, P., Deep, K., Pant, M.: Novel inertia weight strategies for particle swarm optimization. Memetic Comput. **5** (2013). https://doi.org/10.1007/s12293-013-0111-9

5. Chen, G., Huang, X., Jia, J., Min, Z.: Natural exponential inertia weight strategy in particle swarm optimization. In: 2006 6th World Congress on Intelligent Control and Automation, vol. 1, pp. 3672–3675 (2006). https://doi.org/10.1109/WCICA.2006.1713055

6. Clerc, M., Kennedy, J.: The particle swarm-explosion, stability, and convergence in a multi-dimensional complex space. IEEE Trans. Evol. Comput. **6**(1), 58–73 (2002). https://doi.org/10.1109/4235.985692

7. Feng, Y., Teng, G.F., Wang, A.X., Yao, Y.M.: Chaotic inertia weight in particle swarm optimization. In: Second International Conference on Innovative Computing, Informatio and Control (ICICIC 2007), pp. 475–475 (2007). https://doi.org/10.1109/ICICIC.2007.209

8. Harrison, K.R., Engelbrecht, A.P., Ombuki-Berman, B.M.: Inertia weight control strategies for particle swarm optimization. Swarm Intell. **10**(4), 267–305 (2016). https://doi.org/10.1007/s11721-016-0128-z

9. Jiao, B., Lian, Z., Gu, X.: A dynamic inertia weight particle swarm optimization algorithm. Chaos, Solitons & Fractals **37**(3), 698–705 (2008). https://doi.org/10.1016/j.chaos.2006.09.063, https://www.sciencedirect.com/science/article/pii/S0960077906009131

10. Kennedy, J., Eberhart, R.C.: Particle swarm optimization. In: Proceedings of ICNN'95 - International Conference on Neural Networks, pp. 1942–1948. IEEE, Piscataway, NJ (1995). https://doi.org/10.1109/icnn.1995.488968

11. Kulpa, T., Trojanowski, K., Wójcik, K.: Stasis type particle stability in a stochastic model of particle swarm optimization. In: Proceedings of the 2021 Genetic and Evolutionary Computation Conference GECCO 2021, ACM Press (2021). https://doi.org/10.1145/3449639.3459405

12. Nickabadi, A., Ebadzadeh, M.M., Safabakhsh, R.: A novel particle swarm optimization algorithm with adaptive inertia weight. Appl. Soft Comput. **11**(4), 3658–3670 (2011). https://doi.org/10.1016/j.asoc.2011.01.037, https://www.sciencedirect.com/science/article/pii/S156849461100055X

13. Schmitt, M., Wanka, R.: Particle swarm optimization almost surely finds local optima. In: Proceeding of the Fifteenth Annual Conference on Genetic and Evolutionary Computation - GECCO 2013, pp. 1629–1636. GECCO 2013, ACM Press, New York (2013). https://doi.org/10.1145/2463372.2463563

14. Schmitt, M., Wanka, R.: Particle swarm optimization almost surely finds local optima. Theor. Comput. Sci. **561**, 57–72 (2015)

15. Shi, Y., Eberhart, R.C.: A modified particle swarm optimizer. In: Proceedings of the 1998 IEEE International Conference on Evolutionary Computation, IEEE World Congress on Computational Intelligence, pp. 69–73. IEEE (1998). https://doi.org/10.1109/icec.1998.699146

16. Tanweer, M., Suresh, S., Sundararajan, N.: Self regulating particle swarm optimization algorithm. Inf. Sci. **294**, 182–202 (2015). https://doi.org/10.1016/j.ins.2014.09.053, https://www.sciencedirect.com/science/article/pii/S0020025514009657, innovative Applications of Artificial Neural Networks in Engineering

17. Wagdy, A., Hadi, A.A., Mohamed, A.K., Agrawal, P., Kumar, A., Suganthan, P.N.: Problem definitions and evaluation criteria for the CEC 2021 special session and competition on single objective bound constrained numerical optimization (2020)

18. Yang, C., Gao, W., Liu, N., Song, C.: Low-discrepancy sequence initialized particle swarm optimization algorithm with high-order nonlinear time-varying inertia weight. Appl. Soft Comput. **29**, 386–394 (2015). https://doi.org/10.1016/j.asoc.2015.01.004, https://www.sciencedirect.com/science/article/pii/S1568494615000058

LQ-R-SHADE: R-SHADE with Quadratic Surrogate Model

Mateusz Zaborski$^{(\boxtimes)}$ and Jacek Mańdziuk

Faculty of Mathematics and Information Science, Warsaw University of Technology,
Koszykowa 75, 00-662 Warsaw, Poland
{M.Zaborski,J.Mandziuk}@mini.pw.edu.pl

Abstract. The application of evolutionary algorithms in continuous optimization is a well-studied area of research. Nevertheless, recently there have been numerous works associated with surrogate-assisted approaches. This paper introduces LQ-R-SHADE: R-SHADE extended with a quadratic surrogate model. The principles of LQ-R-SHADE and its enhancements over the base R-SHADE are discussed in detail. The extension consists of the three main components: an archive of samples, a prescreening meta-model, and an initialization supported by the meta-model. In order to take advantage of the meta-model utilization as early as possible, a cascade of models is proposed: linear, quadratic, and quadratic with interactions. The proposed algorithm relies on multiple generation of mutated versions of each individual. The prescreening meta-model is then applied to select the most promising candidates for further evaluation with the use of a (costly) true fitness function. The performance of LQ-R-SHADE is evaluated on the well-known COCO BBOB benchmark and compared with the baseline R-SHADE method and its extension SHADE-LM, showing the advantage of the proposed algorithm. Besides numerical assessment, the impact of particular meta-model components on the obtained results is examined.

Keywords: Metaheuristics · R-SHADE · Quadratic model

1 Introduction

Unconstrained continuous global optimization in a black-box scenario is an intensively researched but still under-explored area. While many optimization approaches and experimental designs are task-specific, there is still a lot of interest in the application of general-purpose algorithms, due to their flexibility and ease of adaptation. Evolutionary algorithms are a good example of such methods – they are both simple in their design and easy to hybridize with various optimization approaches.

The process of natural selection inspired a set of metaheuristics that consist of the following three phases: mutation, crossover, and selection. Genetic Algorithm (GA) [11] is the first and the best-known representative of this group.

However, Differential Evolution (DE) algorithm [16] with its various extensions (e.g. external memory or parameter adaptation) has proven more useful in continuous optimization. The above-mentioned DE enhancements have been incorporated into two successive algorithms: Adaptive Differential Evolution with Optional External Archive (JADE) [23] and Success-History Based Parameter Adaptation for Differential Evolution (SHADE) [17]. In addition, a restart mechanism has been applied in SHADE leading to Success-History Based Parameter Adaptation for Differential Evolution with Restart (R-SHADE) [18].

Another group of powerful evolutionary metaheuristics in the continuous optimization domain utilize the Covariance Matrix Adaptation (CMA). The baseline Covariance Matrix Adaptation Evolution Strategy (CMA-ES), like DE, has several well-known extensions, primarily referring to the population size and the restart management (e.g. IPOP-CMA-ES [2], BIPOP-CMA-ES [8], KL-BIPOP-CMA-ES [20]).

In this work we intend to integrate a global surrogate model into R-SHADE. Furthermore, we require the integration to be straightforward and not increase the computational cost considerably. We propose the LQ-R-SHADE algorithm that extends R-SHADE with an archive of samples and a linear-quadratic global surrogate model. Briefly, the extension consists of the following two steps. Firstly, the number of offspring generated by each individual in the mutation phase is greater than the default value of 1. Secondly, the surrogate model prescreens the offspring (trial vectors) to evaluate only the best candidates in the selection phase. Since the number of samples used for model estimation is limited, the computational and memory cost per iteration is constant during the whole optimization run.

The remainder of the paper is structured as follows. Section 2 presents the related literature, including a description of the original R-SHADE. Section 3 introduces the principles of the LQ-R-SHADE algorithm. Experimental results utilizing the BBOB testbed are presented in Sect. 4. Finally, conclusions and plans for future work are discussed in Sect. 5.

2   Related Work

Even though it is hardly possible to determine *a priori* whether utilization of surrogate extensions would become profitable for a given problem, numerous works concerning surrogate-assisted optimization have been proposed recently. Predominantly, the utilization of meta-models assumes an expensive optimization scenario, in which the number of function evaluations is relatively small (e.g. [5,6,13]). At the same time, many low-complexity algorithms are extended with selected surrogate approaches, however, the resulting increase in complexity is generally noticeable.

In particular, the CMA-ES family of algorithms has been extended with complex meta-models (e.g. Gaussian Process [4]), as well as less computationally intensive linear and quadratic models [3,12]. Nevertheless, frequent parameter values estimation of, even low-complexity, meta-model can visibly increase computational load, and therefore the recently proposed LQ-CMA-ES method [9] utilizes global surrogate meta-models of simple forms: linear, quadratic, or quadratic with interaction. This approach has inspired us in designing LQ-R-SHADE.

Likewise, the family of DE-like algorithms can be augmented with meta-models (e.g. Gaussian Process [15]). M-GAPSO [22] hybridizes PSO, DE, and meta-models. Moreover, M-GAPSO utilizes modified surrogate-assisted initial sampling, which has also been incorporated into the LQ-R-SHADE method proposed in this paper. Another algorithm that directly integrates R-SHADE and meta-models is computationally-efficient SHADE-LM [14], which uses a linear model and does not apply any prescreening component.

The related work [21] presents an in-depth analysis of psLSHADE – another LSHADE extension that utilizes a pre-screening mechanism whose initialization in not supported by the meta-model, and which is not in the form of a model cascade. Furthermore, the meta-model employed in psLSHADE's pre-screening mechanism is more complex than that of LQ-R-SHADE and includes inverse transformations of variables.

2.1 R-SHADE

R-SHADE is a population-based optimization method that extends the SHADE algorithm with the restart mechanism. R-SHADE, like SHADE, adapts its control parameters during the optimization run and, similarly to its predecessor JADE [23], utilizes an external archive. R-SHADE has multiple variants differing mainly in the mutation phase implementation. Its version presented in this section refers to *current-to-pbest/1* variant.

In each iteration g, R-SHADE maintains a population $P^g = [\boldsymbol{x}_i^g, \ldots, \boldsymbol{x}_N^g]$ of N individuals. Each individual $\boldsymbol{x}_i^g = [x_{i,1}^g, \ldots, x_{i,D}^g]$ is represented by a vector of D coordinates. The external archive A preserves parent vectors that were replaced with successful trial vectors. The archive's size $|A|$ equals $a \cdot N$ (a is a parameter). A randomly selected element is removed from A if the size is exceeded.

In the mutation phase, each individual $\boldsymbol{x}_i^g$ is transformed to $\boldsymbol{v}_i^g$ according to the following equation:

$$\boldsymbol{v}_i^g = \boldsymbol{x}_i^g + F_i^g(\boldsymbol{x}_{pbest_i}^g - \boldsymbol{x}_i^g) + F_i^g(\boldsymbol{x}_{r1_i}^g - \boldsymbol{x}_{r2_i}^g) \qquad (1)$$

where $\boldsymbol{x}_{pbest_i}^g$ is selected randomly from the best $N \cdot p$ individuals in the current iteration g (p is a parameter). F_i^g is a scaling factor designated independently for each individual i, in each iteration g, using Couchy distribution (4). Indices $r1_i \in \{1, \ldots, N\}$ and $r2_i \in \{1, \ldots, N + |A|\}$ are selected randomly assuming that $r1_i \neq r2_i$ ($r1_i$ corresponds to an individual from the population P, $r2_i$ to an individual from the union of population P and external archive A). Both $r1_i$ and $r2_i$ are designated independently in each iteration g, for each individual i.

The crossover phase is responsible for recombining the mutated vector $\boldsymbol{v}_i^g$ with the parent vector $\boldsymbol{x}_i^g$ to obtain a trial vector $\boldsymbol{u}_i^g$. Each coordinate $d \in \{1, \ldots, D\}$ in the trial vector is randomly chosen from either the mutated individual or the parent one. Crossover rate (CR_i^g) denotes the probability of crossover and is designated independently for each individual i in each iteration g, using Normal distribution (5). Moreover, one coordinate $d_{rand} \in \{1, \ldots, D\}$ is randomly selected for crossover with probability 1. Precisely, each coordinate of the trial vector $\boldsymbol{u}_i^g = [u_{i,1}^g, \ldots, u_{i,D}^g]$ is described by the following equation:

$$u_{i,d}^g = \begin{cases} v_{i,d}^g, & \text{if } rand(0,1) \leq CR_i^g \text{ or } d = d_{rand} \\ x_{i,d}^g, & \text{otherwise} \end{cases} \tag{2}$$

Finally, the trial vector $\boldsymbol{u}_i^g$ is evaluated by the fitness function f, and its value $f(\boldsymbol{u}_i^g)$ is compared with the corresponding value $f(\boldsymbol{x}_i^g)$ of the parent vector $\boldsymbol{x}_i^g$ to select the next generation population:

$$\boldsymbol{x}_i^{g+1} = \begin{cases} \boldsymbol{u}_i^g, & \text{if } f(\boldsymbol{u}_i^g) < f(\boldsymbol{x}_i^g) \\ \boldsymbol{x}_i^g, & \text{otherwise} \end{cases} \tag{3}$$

Parameter adaptation in R-SHADE promotes scaling factor and crossover rate values that have led to solution improvement in previous iterations. The memory consists of H scaling factor entries $M_{F,m}^g$ and H crossover rate entries $M_{CR,m}^g$ ($m \in \{1, \ldots, H\}$) [17].

According to Eq. (1) and Eq. (2), each individual i, in each iteration g has its own parameter values F_i^g and CR_i^g assigned. Both of them are generated randomly (using Cauchy distribution and Normal distribution, resp.). Distribution parameters are randomly sampled (from memory) pair $(M_{F,r_i}^g, M_{CR,r_i}^g)$, where $r_i \in \{1, \ldots, H\}$ (cf. Eqs. (4)–(5)). Besides, if the obtained value of F_i^g is smaller than 0, the random generation is repeated. If it is greater than 1, it is truncated to 1. The value of CR_i^g is truncated from both sides to the range $[0, 1]$.

$$F_i^g = rand_{Cauchy}(M_{F,r_i}^g, 0.1) \tag{4}$$

$$CR_i^g = rand_{Normal}(M_{CR,r_i}^g, 0.1) \tag{5}$$

In each iteration, all pairs of (F_k^g, CR_k^g), $k \in S$, $S = \{1, \ldots, N\}$ that succeeded in the trial vectors generation are recorded. Then, the weighted Lehmer mean (6) of these values is determined and placed in memory. S_k denotes the recorded value of either F_k^g or CR_k^g. The improvement Δf_k (or Δf_l) is understood as $|f(\boldsymbol{u}_k^g) - f(\boldsymbol{x}_k^g)|$. A round-robin algorithm is applied to determine the pair of elements in memory to be replaced by a new pair.

$$mean_{W_L}(S) = \frac{\sum_{k=1}^{|S|} w_k S_k^2}{\sum_{k=1}^{|S|} w_k S_k}, \qquad w_k = \frac{\Delta f_k}{\sum_{l=1}^{|S|} \Delta f_l} \tag{6}$$

R-SHADE is restarted when at least one of the following three conditions occurs: (1) population convergence, (2) population values convergence, (3) stagnation of the best-so-far solution. The first condition takes place if there exists dimension $d' \in \{1, \ldots, D\}$ for which the following inequality is fulfilled $(\epsilon_x = 10^{-12})$:

$$\max_{i=1,\ldots,N}\{x_{i,d'}^g\} - \min_{i=1,\ldots,N}\{x_{i,d'}^g\} < \epsilon_x \max_{i=1,\ldots,N}\{|x_{i,d'}^g|\} \qquad (7)$$

The second one occurs when the following inequality is fulfilled $(\epsilon_f = 10^{-12})$:

$$\max_{i=1,\ldots,N}\{f(\boldsymbol{x}_i^g)\} - \min_{i=1,\ldots,N}\{f(\boldsymbol{x}_i^g)\} < \epsilon_f \max_{i=1,\ldots,N}\{|f(\boldsymbol{x}_i^g)|\} \qquad (8)$$

The third condition is fulfilled if the best-so-far solution has not been updated in the last $500 \cdot D$ fitness function evaluations.

2.2   Contribution

The main contribution of the paper is the proposition of a novel continuous optimization algorithm LQ-R-SHADE that improves R-SHADE performance on expensive optimization problems.

There are three claims underlying the LQ-R-SHADE design: (1) the computational cost of the meta-model usage should be nearly constant during the entire optimization run; (2) the meta-model should be relatively simple and not over-parameterized; (3) the meta-model should not affect the baseline R-SHADE parameter adaptation scheme.

3   LQ-R-SHADE: Surrogate Assisted R-SHADE

LQ-R-SHADE[1] (Algorithm 1) is an extension of the R-SHADE algorithm presented in Sect. 2.1. It utilizes a meta-model that operates in the background and prescreens trial vectors, so as only those with the best surrogate values are further evaluated. Additionally, a meta-model supported initialization is incorporated into LQ-R-SHADE to guide the population to promising regions in the first iteration. Both meta-model related components: supporting initialization and prescreening require the use of an archive of previously evaluated samples. These three key components of the proposed solution are further described in Sects. 3.1, 3.2, and 3.3.

3.1   Archive of Samples

An archive of samples stores already evaluated samples, starting right from the beginning of the algorithm's execution. Coordinates $\boldsymbol{x}_i^g$ and the respective function values $f(\boldsymbol{x}_i^g)$ are inserted into the archive after each fitness function (f.f.) evaluation.

If the archive reaches its maximum capacity, the worst sample (in terms of f.f. value) is removed. If the currently considered (just evaluated) sample is worse

---

[1] LQ-R-RSHADE is maintained as an open source project available at: https://bitbucket.org/mateuszzaborski/lqrshade/.

Algorithm 1. LQ-R-SHADE high–level pseudocode

1: Set all parameter values $N, M_F, M_{CR}, p, a, H, N_a, N_s$ (see Table 2)
2: **while** evaluation budget left **do**
3:     Initialize $M_{F,m}^g$ and $M_{CR,m}^g$ memory entries with default values of M_F and M_{CR}
4:     $P^0 = [\boldsymbol{x}_1^0, \ldots, \boldsymbol{x}_N^0]$ ▷ Population initialization using Algorithm 2
5:     $g = 1$
6:     **while** evaluation budget left **do**
7:         Generate $N \cdot N_s$ mutated vectors $\boldsymbol{v}_i^{g,j}$ using Eq. (9)
8:         Generate $N \cdot N_s$ trial vectors $\boldsymbol{u}_i^{g,j}$ using Eq. (10)
9:         Estimate meta-model parameter values (see Table 1)
10:         Calculate $N \cdot N_s$ surrogate values $f^{surr}(\boldsymbol{u}_i^{g,j})$
11:         For each individual i designate the best trial vector $\boldsymbol{u}_i^{g,best}$
12:         **for** i = 1 to N **do**
13:             Do selection of $\boldsymbol{u}_i^{g,best}$ using Eq. (11)
14:             Add $\boldsymbol{u}_i^{g,best}$ and $f(\boldsymbol{u}_i^{g,best})$ to the archive using rules from Sec. 3.1
15:         **end for**
16:         Update memory with $M_{F,m}^g$ and $M_{CR,m}^g$ using Eq. (6)
17:         **if** restart required **then break**
18:         **end if**
19:         $g = g + 1$
20:     **end while**
21: **end while**

---

than the worst one in the archive, it is not inserted. In addition, the insertion procedure checks whether a sample with exactly the same coordinates or f.f. value already exists in the archive. In either case, the currently considered sample is not inserted.

3.2 Prescreening Meta-model

The prescreening meta-model is a cascade structure composed of the three models, summarized in Table 1, whose parameter values are estimated using Ordinary Least Squares [19]. The simplest model is linear and requires at least df_{lin} samples to be estimated. Then, a quadratic model without interactions is used if the number of samples is equal to at least df_{quad}. The most complex one is a full quadratic model with interactions that requires at least df_{full} samples to be constructed.

Table 1. A description of the prescreening meta-model cascade.

Name	Form	Degrees of freedom
Linear	$X_{lin} = [1, x_1, \ldots, x_D]$	$df_{lin} = D + 1$
Quadratic	$X_{quad} = [X_{lin}, x_1^2, \ldots, x_D^2]$	$df_{quad} = 2D + 1$
Full quadratic	$X_{full} = [X_{quad}, x_1x_2, x_1x_3, \ldots, x_{D-1}x_D]$	$df_{full} = 2D + \frac{D(D-1)}{2} + 1$

In LQ-R-SHADE the mutation phase is altered compared to R-SHADE. Each individual i, randomly generates N_s mutated vectors $\boldsymbol{v}_i^{g,j}$, $j \in \{1, \ldots, N_s\}$ according to Eq. (4) but with different $F_i^{g,j}$, $r1_i$, $r2_i$, and $pbest_i$ values, i.e.

$$\boldsymbol{v}_i^{g,j} = \boldsymbol{x}_i^g + F_i^{g,j}(\boldsymbol{x}_{pbest_i}^{g,j} - \boldsymbol{x}_i^g) + F_i^{g,j}(\boldsymbol{x}_{r1_i}^{g,j} - \boldsymbol{x}_{r2_i}^{g,j}) \tag{9}$$

Then, in each iteration g, each of the N_s mutated vectors $\boldsymbol{v}_i^{g,j}$ coming from a given individual i uses the same CR_i^g value in the crossover phase. Finally, the trial vector $\boldsymbol{u}_i^{g,j} = [u_{i,1}^{g,j}, \ldots, u_{i,D}^{g,j}]$ is obtained as:

$$u_{i,d}^{g,j} = \begin{cases} v_{i,d}^{g,j}, & \text{if } rand(0,1) \leq CR_i^g \text{ or } d = d_{rand} \\ x_{i,d}^g, & \text{otherwise} \end{cases} \tag{10}$$

where $\forall_{j \in \{1, \ldots, N_s\}} \, rand(0,1) = const.$ and $d_{rand} = const.$

A meta-model is built just before the selection phase. For each individual i, the surrogate values $f^{surr}(\boldsymbol{u}_i^{g,j})$ of all N_s trial vectors $\boldsymbol{u}_i^{g,j}$ are determined and only the best trial vector $\boldsymbol{u}_i^{g,best}$ in terms of surrogate function value $f^{surr}(\boldsymbol{u}_i^{g,j})$ is evaluated using a true f.f., what boils down to the following selection procedure:

$$\boldsymbol{x}_i^{g+1} = \begin{cases} \boldsymbol{u}_i^{g,best}, & \text{if } f(\boldsymbol{u}_i^{g,best}) < f(\boldsymbol{x}_i^g) \\ \boldsymbol{x}_i^g, & \text{otherwise} \end{cases} \tag{11}$$

Utilization of the prescreening meta-model does not affect the scaling factor and the crossover rate adaptation procedures. $F_i^{g,best}$ and CR_i^g values associated with the trial vector $\boldsymbol{u}_i^{g,best}$ are treated in the same manner as those associated with $\boldsymbol{u}_i^g$ in R-SHADE. Therefore, the solution improvement in the selection phase requires exactly the same actions to be performed as in the R-SHADE algorithm.

3.3   Initialization Supported by the Meta-model

Initial sampling is not purely random. In the first iteration ($g = 0$), two individuals' coordinates ($\boldsymbol{x}_{df_{lin}+1}^0$ and $\boldsymbol{x}_{df_{quad}+1}^0$) are replaced with the meta-model optima (linear and quadratic). Parameter estimation is performed in the same way as in the prescreening meta-model. In the case of a linear model, each coordinate takes the value of either the lower or the upper search boundary. In the case of a quadratic model, the search boundary points and a parabola peak are considered. Algorithm 2 describes the modified initial sampling procedure.

4   Experimental Results

The performance of LQ-R-SHADE is evaluated using the well-known Black-Box Optimization Benchmarks (BBOB) testbed from the COmparing Continuous Optimizers (COCO) platform [10]. The benchmark contains 24 continuous noiseless functions for $D = \{2, 3, 5, 10, 20, 40\}$. The assumed search space is $[-5, 5]^D$ for all problems. Functions are divided into 5 classes based on their properties: (1) separable functions, (2) functions with low or moderate conditioning,

Algorithm 2. LQ-R-SHADE model initialization

1: $g = 0$
2: $P^0 = [\boldsymbol{x}_1^0, \ldots, \boldsymbol{x}_N^0], \boldsymbol{x}_i^0 \sim \mathcal{U}(x_{min}, x_{max})$
3: **for** i $= 1$ to N **do**
4:    **if** $i == df_{lin} + 1$ **then**
5:       Build linear model (using samples from the archive)
6:       Set $\boldsymbol{x}_i^0$ to linear model optimum, $x_{lin,d}^0 \in \{x_{min}, x_{max}\}$
7:    **else if** $i == df_{quad} + 1$ **then**
8:       Build quadratic model (using samples from the archive)
9:       Set $\boldsymbol{x}_i^0$ to the quadratic model optimum, $x_{quad,d}^0 \in \{x_{min}, x_{max}, x_{peak}\}$
10:    **end if**
11:    Evaluate $\boldsymbol{x}_i^0$ using the fitness function
12:    Add $\boldsymbol{x}_i^0$ and $f(\boldsymbol{x}_i^0)$ to the archive
13: **end for**

(3) functions with high conditioning and unimodal, (4) multi-modal functions with adequate global structure, and (5) multi-modal functions with weak global structure. All test problems are described in detail in [7].

In the experiments, LQ-R-SHADE is compared with the baseline R-SHADE and SHADE-LM. To ensure a meaningful comparison, all parameter values not related to the surrogate model are the same in both algorithms – see Table 2, and follow the original R-SHADE parameterization [18].

Expected Running Time (ERT) is the performance measure used by default for this benchmark. ERT is understood as the time needed to reach the target fitness function value for the first time. The time is expressed in the number of fitness function evaluations. For a better presentation, the final number of evaluations is divided by the problem dimension D and then logarithmized. The target fitness function value is defined using the absolute difference Δf from the fitness function optimum f^{opt}. The lower Δf is, the more difficult the target $f^{target} = f^{opt} \pm \Delta f$ is to achieve.

Table 2. R-SHADE and LQ-R-SHADE parameters. $r(\cdot)$ returns the argument rounded to the nearest integer.

R-SHADE & LQ-R-SHADE		LQ-R-SHADE only	
Population size N	$r(3.96 \cdot D)$	Archive size N_a	$\max(N, 2df_{full})$
Initial M_F	0.38	Surrogates per individual N_s	10
Initial M_{CR}	0.94		
Best rate p	0.09		
Archive rate a	0.12		
Memory size H	11		

Experimental results for all 24 functions $f1 - f24$ are jointly presented in Fig. 1. The plots show empirical cumulative distribution functions (ECDF) of

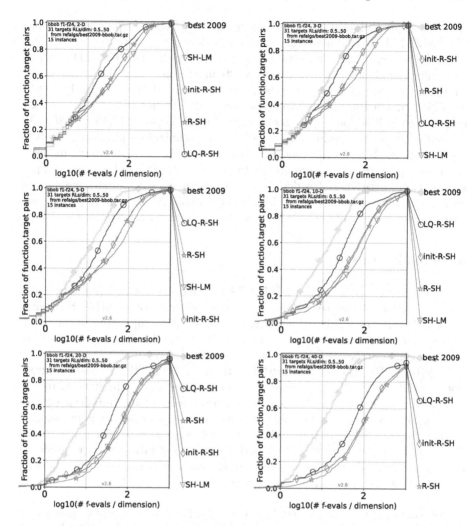

Fig. 1. Cumulative results of LQ-R-SHADE (denoted as LQ-R-SH) and init-R-SHADE (denoted as init-R-SH) versus R-SHADE (denoted as R-SH) and SHADE-LM (denoted as SH-LM) for all 24 functions on 2D, 3D, 5D, 10D, 20D and 40D, resp., with $10^3 \cdot D$ optimization budget.

ERT. ECDF value of 1 means that all targets f^{target} have been achieved at the given ERT level. The *best 2009 line* represents the best ERT observed during BBOB 2009 competition(*best 2009* is a baseline result used in the current COCO version 2.4). It is computed for each selected target independently. The evaluation was made using an expensive scenario, i.e. evaluation budget was limited to $10^3 \cdot D$ evaluations.

In addition, the performance of LQ-R-SHADE with initialization supported with the meta-model but without the use of the prescreening meta-model is checked. This configuration is marked as init-R-SHADE. Experimental results of SHADE-LM are obtained from the COCO BBOB web data archive [1].

In the context of the entire evaluation budget ($10^3 \cdot D$), LQ-R-SHADE outperforms all other algorithms in 5, 10, 20 and 40 dimensions. For 2 and 3 dimensions, all algorithms end up with comparable performance. More importantly, LQ-R-SHADE is noticeably more efficient between $10^1 \cdot D$ and $10^2 \cdot D$ optimization budgets in all dimensions. The meta-model initialization alone (init-R-SHADE results) improves the performance in the early optimization phase compared to the original R-SHADE. However, incorporating the prescreening meta-model is crucial for performance improvement from the perspective of the entire optimization run. SHADE-LM is designed for experiments with greater numbers of evaluations (non-expensive scenarios), what explains its noticeably poorer performance compared to LQ-R-SHADE under the assumed experiment conditions.

5 Conclusions

In this work, we propose the LQ-R-SHADE algorithm which is a surrogate-assisted version of the well-known R-SHADE method. LQ-R-SHADE extends R-SHADE with the archive of samples, the initialization supported by the meta-model, and the prescreening meta-model. The method scales quasi-linearly in time with the number of fitness function evaluations.

LQ-R-SHADE distinctly outperforms the baseline R-SHADE and its extension SHADE-LM in the expensive scenario. Furthermore, experimental results proved that both meta-model components (prescreening and initialization) are relevant, but the prescreening part is essential for the performance improvement.

The future work concerns finding more effective meta-model parametrizations, including the archive size (N_a) and the number of surrogates per individual (N_s). Temporary deactivation of meta-model in case of inaccurate predictions seems to be a promising direction, as well.

Acknowledgments. Studies were funded by BIOTECHMED-1 project granted by Warsaw University of Technology under the program Excellence Initiative: Research University (ID-UB).

References

1. COCO data archives. https://numbbo.github.io/data-archive/. Accessed 14 Jan 2021
2. Auger, A., Hansen, N.: A restart CMA evolution strategy with increasing population size. In: 2005 IEEE Congress on Evolutionary Computation, vol. 2, pp. 1769–1776. IEEE (2005)
3. Auger, A., Schoenauer, M., Vanhaecke, N.: LS-CMA-ES: a second-order algorithm for covariance matrix adaptation. In: Yao, X., et al. (eds.) PPSN 2004. LNCS, vol. 3242, pp. 182–191. Springer, Heidelberg (2004). https://doi.org/10.1007/978-3-540-30217-9_19

4. Bajer, L., Pitra, Z., Repický, J., Holeňa, M.: Gaussian process surrogate models for the CMA evolution strategy. Evol. Comput. **27**(4), 665–697 (2019)
5. Can, B., Heavey, C.: A comparison of genetic programming and artificial neural networks in metamodeling of discrete-event simulation models. Comput. Oper. Res. **39**(2), 424–436 (2012)
6. Fang, H., Horstemeyer, M.F.: Global response approximation with radial basis functions. Eng. Optim. **38**(04), 407–424 (2006)
7. Finck, S., Hansen, N., Ros, R., Auger, A.: Real-parameter black-box optimization benchmarking 2009: Presentation of the noiseless functions. Technical Report, Citeseer (2010)
8. Hansen, N.: Benchmarking a BI-population CMA-ES on the BBOB-2009 function testbed. In: Proceedings of the 11th Annual Conference Companion on Genetic and Evolutionary Computation Conference: Late Breaking Papers, pp. 2389–2396 (2009)
9. Hansen, N.: A global surrogate assisted CMA-ES. In: Proceedings of the Genetic and Evolutionary Computation Conference, pp. 664–672 (2019)
10. Hansen, N., Auger, A., Ros, R., Mersmann, O., Tušar, T., Brockhoff, D.: COCO: a platform for comparing continuous optimizers in a black-box setting. Optim. Methods Softw. **36**(1), 114–144 (2021)
11. Holland, J.H., et al.: Adaptation in Natural and Artificial Systems: an Introductory Analysis with Applications to Biology, Control, and Artificial Intelligence. MIT Press, Cambridge (1992)
12. Kern, S., Hansen, N., Koumoutsakos, P.: Local meta-models for optimization using evolution strategies. In: Runarsson, T.P., Beyer, H.-G., Burke, E., Merelo-Guervós, J.J., Whitley, L.D., Yao, X. (eds.) PPSN 2006. LNCS, vol. 4193, pp. 939–948. Springer, Heidelberg (2006). https://doi.org/10.1007/11844297_95
13. Kleijnen, J.P.: Kriging metamodeling in simulation: a review. Eur. J. Oper. Res. **192**(3), 707–716 (2009)
14. Okulewicz, M., Zaborski, M.: Benchmarking SHADE algorithm enhanced with model based optimization on the BBOB noiseless testbed. In: Proceedings of the Genetic and Evolutionary Computation Conference Companion, pp. 1259–1266 (2021)
15. Ren, Z., et al.: Surrogate model assisted cooperative coevolution for large scale optimization. Appl. Intell. **49**(2), 513–531 (2019)
16. Storn, R., Price, K.: Differential evolution-a simple and efficient heuristic for global optimization over continuous spaces. J. Global Optim. **11**(4), 341–359 (1997)
17. Tanabe, R., Fukunaga, A.: Success-history based parameter adaptation for differential evolution. In: 2013 IEEE congress on evolutionary computation, pp. 71–78. IEEE (2013)
18. Tanabe, R., Fukunaga, A.: Tuning differential evolution for cheap, medium, and expensive computational budgets. In: 2015 IEEE Congress on Evolutionary Computation (CEC), pp. 2018–2025. IEEE (2015)
19. Weisberg, S.: Applied Linear Regression. Wiley, New York (2013)
20. Yamaguchi, T., Akimoto, Y.: Benchmarking the novel CMA-ES restart strategy using the search history on the BBOB noiseless testbed. In: Proceedings of the Genetic and Evolutionary Computation Conference Companion, pp. 1780–1787 (2017)
21. Zaborski, M., Mańdziuk, J.: Improving LSHADE by means of a pre-screening mechanism. In: Proceedings of the Genetic and Evolutionary Computation Conference, GECCO 2022, pp. 884–892. Association for Computing Machinery, New York, USA (2022). ISBN 9781450392372. https://doi.org/10.1145/3512290.3528805

22. Zaborski, M., Okulewicz, M., Mańdziuk, J.: Analysis of statistical model-based optimization enhancements in generalized self-adapting particle swarm optimization framework. Found. of Comput. Decis. Sci. **45**(3), 233–254 (2020)
23. Zhang, J., Sanderson, A.C.: JADE: adaptive differential evolution with optional external archive. IEEE Trans. Evol. Comput. **13**(5), 945–958 (2009)

Pattern Classification

Overload Monitoring System Using Sound Analysis for Electrical Machines

Nguyen Cong-Phuong[✉] and Nguyen Tuan Ninh

School of Electrical and Electronic Engineering, Hanoi University of Science and Technology, Hanoi, Vietnam
phuong.nguyencong@hust.edu.vn

Abstract. Overloading is one of the faults that occur very often in the operation of electrical machines. Therefore, a continuous monitoring and diagnosis for this is necessary in safety-critical applications. This paper presents a sound analysis system used for detecting and classifying induction motor and power transformer overload levels with a microphone. Three acoustic features and six classification models are evaluated. The obtained results show that this is a promising way to monitor electrical machines overload.

Keywords: Sound analysis · Electrical machine overload · Machine learning

1 Introduction

Electrical machines, including electric motors, generators, and transformers, can be found from homes to industries, transportation, agriculture, etc. Because of their wide diffusion and popularity, electrical machines' stable and smooth operations are very essential. For example, the total cost of the Hydro-Quebec incidents is estimated to be USD 6 billion [1], or an eight-hour interruption can result in an average loss of nearly USD 94,000 for medium and large industrial manufactures in the United States [2].

Unfortunately, there are many factors causing the failures of transformers, such as electrical breakdown (caused mainly by contaminated oil, thermal ageing, repetitive excessive voltage, mechanical deformation), lightning, insulation, loose connection, improper maintenance, moisture, overload, and other. For induction motors, they are bearing (e.g., wear out of bearings), stator (insulation damages, for instance), rotor (broken rotor bars or cracked rotor end-rings), and other faults (eccentricity, for example). Readers can refer to [3] for a literature review on methods for assessing the condition of power transformer. Another literature review about condition monitoring of induction motors can be found in [4].

Among power transformer and induction motor failures, overloading can have serious consequences. Power transformer overload can result in overheating for the oil and the core which in turn accelerate the aging process, cause internal damage, or even lead to the outage. When a motor is overloaded, it can draw more current, causing excessive temperatures. A too high temperature may burn motors. Besides, overload can result in tooth breakage, or wear in roller bearings and gears.

L. Rutkowski et al. (Eds.): ICAISC 2022, LNAI 13588, pp. 279–287, 2023.
https://doi.org/10.1007/978-3-031-23492-7_24

An overload monitoring system for induction motors was proposed in [4] and another one for power transformers in [3]. Both are based on sound analysis. In this research, the two systems are combined into a unified system. This new one is to detect and classify levels of power transformer overload and induction motor overload. The advantage of the combination is that we do not have to switch manually to electric motor mode or transformer mode, especially if a telemonitoring is needed. The levels to be classified of power transformer are 90% of load (underload), 100% of load (full load), and 110% of load (10% overload). Those of induction motor are 100%, 110%, 120%, 150%, and 200%. It should be mentioned that an induction motor with 10% overload can still run about 30 min, while a 100% overload can last 10 min only. The approach of using sound analysis has three advantages. Firstly, sounds of overload appear earlier than its consequences (excessive temperatures, tooth breakage, etc.), so detecting an abnormal sound can prevent these consequences. Secondly, a sound sensor and its installation and maintenance are inexpensive compared to other sensors, such as dissolved gas sensor. And finally, it does not need to stop the motor during the detection, therefore, a continuous and online monitoring system is available.

The organization of the paper is as follows. Section 2 informs the studies recently presented in the literature that refer to induction motor and power transformer overload detection methods. Section 3 is about the corpus used in this research. Section 4 describes in detail the proposed method, experiments, and results. Section 5 concludes the paper and presents future developments.

2   Related Works

Sounds can be used to detect some mechanical and electrical problems of electric motors, such as unbalance, bearing, broken rotor bars, eccentricity, soft foot, shorted rotor coils, tooth damage in the gearbox. Transformer overload can be detected by monitoring current, temperature, water content of oil, vibration signal. These signals are usually processed by artificial intelligence. For more details on published works concerning induction motor overload and power transformer overload, readers can refer to [4] and [3], respectively.

To our best knowledge, no study of using sounds to recognize induction motor and power transformer overload levels has been published so far. Sample Heading (Third Level). Only two levels of headings should be numbered. Lower level headings remain unnumbered; they are formatted as run-in headings.

3   The Database

The object of our study is a 63MVA power transformer and a 4kW three-phase induction motor. Sounds of underload, full load and overload operations are recorded by a microphone. This microphone is connected directly to the audio input of a laptop, so the recorder is the laptop's sound card. The distance between the electrical machine and the microphone is about 2 m. Nonstop soundtracks are acquired with the following parameters: sampling frequency is 44.1 kHz, bit resolution is 16, and mono channel. Levels of overload and durations of recorded sounds are in Table 1 and Table 2.

Table 1. Power transformer overload sound corpus.

Overload level	90% (Underload)	100% (Full load)	110% (10% overload)
Duration (s)	292	306	305

Table 2. Induction motor overload sound corpus.

Overload level	100% (Full load)	110% (10% overload)	120% (20% overload)	150% (50% overload)	200% (100% overload)
Duration (s)	137	131	142	146	112

4   The Experiments

4.1   The 8-Class Classifier

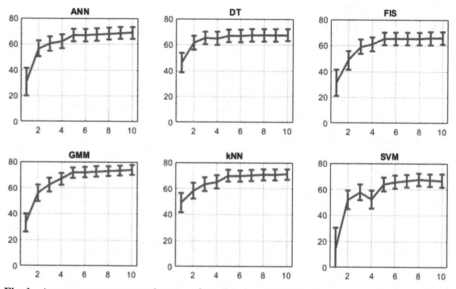

Fig. 1. Average accuracy rate plots (n-o-f varying from 1 to 10) of six models. Horizontal axes are n-o-f. Vertical axes are accuracy (in %). Vertical bars are standard deviations.

In the first phase, we try to categorize sound signals directly into 8 classes mentioned in Sect. 3. Therefore, an eight-class classifier should be used. It includes a set of discriminant features and a classification model. Our discriminant features consist of F0, Mel-Frequency Cepstral Coefficients (MFCC, 12 coefficients, [5]), and Band Energy Ratio (BER, 4 bands, [6]). These very popular features in audio signal processing form a starting feature set of 17 elements. Then the Principal Component Analysis (PCA,

[7]) is applied to reduce the dimension of our feature vector without losing too much information. Essentially, the PCA extracts the important information from the original feature set to rebuild them as a set of new orthogonal features (principal components), and hence to gain a better representation of classes by reducing the number of features (n-o-f). To find the appropriate number of new features, we test six classification models for frame classification: artificial neural network (ANN [8]), decision tree (DT [9]), fuzzy inference system (FIS [10]), Gaussian mixture model (GMM [11]), k-nearest neighbors (kNN [12]), and support vector machine (SVM [13]).

Table 3. Parameters of classification models.

Model	Parameter(s)
ANN	20 hidden neurons, feedforward network, gradient descent training algorithm
DT	At least 10 observations
FIS	Backpropagation
GMM	2 mixtures, diagonal covariance matrices
kNN	10 neighbors
SVM	Error – correcting output codes

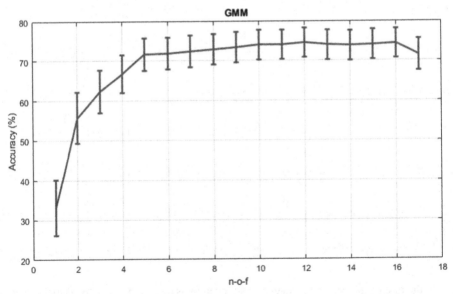

Fig. 2. Average accuracy rate plot of GMM. Vertical bars are standard deviations.

For each classification model, the n-o-f varies from 1 to 10, and the 10-fold cross validation is applied to find the best n-o-f for that model. The accuracy rate is employed in model evaluation for 1024-sample frames. This criterion is also used to select best parameter(s), if any, for each model. The chosen parameters are the ones resulting in highest accuracy rates. Those parameters are in Table 3. Plots of average accuracy rates depending on n-o-f are in Fig. 1. It is noticeable that these plots are of the same trend: start at a very low rate (corresponding to one feature), then rise rapidly when the n-o-f increases to 5, and finally go nearly horizontally. The operations of these models are stable, because the standard deviations are rather small for each n-o-f. Of these models, GMM performs the best, so we keep experimenting with it by increasing n-o-f.

Table 4. Confusion matrix of GMM with 12 features.

			100%	110%	120%	150%	200%	90%	100%	110%
		100%	56.73	29.69	9.84	4.09	0.062	0	0	0
	Induction motor	110%	25.73	53.06	10.28	8.23	0.25	0	0	0
True class		120%	29.70	30.74	21.80	19.27	1.21	0	0	0
		150%	3.20	7.09	8.35	76.22	8.25	0	0	0
		200%	0	0.13	0.80	8.01	88.35	0	0	0
	Pow. trans.	90%	0	0.017	0.057	0.024	0	99.88	0.072	0
		100%	0	0	0	0	0	0.008	99.99	0
		110%	0	0	0	0	0	0	0	100

(columns grouped as Induction motor: 100% 110% 120% 150% 200%; Power transformer: 90% 100% 110%. Predicted class.)

Table 5. Standard deviation.

			100%	110%	120%	150%	200%	90%	100%	110%
		100%	1.54	2.20	1.01	0.37	0.10	0	0	0
	Induction motor	110%	1.26	1.11	0.86	0.44	0.18	0	0	0
True class		120%	1.09	0.93	0.71	1.06	0.36	0	0	0
		150%	0.54	1.19	0.91	1.65	1.43	0	0	0
		200%	0	0.14	0.32	0.45	0.96	0	0	0
	Pow. trans.	90%	0	0.036	0.055	0.039	0	0.051	0.052	0
		100%	0	0	0	0	0	0.017	0.016	0
		110%	0	0	0	0	0	0	0	0

(columns grouped as Induction motor: 100% 110% 120% 150% 200%; Power transformer: 90% 100% 110%. Predicted class.)

To improve the performance of GMM, the evaluation is conducted with n-o-f varying from 11 to 17. The outcomes (Fig. 2) show that the best n-o-f is 12, resulting in an accuracy of 74.50%, although the effect of this value is not so outstanding compared to others. But this is just the average accuracy of 8 classes. To better understand the efficiency of this model for each class, we consider the confusion matrix (Table 4) corresponding to the n-o-f of 12. The standard deviations of elements in Table 4 are shown in Table 5. The first thing we notice is that this classifier works very well for power transformer. In particular, the accuracy of the 110% class of this electrical machine is 100% with zero standard deviation. Out of 36 off-diagonal elements of power transformer, 31 are equal to 0. As for induction motor, the story is not so good. Its diagonal elements vary from 21.80 to 88.35. The greatest off-diagonal element reaches 30.74 (the 110–120 element). It can be seen that the accuracy of power transformer's signals contributes mainly to the average accuracy of this 8-class classifier. In short, this model is not suitable for the task of directly categorizing 8 sound classes.

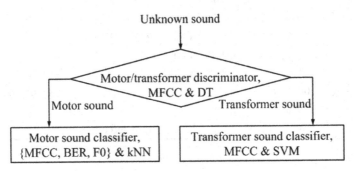

Fig. 3. The two-stage classifier.

A new classifier is needed to get better results. Because power transformer sounds are classified very well, the classification should be divided into two stages (illustrated in Fig. 3). In the first stage, the unknown sound is classified as induction motor or power transformer. In the second stage, if it comes from induction motor, it is further classified by the motor sound classifier in [4]. Otherwise, the transformer sound classifier in [3] is employed. The first stage therefore needs a motor/transformer discriminator.

4.2 The Motor/Transformer Discriminator

A discriminator is actually a two – class classifier. Because the motor sound classifier [4] and the transformer sound classifier [3] do not use PCA, we do not apply PCA to this discriminator either. The full set of discriminant features (MFCC, BER, and F0) is evaluated first. Then each feature is evaluated separately. Among the six classification models mentioned in Sect. 4.1, DT is the simplest one in terms of testing. So, this model (with at least 10 observations) is tested first using the 10-fold cross validation. The obtained results are provided in Table 6.

Table 6. DT-based discriminator's performance.

Feature set	MFCC, BER, & F0	MFCC	BER	F0
Accuracy (%)	100	100	93.65	82.18
Standard deviation (%)	0	0	3.22	2.40

If all three discriminant features are used, the accuracy is 100% and the accompanying standard deviation is 0%. These results are still the same if only MFCC is used. If less features are used (BER only or F0 only), the discriminant efficiency reduces. And the corresponding standard deviations increase, but are still quite small, without affecting the stability of this discriminator. Since 12 MFCC features are less than 17 features of the full set (MFCC, BER, and F0), while their performance is the same, only MFCC is chosen for the discriminator. It should be noted that fewer features make the discriminator simpler.

Since the results obtained with DT are already good enough, the other five models do not need to be evaluated anymore. Finally, the chosen motor/transformer discriminator is based on MFCC and DT.

4.3 The Overload Monitoring System

This system is expected to monitor electrical machines every one second, so the classification of one-second segments is evaluated in the final phase. The number of 1-s segments is fewer than that of frames of 1024 samples, so in this stage, the "leave-one-out" cross validation is applied to classify 1-s segments. Because the sampling frequency is 44100 Hz and the overlap is 512 samples, each 1-s segment includes 86 frames of 1024 samples. To categorize a 1-s segment, "the winner takes it all" tactic is applied. For example, if a 1-s segment of a motor sound has 30 frames of 100%, 10 of 110%, 25 of 120%, 15 of 150%, and 6 of 200%, it will be recognized as an 100% one.

The confusion matrix of classification performed by our system (Fig. 3) is presented in Fig. 4. It is obvious that the system gains an error rate of 0: all off-diagonal entries are zero. Based on this excellent performance, the classifier in Fig. 3 is selected for our overload monitoring system. To classify overloads, sounds of the electrical machines are collected by a microphone. For each sound signal, a Hanning window is applied to frames of 1024 samples, the overlap is 512 samples, then F0, MFCC, and BER are computed from each frame and fed to the input of the proposed system. One second of sound takes our algorithm (installed in a laptop with Intel Core i5 and 8 GB of RAM) about 0.1 s to process. The output of this system will tell that the recorded signal came from one of eight levels of overload of electrical machines. By detecting and classifying overload condition, this system can come to the root of some problems of electrical machines, such as tooth breakage, overheating or internal damage. It is difficult to compare the accuracy of our approach to the other ones, because we cannot find any published reports of accuracy of electrical machine overload sound classification so far.

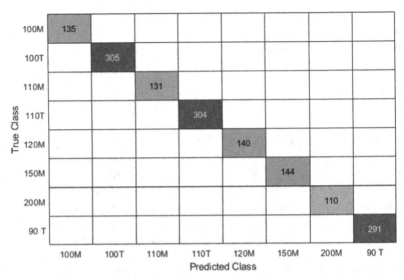

Fig. 4. Confusion matrix of the overload monitoring system. M: motor; T: transformer.

5   Conclusions

This paper presents a method for classifying sounds of electrical machine overload using sound analysis. Audio signals are recorded by a microphone placed near an electrical machine to monitor its overload. A feature set (including F0, MFCC, and BER) and six classification models are evaluated. Experiments prove that DT, kNN, and SVM fit our two-stage classifier. Our proposed system can be an online monitoring method because it does not need to stop electrical machines. This system requires a microphone, so it is inexpensive. It is also flexible, meaning that if we can collect sounds of other faults, we can upgrade it by retraining it. Filtering techniques should be applied if this method is moved to industrial environment to reduce noises. Future developments can be related to other levels of overload and other faults, such as eccentricity, bearing, rotor bars, etc.

References

1. OECD: OECD Reviews of Risk Management Policies. 2011. Future Global Shocks: Improving Risk Governance, p. 36 (2011)
2. Hodge, F.: Energy risks–the dangers of power cuts and blackouts. In: INSIGHTS Special topic–Emerging risks, Allianz Global Corporate & Specialty, AGCS ALLIANZ, pp. 28–33 (2015)
3. Nguyen, C.P.: Large power transformer overload detection using sound analysis. In: SPML 2021, pp. 135-141 (2021)
4. Nguyen, C.P.: A simple and effective sound-based five-class classifier for induction motor overload. In: ICAAI 2021, pp. 170-175 (2021)
5. Carey M.J., Parris, E.S., Lloyd – Thomas, H.: A comparison of features for speech, music discrimination. In: ICASSP'99, pp. 149–152 (1999)

6. McCowan, I., Gatica-Perez, D., Bengio, S., Lathoud, G., Barnard, M., Zhang, D.: Automatic analysis of multimodal group actions in meetings. IEEE Trans. Pattern Anal. Mach. Intell. **27**(3), 305–317 (2005)
7. Jolliffe, I.: Principal Component Analysis. Springer-Verlag, New York (1986)
8. Bishop, C.M.: Neural Networks for Pattern Recognition, Oxford University Press (1995)
9. Breiman, L., Friedman, J., Olshen, R., Stone, C.: Classification and Regression Trees, Belmont, USA (1984)
10. Jang, J.-S. R.: ANFIS: adaptive-network-based fuzzy inference systems. IEEE Trans. Syst. Man Cybern. **23**(3), 665–685 (1993)
11. McLachlan, G., Peel, D.: Finite Mixture Models. John Wiley & Sons Inc, Hoboken, NJ (2000)
12. Scheirer, E., Slaney, M.: Construction and evaluation of a robust multifeature music/speech discriminator. In: ICASSP 1997, vol. II, pp. 1331–1334 (1997)
13. Vapnik, V.N.: Statistical Learning Theory. Wiley, N.Y. (1998)

Employing Convolutional Neural Networks for Continual Learning

Marcin Jasiński and Michał Woźniak[✉][iD]

Department of Systems and Computer Networks, Wrocław University of Science
and Technology, Wybrzeże Wyspiańskiego 27, 50-370 Wrocław, Poland
marcin.jasinski1997@gmail.com, michal.wozniak@pwr.edu.pl

Abstract. The main motivation for the presented research was to investigate the behavior of different convolutional neural network architectures in the analysis of non-stationary data streams. Learning a model on continuously incoming data is different from learning where a complete learning set is immediately available. However, streaming data is definitely closer to reality, as nowadays, most data needs to be analyzed as soon as it arrives (e.g., in the case of anti-fraud systems, cybersecurity, and analysis of images from on-board cameras and other sensors). Besides the vital aspect related to the limitations of computational and memory resources that the proposed algorithms must consider, one of the critical difficulties is the possibility of *concept drift*. This phenomenon means that the probabilistic characteristics of the considered task change, and this, in consequence, may lead to a significant decrease in classification accuracy. This paper pays special attention to models of convolutional neural networks based on probabilistic methods: Monte Carlo dropout and Bayesian convolutional neural networks. Of particular interest was the aspect related to the uncertainty of predictions returned by the model. Such a situation may occur mainly during the classification of drifting data streams. Under such conditions, the prediction system should be able to return information about the high uncertainty of predictions and the need to take action to update the model used. This paper aims to study the behavior of the network of the models mentioned above in the task of classification of non-stationary data streams and to determine the impact of the occurrence of a sudden drift on the accuracy and uncertainty of the predictions.

Keywords: Continous learning · Concept drift · Bayesian convolutional neural network · Deep learning

1 Introduction

One of the still current problems of continual learning is the classification of non-stationary data streams [3]. Methods dedicated to their analysis require efficient classifiers, which will identify the occurrence of the so-called *concept drift* on the

L. Rutkowski et al. (Eds.): ICAISC 2022, LNAI 13588, pp. 288–297, 2023.
https://doi.org/10.1007/978-3-031-23492-7_25

one hand, and on the other hand, correctly respond to emerging changes in probabilistic characteristics of the classification task [2]. It is also worth mentioning the need for restoration analysis [13], which aims to assess the speed of model adjustment to the new probability characteristics - *restoration time* and *maximum performance loss*, which measures the prediction performance degradation after the drift. This paper focuses on analyzing the behavior of convolutional neural networks based on probabilistic methods (Monte Carlo dropout and Bayesian convolutional neural networks) in case of *concept drift* occurrence.

Let us first briefly characterize the techniques used in this work. The Bayesian neural network describes a model in which probability distribution parameters replace the real-number weights of individual neurons. In the traditional view, a neural network model N represents some function $y = N(x)$, where y is the output of the model's prediction based on input x. Each layer of the network performs a linear transformation of the previous layer's output using the weights in that layer and then a nonlinear transformation of the result of this operation using an activation function. The process of learning the network involves matching weights and biases in the layers using a backpropagation algorithm to minimize the cost function [11]. From a statistical point of view, this process corresponds to the use of the point-estimate method. A Bayesian neural network is a network in which, instead of point-estimate, interval estimation of model parameters is performed, and point values of parameters are replaced by parameters of probability distribution [10]. The process of learning such a network takes place using the rules of Bayesian inference. The method of learning Bayesian neural networks known as *Bayes by Backprop* was proposed by Charles Blundell in [1]. Like a neural network using Monte Carlo dropout, inference using a Bayesian neural network can be treated as inference using an ensemble of classifiers. However, a single model is trained. By replacing point values with distribution parameters, each prediction made with a Bayesian model will result from a model with different weight values in the layers, again allowing a large number of predictions to be collected and acted upon. The need for regularization mechanisms in the learning process of neural networks arises from their tendency to overfit the training data. To this end, many methods are used to prevent networks from overfitting, one of the most popular of which is a dropout. G. E. Hinton proposed this technique in [6]. Dropout also finds application in aspects of deep learning other than preventing overfitting. It has been successfully used in classifying data streams with recursive drift [5]. Monte Carlo dropout is a method proposed by Yarin Gal and Zoubin Ghahramani in [4]. According to the original concept, dropout is based on randomly switching off neurons in selected network layers during the model learning process on training data. The method proposed by the authors assumes leaving the dropout active also during the classification of new samples. The paper's authors demonstrated that the method "can be interpreted as a Bayesian approximation of another well-known probabilistic model: the deep Gaussian process." Leaving neurons to drop out when inferring new, previously unseen samples actively allows multiple, potentially different prediction results to be obtained with a single model without explicitly modifying it-performing a prediction on given sample dozens or even hundreds of times using such neural network results in each prediction being the result of a model with a different internal structure (with a different set of active neurons), which corresponds to a situation where a set of different neural networks makes the predictions. This method provides information on the uncertainty of the predictions without making trade-offs between computational efficiency and time and classification quality. Bayesian machine learning approaches seem to be particularly relevant in image processing and image classification tasks. In medicine, for example, information about the uncertainty of model prediction results is crucial. However, due to its mode of operation, Bayesian inference in neural networks leads to at least a doubling of the number of model parameters (replacing the point value with the mean and variance or standard deviation of the distribution). This resulted in the need for a new approach to model training and a backpropagation algorithm that considers the problem of a very large number of parameters.

This paper aims to study the behavior of conventional neural networks in the task of classifying nonstationary data streams and to determine the effect of the occurrence of a sudden drift on the accuracy and uncertainty of the predictions. It is particularly important to evaluate the decision uncertainty, which can be helpful when designing *concept drift* detectors for the mentioned models.

2   Experimental Evaluation

The study was conducted to investigate the behavior of conventional neural networks in classifying non-stationary data streams and determine the effect of the occurrence of a sudden drift in the stream on the accuracy and uncertainty of the predictions. In particular, the experiments are designed to answer the following research questions:

RQ1: What effect does a sudden drift have on convolutional networks' classification accuracy and behavior?

RQ2: How high is the prediction uncertainty, as determined by the measured prediction entropy, mean value and variance of the softmax function response?

RQ3: What effect does the choice of optimizer have on model learning and the consequences of drift?

RQ4: What is the maximum decrease in classification accuracy, and how long does it take for the model to reach prediction accuracy before drift occurs?

2.1   Setup

Data Streams. In order to perform the experimental evaluation two datasets were used for the experiments to form data streams: CIFAR10 [9] and MNIST [12]. Each stream consists of 6000 samples. Drift in the data stream produced from the CIFAR10 dataset was induced by switching samples from the automobile and deer classes to samples from the truck and horse classes at a given point in time (with switching back to the first set of classes for recurrent drift), while drift in the data stream from the MNIST dataset was simulated by introducing significant random noise in the samples.

Analyzed Methods. During the experiments, the following neural network configurations were compared:

- CNN with SGD and Adam optimizers (Normal CNN)
- CNN using Monte Carlo dropout with SGD and Adam optimizers (Monte Carlo CNN)
- Bayesian CNN with SGD and Adam optimizers (Bayesian CNN)

Hyperparameters of Models and Experimental Environment. The following values were set for the hyperparameters of the models under study and the testing environment itself: the group size of samples taken from the stream was set to 32, while the number of model learning epochs on a single group was set to 50. These determinations were made keeping in mind the characteristics of learning from stream data regarding the speed of data processing by the model. SGD and Adam methods were chosen as optimizers for the convolutional networks. The default value of the optimizers parameter of 0.001 was used during the experiments. The ReLU function was used as the activation function in the convolutional layers.

Reproducibility. The research was conducted in the Python environment version 3.6.5, using the TensorFlow (v. 1.15.0) and Keras (v. 2.3.1) libraries. The Bayesian convolutional network was implemented using the TensorFlow Probability library (v. 0.8.0). Code written by Robert Romijnders was used to implement metrics related to the uncertainty of the network predictions: entropy, variance, and mean response of the softmax function[1]. All experiments presented in this article can be replicated independently using the code available in the *GitHub* repository[2].

Experimental Protocol. All experiments were based on the *Test-Than-Train* [8] evaluation protocol. The estimation of the classification accuracy and the parameters determining the prediction uncertainty (entropy, mean value, and variance of the softmax function response) consisted of performing the prediction with the tested model 100 times on the current group of samples. The tested parameters were then calculated for each sample in the group, and in the next step the results were averaged over the entire group. This approach, based on performing the prediction multiple times, made it possible to study the behavior of the models taking into account the properties of Monte Carlo dropout-based and Bayesian neural networks.

2.2   Results

Three experiments were conducted during the study. In the first, a data stream generated from the CIFAR-10 database with one injected sudden drift was analyzed. In the second, two sudden drifts were added. The third one analyzed

---

[1] https://github.com/RobRomijnders/weight_uncertainty.
[2] https://github.com/marcin-jasinski/cnn_data_streams.

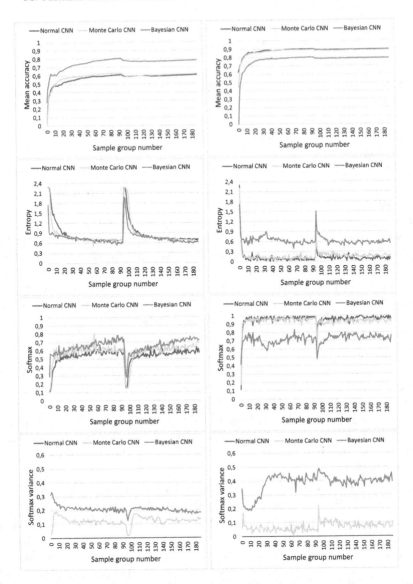

Fig. 1. Results for CIFAR-10 with single drift. From top to botom: average prediction accuracy, prediction entropy, average softmax response values, softmax variance for SGD (left) and Adam (right) optimizers.

the data stream generated from the MNIST database. For each experiment, the results of mean accuracy, entropy, mean value, and variance of the softmax function response for two Bayesian network models with Stochastic Gradient Descent (SGD) [14], and Adam [7] optimizers are presented. The detailed results of the

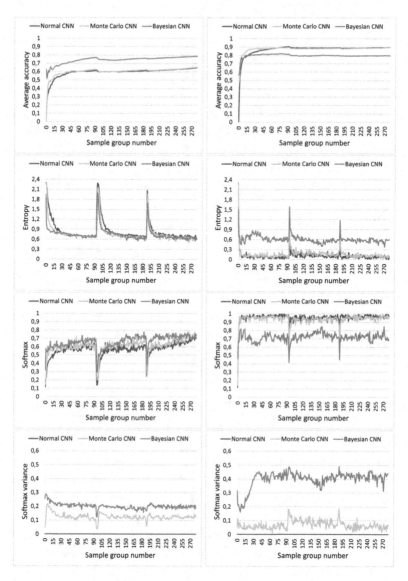

Fig. 2. Results for CIFAR-10 with double drift. From top to botom: average prediction accuracy, prediction entropy, average softmax response values, softmax variance for SGD (left) and Adam (right) optimizers.

experiments are included in Fig. 1, 2 and 3, while the aggregated results are summarized in Tab. 1.

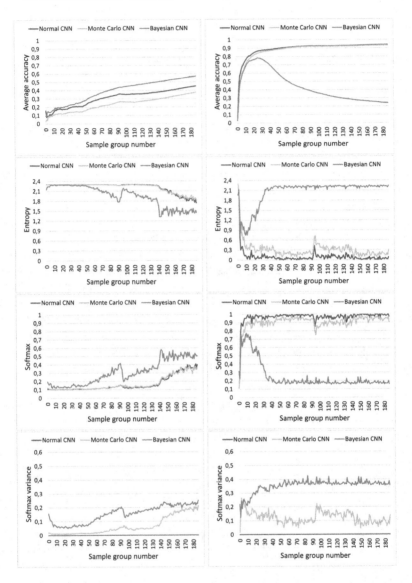

Fig. 3. Results for MNIST dataset. From top to botom: average prediction accuracy, prediction entropy, average softmax response values, softmax variance for SGD (left) and Adam (right) optimizers.

2.3 Lessons Learned

Based on the results of the experiments, let's try to answer the formulated research questions.

RQ1: What effect does a sudden drift have on convolutional networks' classification accuracy and behavior?

Table 1. Maximum loss and restoration time [%]

	Maximum loss [%]	Restoration time (s)
Bayesian (Adam)	93,75	0,0909
Bayesian (SGD)	100,00	0,1230
Monte Carlo (Adam)	84,38	0,0053
Monte Carlo (SGD)	87,50	0,1337
Normal (Adam)	75,00	0,5508
Normal (SGD)	84,38	0,1390

The sudden drift resulted in a rapid decrease in classification accuracy and an equally rapid increase in prediction uncertainty. However, the models tested were able to readjust to the new probabilistic characteristics quickly for both single and double drift, with the jumps in accuracy and increase in model uncertainty being smaller for the recursive data stream.

RQ2: How high is the prediction uncertainty, as determined by the measured prediction entropy, mean value, and variance of the softmax function response?

The *concept drift* caused a spike in the uncertainty of the predictions provided by the models. Conclusions regarding prediction uncertainty should be based on a synthesis of information from all three parameters discussed. The results of the research indicate a much higher prediction uncertainty of the Bayesian convolutional network model compared to the network based on Monte Carlo dropout. It results from the fact that the network learning process is approached differently, and twice as many parameters have to be fitted under the same learning conditions on streaming data. The predictions were very accurate with low uncertainty for the Monte Carlo network using the Adam optimizer. The sudden change in entropy and the response values of the softmax function when drift occurs indicate a high potential for their use in constructing drift detectors for models operating on streaming data.

RQ3: What effect does the choice of optimizer have on model learning and the consequences of drift?

The selection of an appropriate optimizer has a key impact on the learning behavior of the model on dynamic streaming data. Convolutional networks using the Adam optimizer had higher classification accuracy and lower prediction uncertainty. It should be noted that this effect is less pronounced in the case of the Bayesian network, where changing the optimizer during learning on the set of CIFAR10 did not significantly affect the classification accuracy. However, the model behaved more stable and reacted faster to changes resulting from the occurrence of drift. An exception to this is a study conducted on the MNIST dataset, during which the Bayesian network quickly overfitted. This indicates the likely need for additional model regularization methods in this application. In addition, the models using the SGD optimizer underperformed when learned on the data stream from the MNIST set. A possible improvement in the performance of these models could be achieved by increasing the number of samples in the group or increasing the number of model learning epochs.

RQ4: What is the maximum decrease in classification accuracy, and how long does it take for the model to reach prediction accuracy before drift occurs?

A sudden drift led to a rapid drop in the classification accuracy of the models, whose measured values for a single drift ranged from 75% for a simple convolutional network, about 85% for a Monte Carlo network to as much as 100% for a Bayesian network. When drift occurred, the models almost completely lost their ability to generate correct predictions about new data. At the same time, this effect was very short-lived – the models under study were able to adjust to the unknown samples extremely quickly, ultimately causing the moment of drift occurrence to show up as a slight drop in the average prediction accuracy plot of the models, with no significant effect on the overall accuracy of the models, again – this effect is smaller for the network with the Adam optimizer.

3 Conclusion

This article aimed to investigate conventional neural networks' behavior in classifying non-stationary data streams. We would also determine the impact of a sudden drift on the accuracy and uncertainty of the predictions. The experiments conducted allowed us to analyze the behavior of neural networks in the classification of non-stationary data streams, both with single and recursive drift. Based on the measured entropy of the predictions, the mean value, and the variance of the softmax function response, the level of uncertainty in the predictions generated by the models was determined, and the maximum decrease in classification accuracy and the time required to restore the model were estimated. The behavior of networks using different optimizers was also compared. The analysis of the obtained results indicates that the study of the obtained characteristics could provide a solid basis for proposing a new unsupervised drift detector, which would consider prediction entropy or softmax response for the analysis in terms of drift occurrence. The authors identified this issue as the most important direction for further work.

However, realizing the study's limited scope, it may be pointed out that convolutional networks using Monte Carlo dropout with Adam optimizer may be the best model for classifying non-stationary data streams. The average classification accuracy obtained by this model was higher than that of the Bayesian convolutional network model with the same optimizer, and the obtained predictions were subject to much lower uncertainty. At the same time, due to the way the model was constructed, it had a lower cost and speed of data processing (half the number of parameters compared to the Bayesian network), which is an additional advantage in applications for stream data analysis.

Acknowledgements. This work is supported in part by the CEUS-UNISONO programme, which has received funding from the National Science Centre, Poland under grant agreement No. 2020/02/Y/ST6/00037 and by the Research Fund of Department of Systems and Computer Networks, Faculty of ICT, Wroclaw University of Science and Technology.

References

1. Blundell, C., Cornebise, J., Kavukcuoglu, K., Wierstra, D.: Weight uncertainty in neural networks (2015)
2. Delange, M., et al.: A continual learning survey: defying forgetting in classification tasks. IEEE Trans. Pattern Anal. Mach. Intell. 1 (2021). https://doi.org/10.1109/tpami.2021.3057446, https://doi.org/10.1109%2Ftpami.2021.3057446
3. Duda, P., Jaworski, M., Cader, A., Wang, L.: On training deep neural networks using a streaming approach. J. Artif. Intell. Soft Comput. Res. **10**(1), 15–26 (2020). https://doi.org/10.2478/jaiscr-2020-0002
4. Gal, Y., Ghahramani, Z.: Dropout as a Bayesian approximation: representing model uncertainty in deep learning. In: 33rd International Conference on Machine Learning, ICML 2016, vol. 3, pp. 1651–1660 (2016)
5. Guzy, F., Wozniak, M.: Employing dropout regularization to classify recurring drifted data streams. In: Proceedings of the International Joint Conference on Neural Networks (2020). https://doi.org/10.1109/IJCNN48605.2020.9207266
6. Hinton, G.E., Srivastava, N., Krizhevsky, A., Sutskever, I., Salakhutdinov, R.R.: Improving neural networks by preventing co-adaptation of feature detectors (2012)
7. Kingma, D.P., Ba, J.: Adam: a method for stochastic optimization (2014)
8. Krawczyk, B., Minku, L.L., Gama, J., Stefanowski, J., Woźniak, M.: Ensemble learning for data stream analysis: a survey. Inf. Fusion **37**, 132–156 (2017)
9. Krizhevsky, A., Nair, V., Hinton, G.: Cifar-10 (Canadian institute for advanced research). http://www.cs.toronto.edu/kriz/cifar.html
10. Jospin, L.V., Laga, H., Boussaid, F., Buntine, W., Bennamoun, M.: Hands-on bayesian neural networks - a tutorial for deep learning users. CoRR abs/2007.06823 (2020). https://arxiv.org/abs/2007.06823
11. LeCun, Y.: Backpropagation applied to handwritten zip code recognition. Neural Comput. **1**, 541–551 (1989). https://doi.org/10.1162/neco.1989.1.4.541
12. LeCun, Y., Cortes, C.: MNIST handwritten digit database (2010). http://yann.lecun.com/exdb/mnist/
13. Shaker, A., Hüllermeier, E.: Recovery analysis for adaptive learning from non-stationary data streams: Experimental design and case study. Neurocomputing **150**, 250–264 (2015)
14. Shamir, O., Zhang, T.: Stochastic gradient descent for non-smooth optimization: convergence results and optimal averaging schemes. In: Dasgupta, S., McAllester, D. (eds.) Proceedings of the 30th International Conference on Machine Learning. Proceedings of Machine Learning Research, vol. 28, pp. 71–79. PMLR, Atlanta, Georgia, USA, 17–19 June 2013

Explaining Machine Learning Predictions in Botnet Detection

Sean Miller[(✉)] and Curtis Busby-Earle

The University of the West Indies Mona Kingston, Kingston, Jamaica
{sean.miller02,curtis.busbyearle}@uwimona.edu.jm

Abstract. As the war against cybercrime continues, security experts and attackers improve their skills and tools to accomplish their desired goal. Botnets have been the method of choice for a majority of cyber attacks, from phishing and spam campaigns to full scale distributed denial of service (DDoS) attacks. The application of machine learning methods to help fight cybercrime has shown promising results. Unlike other domains where machine learning (ML) has been applied, there is another human being capable of using the same tools to counter the security experts. ML models are more effective at analyzing a large amount of traffic than human beings, but providing a black box response is not sustainable against dynamic human opponents. In this paper we begin to describe a method that presents the underlying patterns discovered by ML methods that allow researchers and experts to gain the insight needed to effectively design more robust responses to the ever-changing threats with which they are faced.

Keywords: Machine learning · Botnets · Cyber security

1 Introduction

1.1 Machine Learning and Botnets

Machine learning approaches have been applied to various problems in multiple domains. These applications have been met with varying success [1]. As researchers employ these widely successful models to cybersecurity, their ability to identify patterns in data proved to be quite effective in recognizing threats. However, unlike most other fields where machine learning has been applied, cybersecurity has one unpredictable and resourceful variable, the attacker. Being able to access similar tools, hardware, software and now artificial intelligence, getting a high accuracy on a model is not a solid indicator of whether or not this system will be able to defend against new unseen threats or even withstand updated versions of the malware from the attackers.

Despite their outstanding performances in various fields, machine learning models are quite susceptible to adversarial learners as exemplified in [2]. These adversarial learners can generate examples for which the model will almost

L. Rutkowski et al. (Eds.): ICAISC 2022, LNAI 13588, pp. 298–309, 2023.
https://doi.org/10.1007/978-3-031-23492-7_26

always produce the incorrect classification. For some datasets, the adversarial examples are so close to the actual examples that even the human eye is unable to distinguish between them [3]. There is no exception to this fact when applied to cybersecurity. Classifiers designed to detect malware, are easily fooled by well-crafted adversarial examples [2]. With attackers also having these tools at their disposal and new malware specimens being released into the wild, the effects of adversarial examples have raised questions about the applicability of these algorithms. Researchers in one study suggest that although learning techniques obtain excellent performance on the test set, they are not learning the true underlying concepts that relate to the correct label [3]. In a 2016 study [4], researchers showed how two different high scoring algorithms on the same data set arrive at their decisions. Even though both used the same data set, the set of features selected as the most influential was quite different. Other limitations of developing machine learning models for cybersecurity include data, the length of time and how frequent a model would have to be retrained [5].

Models alone will not be sufficient in fighting cyber threats [6]. Though more effective than some rule-based approaches of the past [7], AI will not provide a silver bullet to solve our security needs. Our approach aims to give security experts and researchers the ability to interact and learn from the algorithms they implement. With the ability for models to explain their results, it is anticipated that security experts will be able to gain insight into the structure and behavior of the data set and this will thereby lead to the creation of stronger and more robust models.

In the sections that follow, Sect. 2 presents related work in the area of explaining machine learning models applied in various fields. In Sect. 3 we propose the theoretical framework for our approach to explaining models related to botnets. Section 3.2 we discuss the work done thus far in implementing our proposed method. Finally in Sect. 4 we conclude with a summary and future work.

2   Related Work

The application of machine learning techniques has proven to be quite successful in several fields over the years. The major limitation to the widespread application of these methods in some fields is as a result of their black-boxed nature, their inability to be easily interpreted [8]. Some researchers have done work to remove this limitation and have identified many benefits to interpretable models. In 2016 Ribeiro et al. [8] proposed a technique LIME that explains the prediction of any classifier. This method learns an interpretable model based on the one created from the original classifier which is said to be easy to understand and locally faithful.

A decision boundary is learned by a non-linear model, that is the original classifier f. It does this by sampling instances in close proximity to the target instance. It then uses the classifier f to get the predictions of the surrounding instances. The main difference between our proposed method and LIME is the implementation and how explanations are generated. LIME creates an interpretable model, our method creates an ensemble of classifiers. LIME is interpreted based on features and how they relate to the instances sampled. Our method explains a model based on the impact different features have on a given instance.

Robnik-Šikonja and Kononenko [10] suggested two levels of explanation, the domain level, and the model level. The domain level explanation was not achievable unless true causal relationships and probability distributions were known in advance. The EXPLAIN method takes the i-th input variable and calculates its importance, by comparing the difference in the model's prediction with and without that variable. The size of the difference indicates the importance of that feature. Unlike EXPLAIN, IME considers all subsets of the selected attribute. This results in an infeasible time complexity requiring 2^a steps increasing exponentially. However, this can be practically implemented by sampling the feature space.

Bohanec et al. [9] applied two different model interpretation techniques to a real-world business problem. This study shows just how useful interpretable models are to help make decisions. With this, human experts were able to evaluate and update their beliefs as they interact with these models. EXPLAIN and IME were the two models used in this study originally presented by [10] in 2008. Both methods determine the significance of an attribute in similar ways. They did this by simulating the absence of the target attribute after which estimation is made based on the results of the simulation. The result of the method is a break down of the contribution of each attribute.

In 2016 Lou et al. [11] proposed an explanation method for automatic explanation without losing prediction accuracy. They used this explanation method for predicting whether or not patients will develop type 2 diabetes within the next year. The authors used a champion machine learning predictive model, from the practice fusion diabetes classification competition. This model was formed by combining eight boosted regression trees and four random forests using a generalized additive model with cubic splines. Using the model produced by this setup and the diabetes patents dataset, the authors then applied an associative rule classifier to learn associations between variables, based on the output of the model. The association classifier did this without affecting the accuracy of the model and without consideration for its accuracy. This method creates an explanatory method from an existing non-explanatory method. Building on the MPML method our model is able to explain the output it produced without creating another model.

3   Proposal - Model Explanation via MPML

The foundation of our approach finds its root in multi-perspective machine learning (MPML). MPML is primarily supported by two concepts in the machine learning literature, Ensemble learning, and Multi-view learning. We will build on these concepts to explain the MPML approach and thereby effectively describe its application in explaining predictions.

3.1 Multi-perspective Machine Learning

The perspective is the core concept of MPML. A clear definition of perspectives is required for it to be applied effectively. A perspective is a group of related features, a subset taken from the set of all features. In [11] the authors describe the benefits of what they call a multi-view learning strategy. This learning strategy describes how the learning problem is split. The perspective selected will define the learning strategy for any MPML task.

Let T be a specific learning problem. Where each element t_x is a feature of the learning problem T.

$$T = \{t_1, t_2, t_3...t_n\}$$

Let P be the set of perspectives of problem task T

$$P = \{p_1, p_2, p_3...p_n\}$$

where $p_x \subset T$

for example $p_1 = t_1, t_4, t_7, t_9$

Let L be the set of learning strategies that can be applied to solving problem task T using one or more perspectives.

$$L = \{l_1, l_2, l_3...l_n\}$$

$$l_1 = \{p_1, p_2, p_3...p_n\}$$

$$l_x \subset P$$

These perspectives are selected based on the preference of the researcher (if they are created manually) or the bias of the method describing feature relations (if it is done automatically). Two learning strategies may arise simply based on how the problem is viewed. If, for example, we are looking at botnet detection, one may decide or the algorithm may find correlation between time-related features, such as; t_1 - the number of times a particular server connection is attempted within a given time period and t_2 - the number of times a particular service is requested within a given time period. Another correlation may be found between host-based features, such as; t_3 - number of data bytes transferred from source to destination in a single connection and t_4 - number of data bytes transferred from destination to source in a single connection. For one type of bot it may be that t_1 and t_2 find themselves in the same perspective, simply because of how this bot (bot_x) was designed to operate, the relation between these two features is strong and the same can be said for t_3 and t_4.

However, bots may operate completely differently, we may find that the connection to a particular server is quite in sync with the number of bytes transferred from source to destination. We may then find t_1 and t_3 being in one perspective and t_2 and t_4 in another. Hence the learning strategy defines the foundation of the approach. The aim of this method is to be able to capture and explain these differences.

3.2 Proposed Method

The structure of the proposed method is seen in Fig. 1. A typical MPML setup with the explanation being possible at each level. For any given instance the system is able to explain why the given prediction was made, by identifying the perspective with the greatest impact score and by suggesting a combination of features that had the highest individual impact score on each perspective. Each perspective focuses on a single aspect of the learning problem. Understanding how each perspective affects the prediction y gives a general explanation of that particular instance. For example (see Fig. 2) if perspective one (p_1) is the most influential perspective for predicting a particular type bot and this perspective (p_1) is made up of time-related features, then this would provide good insight into the core function or behavioral pattern for that bot.

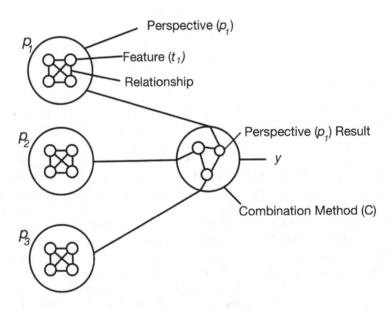

Fig. 1. Model overview

By going deeper than the perspective level we can identify what features are most influential inside the most influential perspective. This now provides deeper insight into what type of time-based feature is most suited for predicting this threat. Understanding how the individual features relate to each other is crucial to understanding the underlying behavior of the threat.

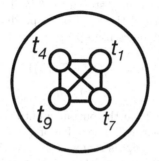

Fig. 2. A single Perspective p_1

P is the set of all perspectives for a given learning Task T .

$$P = \{p_1, p_2, p_3...p_n\}$$

Each perspective p_x contains a subset of features t_x from the learning task T.

$$p_x = \{t_1, t_2, t_3...t_n\}$$

The model generated by applying a learning algorithm S to any perspective p_x is represented as:

$$S_x(p_x)$$

The set of all models S produced from each perspective in P is denoted by:

$$Q = \{S_1(p_1), S_2(p_2), S_3(p_3)...S_n(p_n)\}$$

These models are then combined using a combination method C and the final result (the prediction) is represented by y

$$y = C(S_1(p_1), S_2(p_2), S_3(p_3)...S_n(p_n))$$

$$y = C(Q)$$

We then attempt to explain y using a technique similar to the EXPLAIN method by identifying which perspective, when removed, produces the greatest change in y. To calculate the change in y we use the confidence of the model for the predicted value of y. For instance, in the case of botnet detection the result y can be either BOT or NOT. We record the confidence of the model for each class. If the model (with all perspectives) produces a y with 90% confidence that it is a bot (BOT) and 10% that it is not (NOT) the confidence of the correct class is noted, in this case 90%. The result after removing a perspective $(S_x(p_x))$ would be stored in $\hat{y}$. If the confidence of $\hat{y}$ is 70% that it is a bot and 30% it is not a bot, then the change in y stored in d would be 90 - 70 = 20. $S_x(p_x)$ has the greatest impact on y if the resulting $\hat{y}$ produced with the exception of $S_x(p_x)$ has the greatest difference from y $\forall p_x \in P$.

$$\hat{y} = C(Q - \{S_x(p_x)\})$$

$$y - \hat{y} = d$$

This is done recursively from the output y until the strongest feature in the strongest perspective is identified. The value of d also denotes the direction the model is heading in when $S_x(p_x)$ is removed. If removing $S_x(p_x)$ brings $\hat{y}$ closer to the correct prediction, then $S_x(p_x)$ has a negative impact on the result y for that particular instance, if the opposite is true then $S_x(p_x)$ has a positive impact on y. Both impacts are useful in guiding a user to a greater understanding of the nature of the problem being studied.

What knowledge can be gained from -ve impact vs +ve impact? Let's look at an actual example. The botnet dataset used was provided by the Canadian Institute for Cybersecurity [12]. The details of how the data was cleaned and assembled can be found in [13]. We start by generating the relationship between the features. The relationship score describes the correlation between two features and is useful in grouping similar features into a perspective. The details of generating perspectives can be found in [14]. The aim was to generate an explanation of the prediction for a single instance.

Table 1. Test sets

Test Set #	No. of Inst	ML Algo	# of Perspectives	Majority Vote Accuracy	Combination ML Accuracy	Perspective Accuracy				Accuracy without Perspectives
						P0	P1	P2	P3	
Testset#1	9,000	GNB	2	82.647%	82.756%	82.756%	58.469%	–	–	81.693%
Testset#2	1,000	GNB	4	75.2%	93.939%	76.969%	54.545%	86.363%	60%	83.636%
Testset#3	1,000	SVM	4	97%	94.545%	93.303%	62.424%	94.545%	76.969%	92.727%
Testset#4	9,000	SVM	2	96.813%	94.231%	94.231	71.311%	–	–	93.776%

In the following series of tests (Table 1), we will run the model with the following variables. The aim of these tests is to see whether or not the conclusions drawn by the model will remain consistent across the various changes in these variables:

- The number of instances used in preparing the model.
- The specific learning algorithm used.
- The particular instance being focused on.
- The number of perspectives generated.
- The method used to combine the perspectives.

The number of instances indicates the number of flows included in the test. With this variable we observed how the model behaved with different numbers of training instances. The number of instances given was split into train and test set, thirty-three percent (33%) was used for testing and the rest (67%) for training. The learning algorithm is an important factor in the learning process with each method having it's own strengths and weakness and specific characteristics that affect how the data is interpreted. The instance being focused on remained in the same position throughout each test. The number of perspectives generated was dependent on the MPML method of grouping features, varying the number of instances will influence how features relate and the effects of these perspectives on the model was an important observation. How the perspectives are combined to obtain the final result was also taken into consideration.

Table 2. Test results from test set 1

Test #	Instance #	Highest +ve Impact Perspective	Highest −ve Impact Perspective	Highest +ve Impact Feature	Highest −ve Impact Feature	Prediction	Class
1	497 (RBot)	P1 (88.964%)	P0 (64.347%)	var_byte (+22.807)	byte_exc (−31.394)	BOT	BOT
2	342 (RBot)	P1 (90.280%)	P0 (0.053%)	var_byte (+20.352)	src_port (−98.241)	NOT	BOT
3	345 (RBot)	P1 (88.810%)	P0 (0.00025%)	var_byte (+23.014)	dst_port (−99.978)	NOT	BOT
4	27 (Virut)	P0 (99.685%)	P1 (90.246%)	var_byte (+28.028)	percent_push (−0.024)	BOT	BOT
5	218 (Virut)	P0 (99.685%)	P1 (90.246%)	var_byte (+28.028)	percent_push (−0.024)	BOT	BOT
6	227 (Virut)	P0 (99.257%)	P1 (90.246%)	var_byte (+28.028)	percent_push (−0.101)	BOT	BOT
7	29 (Neris)	P0 (97.610%)	P1 (90.246%)	var_byte (+28.028)	percent_push (−0.187)	BOT	BOT
8	28 (Neris)	P0 (99.292%)	P1 (90.246%)	var_byte (+28.028)	percent_push (−0.097)	BOT	BOT
9	54 (Neris)	P0 (99.296%)	P1 (90.246%)	var_byte (+28.028)	percent_push (−0.096)	BOT	BOT
10	8 (Normal)	P0 (99.990%)	P1 (9.662%)	src_port (+89.731)	var_byte (−20.346)	NOT	NOT
11	112 (Normal)	P0 (94.405%)	P1 (99.999%)	std_byte (+89.731)	src_port (−11.145)	NOT	NOT
12	620 (Normal)	P1 (10.691%)	P0 (0.521%)	pack_push (+0.023)	var_byte (−22.311)	BOT	NOT

From the list of randomly selected instances in Table 2 we examine instance #29 from test set 1 (Table 1) in detail. By applying the Gaussian Naive Bayes algorithm in a traditional manner the model generated produced an accuracy of 81.693%. Test case 1 used 9,000 instances and produced two perspectives. Using the majority vote method we combined the resulting models from these 2 perspectives which produced an accuracy of 82.647%. Both perspectives are listed below along with their individual accuracy.

Perspective #0 - 82.756%

- Average byte per packet per flow (avg_byte)
- The protocol used in the flow (protocol)

- The number of packets exchanged in the flow (pack_exc)
- Source port (src_port)
- Destination port (dst_port)
- The number of bytes exchanged in flow (byte_exc)
- Packets with the push flag set to 1 (pack_push)
- The percent of packets pushed (percent_push)

Perspective #1 - 58.469%

- Standard deviation of number of bytes per packet in flow(std_byte)
- Variance of bytes per packet per flow (var_byte)

We also record the confidence of the predictor on each class for the instance. The confidence of the combined perspectives are:

- Confidence for Not - 6.071440117785192%
- Confidence for Bot - 93.92855988221498%

By removing one perspective at a time we obtain the impact score d for each perspective. The impact score was determined based on the increase or decrease in the confidence of the entire model for the correct class. This was done by calculating the difference between the confidence level of the model with (y) and without that perspective $(\hat{y})$.

$$y - \hat{y} = d$$

Result without Perspective #0: With both perspectives, the confidence that instance #29 is a bot was - 93.928%, without perspective p_0 (Fig. 3) the confidence fell to - 90.246%.

$$y - \hat{y} = d$$

$$93.928 - 90.2466 = +3.681$$

The impact Perspective 0 (p_0) has on the model is a positive one shown by the impact score of +3.681. The same is done for Prospective 1 (p_1) to obtain its impact score on this instance.

$$y - \hat{y} = d$$

$$93.610 - 97.610 = -4$$

After examining the perspectives the impact score for each was obtained. Perspective #0 - +3.681 and Perspective #1 - (-4)

After the perspectives are examined the features are analyzed to find the most influential feature in each perspective. The confidence of p_0 with all features is, 1.33159871% confident that it is not a bot (NOT) and 98.66840129% confident it is a bot (BOT). The impact of each feature is determined by removing each feature and observing the changes to the confidence level of that perspective for the correct class.

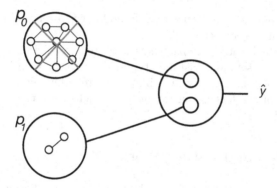

Fig. 3. Examining instance #29 - Removing p_0

We obtained results by running multiple tests using different test sets. With each test set, there were 12 tests, the same 12 (Table 2) instances were used in each test set. A test is carried out on a single instance to obtain the model's explanation for the class predicted. Instances were randomly chosen from different types of bots and also non-bot traffic flows. The aim of this experiment was to observe the results produced by the model across different flows and if any patterns emerge among the explanations.

We designed a tool that traverses the layers of the MPML model reporting on the impact scores of perspectives and the impact scores of features in each perspective for a given instance. Using instance #29 as an example from the test set 1 we saw where the perspective p_0 had a higher impact on prediction for that particular instance. Looking further into that perspective we saw that the most influential features turned out to be source port (src_port) with an impact score of +4.248, destination port (dst_port) with an impact score of +1.004, and the number of packets exchanged in the flow (pack_exc) with an impact score of +1.223.

Examining the other instances that were marked as Neris bots (instance #28 and #54) similar patterns were found for the features that had the most impact on the decision. And when done for bot traffic of similar types such as the ones marked Virut, similar patterns emerged. Observing the impact score for features of multiple instances of the same type, revealed similarities in the impact score, we refer to this as the impact signature for those set of instances. We observed that similar bot types recorded comparable impact signatures, irrespective of the machine learning algorithm used. Examining the results of this tool revealed subtle patterns emerging between the consistency of impact scores for features and the correctness of the predictions made. For instances that were labeled RBot the model was less confident in its prediction and even got some predictions wrong. For these instances, we observed no discernible patterns in the impact scores reported. The consistency of the impact signatures has a relation to the confidence of the prediction, which is important to us because a model with high accuracy and high confidence on the correct labels is more desirable and demonstrates a more stable nature than a model with even higher accuracy but low confidence on the correct label.

Although p_1 had a lower impact on the correct prediction of this instance, across different test sets and instance types, the variations in bytes per packet per flow consistently had a very high impact score. This raises a question as to the true impact of the feature on the entire model versus the impact it has on the perspective it is in.

4 Conclusion and Future Work

This paper proposed a method for explaining botnet prediction using MPML. The results demonstrate the model's ability to identify patterns in specific botnet traffic and the ability to present features that contribute to those specific patterns. Understanding what features greatly impact specific bot traffic can help researchers and security experts alike defend against bots. At this point, the method is able to identify the features with the highest impact on a specific instance. Simply presenting these features would not prove to be fully interpretable or complete, but it is definitely a stepping stone towards these goals. The patterns observed across perspectives within the model demonstrate the potential of this method and the promise of providing truly intelligible results.

There is still much work to be done on this technique. Such work includes: How do the patterns change based on the phase the bot is in. The change in the number of instances affected the number of perspectives created, with perspectives being integral to the process in future work we aim to understand exactly why this occurs. Additionally, more work is needed to establish the best method for scoring feature relations, methods such as correlation matrix will be explored. How does the impact score change with different features in a single perspective? Another aspect for more work is looking into what role the relationship between features play in the interpretability and the completeness of the explanation.

References

1. Sommer, R., Paxson, V.: Outside the closed world: On using machine learning for network intrusion detection. In: 2010 IEEE symposium on security and privacy, pp. 305–316. IEEE (2010), Treatise on Electricity and Magnetism, 3rd edn., vol. 2, pp. 68–73. Clarendon, Oxford (1892)
2. Demetrio, L., Biggio, B., Lagorio, G., Roli, F., Armando, A.: Explaining Vulnerabilities of Deep Learning to Adversarial Malware Binaries. arXiv preprint arXiv:1901.03583 (2019)
3. Goodfellow, I.J., Shlens, J., Szegedy, C.: Explaining and harnessing adversarial examples. arXiv preprint arXiv:1412.6572 (2014)
4. Ribeiro, M.T., Singh, S., Guestrin, C.: Why should I trust you?: Explaining the predictions of any classifier. In: Proceedings of the 22nd ACM SIGKDD International Conference On Knowledge Discovery and Data Mining, pp. 1135–1144. ACM (2016)

5. Buczak, A.L., Guven, E.: A survey of data mining and machine learning methods for cyber security intrusion detection. IEEE Commun. Surv. Tutor. **18**(2), 1153–1176 (2015)
6. Mohanty, S., Sachin V.: Cybersecurity and AI. In: How to Compete in the Age of Artificial Intelligence, pp. 143–153. Apress, Berkeley, CA (2018)
7. Veiga, A.P.: Applications of artificial intelligence to network security. arXiv preprint arXiv:1803.09992 (2018)
8. Montavon, G., Lapuschkin, S., Binder, A., Samek, W., Müller, K.-R.: Explaining nonlinear classification decisions with deep taylor decomposition. Pattern Recogn. **65**, 211–222 (2017)
9. Bohanec, M., Borštnar, M.K., Robnik-Šikonja, M.: Explaining machine learning models in sales predictions. Expert Syst. Appli. **71**, 416–428 (2017)
10. Robnik-Šikonja, M., Kononenko, I.: Explaining classifications for individual instances. IEEE Trans. Knowl. Data Eng. **20**(5), 589–600 (2008)
11. Luo, G.: Automatically explaining machine learning prediction results: a demonstration on type 2 diabetes risk prediction. Health Inf. Sci. Syst. **4**(1), 2 (2016)
12. Beigi, E.B., Jazi, H.H., Stakhanova, N., Ghorbani, A.A.: Towards effective feature selection in machine learning-based botnet detection approaches. In: 2014 IEEE Conference on Communications and Network Security, pp. 247–255. IEEE (2014)
13. Miller, S., Busby-Earle, C.: The impact of different botnet flow feature subsets on prediction accuracy using supervised and unsupervised learning methods. Int. J. Internet Technol. Sec. Trans. (2016)
14. Miller, S.T., Busby-Earle, C.: Multi-perspective Machine Learning (MPML)-A Machine Learning Model for Multi-faceted Learning Problems. In: 2017 International Conference on Computational Science and Computational Intelligence (CSCI), pp. 363–368. IEEE (2017)

Learning Functional Descriptors Based on the Bernstein Polynomials – Preliminary Studies

Wojciech Rafajłowicz[1]([✉]) [iD], Ewaryst Rafajłowicz[1] [iD],
and Jędrzej Więckowski[2] [iD]

[1] Faculty of Information and Comunication Technology, Wroclaw University
of Science and Technology, Wyb. Wyspianskiego 27, 50 370 Wrocław, Poland
{wojciech.rafajlowicz,jedrzej.wieckowski}@pwr.edu.pl
[2] Faculty of Mechanical Engineering, Wroclaw University of Science and Technology,
Wyb. Wyspianskiego 27, 50 370 Wrocław, Poland
ewaryst.rafajlowicz@pwr.edu.pl

Abstract. We propose a new method of constructing and estimating descriptors for classifying functional data. These descriptors are based on Bernstein polynomials and their estimation is based on noisy samples of a function (signal) to be classified.

The next step is to select an appropriate classifier, well suited to these descriptors. Although the result can be case dependent, we provide the methodology of running comparisons. As a vehicle for presenting the results, we choose benchmark data published in [32]. They represent shocks and vibrations of the operator's cabin in a large mechanical structure.

Keywords: Functional data classification · Machine learning · Estimating descriptors · Bernstein polynomials

1 Introduction

Descriptors of functional data, for example signals and curves, are created to extract features from the data to provide high-quality classification. At the same time, the descriptors should provide a significant degree of compression of the functional data, allowing it to be stored in computer memory in a cost-effective manner.

Approaches to creating descriptors can be divided into two large groups. The first includes methods tailored to a specific class of signals. These methods make significant use of specialized knowledge about a particular class of signals and their specific characteristics. A classic example of this class of methods is the recognition of electrocardiogram (ECG) signals based on so-called Q, R, S waveforms. We refer the reader to the following recent papers [1,17] [3] on classifying ECG signals. Specialized methods, dedicated to feature selection from electroencephalogram (EEG) signals, are developed and surveyed in [10,11], while in [2]

L. Rutkowski et al. (Eds.): ICAISC 2022, LNAI 13588, pp. 310–321, 2023.
https://doi.org/10.1007/978-3-031-23492-7_27

one can find the survey on electromyography signals. In [33] a representative artificial intelligence (AI) method applied to acoustic signals is described. A common feature of application-specific feature extraction methods is that they are highly labour intensive, which is justified mainly by applications in sensitive fields such as medicine. The second group of descriptor generation methods aims to significantly automate the feature extraction process for pattern, signal and image recognition. The expected result is a significant reduction in labor intensity, while maintaining satisfactorily high classification quality that is sufficient for applications in less demanding areas, for example, in technology and manufacturing processes.

Descriptors for Functional Data. The first examples of applications of methods from this group date back to the 1960 s s and are related to the development of algorithms known collectively as Fast Fourier Transform (FFT). In recent years, there has been renewed interest in this class of feature extraction methods due to the emergence of functional data classification methods. A special subclass within this group of methods are approaches that require the classifier to be sensitive to the shapes of the functions (signals) being classified. We refer the reader to [12,16,29,34] for more details on such approaches and to [23] for the latest contribution.

In these papers, the primary tool for obtaining the sensitivity of algorithms to the shape of signals is to consider the waveforms of their derivatives.

Advantages of Applying Bernstein Polynomials. In contrast, the approach proposed in this work is based on obtaining shape-sensitive descriptors of signals by comparing them with elements of the function space basis that have shape-preserving features. The best-known basis with these properties is that spanned by Bernstein polynomials. In the theoretical version of the proposed method, this comparison is implemented by computing scalar products between the signal to be classified and successive Bernstein polynomials. These products attain high values when individual signal (function) fragments are well matched to successive Bernstein polynomials and, conversely, the values are small when a given signal fragment is orthogonal to successive polynomial Bernsteins. For this reason, we choose these products (after possible normalization) as descriptors sensitive to signal shapes.

The question of whether to normalize descriptors or to use only non-normalized scalar products has no clear answer. In situations where the signal amplitudes vary considerably between classes, normalization is not advisable. On the other hand, when membership of a signal to a given class is determined only by its shape, the use of normalization will be useful.

Why is Learning Needed? In practice, we usually do not have a signal at all points of the observation interval, but only its samples, taken most often at equidistant moments of time, and observed with random disturbances. For these reasons, the process of learning the features of this signal is needed. In fact, we need to apply descriptor learning in two situations. The first one appears when we extract signal descriptors contained in the training sequence. The second one is needed when – after learning the classifier – we acquire new signals to be classified. In the first case, a learning process can be more accurate, since it is usually performed off-line. In the second one, it can be desirable (or necessary) to learn descriptors on-line.

Assumptions. A common feature of all approaches to the construction of classifiers for functional data is the assumption of statistical repeatability of signals and their (dis-)similarities when they come from the same or different classes. Since the description of probability distributions in function spaces is complex, in this paper we will make the simplifying assumption that we describe the probability distributions of signals of particular classes as finite-dimensional distributions of the coefficients of the expansion of that signal into a series of Bernstein polynomials of given degree $N > 1$. We refer the reader to [23, 24] and [28] for a more advanced model of imposing a probability structure on random functions.

The well-known Weierstrass theorem on the approximation of a continuous function on a finite interval by a polynomial of a sufficiently high degree can serve as a justification of this assumption. Bernstein polynomials form the basis of a constructive proof of this theorem.

We emphasize that knowledge of these probability distributions is not assumed in this paper. On the contrary, we only assume their existence and the complete lack of knowledge about them. Thus, the proposed approach is non-parametric, even though it deals with a finite number $(N + 1)$ of parameters, since this number can be chosen depending on the number of observations n and can grow with it.

Our Approach. In summary, the method proposed in this work to construct classifiers for functional data consists of two steps. In the first one, we learn vectors of Bernstein descriptors for each class, based on the learning sequence. In the second stage, we select a descriptor classification method from among known algorithms in such a way that it gives a satisfactory classification quality for a given application.

Other Approaches Based on Bernstein Polynomials. Another approach to constructing classifiers based on Bernstein polynomials was proposed in [21]. The difference is that in [21] Bernstein polynomials were used to estimate the probability densities of the classes. Classifiers or predictors acting as neural networks based on Bernstein polynomials are constructed in a similar way (cf. [18,30]). Advantages of using Bernstein polynomials occurred to be useful in estimating quantile functions [19]. Recently, an interesting application of Bernstein polynomials to modeling Covid-19 growth was proposed in [25].

Paper Organization. The paper is organized as follows. The next section presents the basic properties of Bernstein polynomials that are needed later in the paper. In Sect. 3, we formulate the assumptions and pose the descriptor learning problem. We present the learning algorithm itself and its elementary properties in Sect. 4. In that section we also describe the interaction of this algorithm with the decision function of the classifier. We then illustrate the selection of the decision function using the example of classification of the acceleration signals of the excavator operator's cab as a function of ground hardness.

2 Descriptors Based on the Bernstein Polynomials

We refer the reader to [5,7,14] for classic and more recent results on Bernstein polynomials (BP) and to [6] for their application to nonparametric estimation of probability density functions (p.d.f.).

It should be emphasized that Bernstein polynomials do not form an orthogonal basis, but many formulas are similar to those typical for nonparametric estimation methods based on orthogonal expanssions (see, e.g., [20,26,27] and the bibliography cited therein).

Definition and Elementary Properties of Bernstein Polynomials

Bernstein polynomials are usually defined on the interval $X = [0,1]$. Further in this paper we will assume that also all other functions considered here are defined on X.

For $x \in X$ k-th of order $N \geq k$ the Bernstein polynomial, denoted as $B_k^{(N)}(x)$, is defined as follows

$$B_k^{(N)}(x) = \binom{N}{k} x^k (1-x)^{N-k}, \quad k = 0, 1, \ldots, N.$$

We extend this definition by setting $B_k^{(N)}(x) \equiv 0$, if $k < 0$ or $k > N$.

We summarize and comment on the following, well-known, properties of the BPs.

$$\forall x \in X \quad \sum_{k=0}^{N} B_k^{(N)}(x) = 1, \quad 0 \leq B_k^{(N)}(x) \leq 1. \tag{BP 1}$$

Observe that (BP 1), being a partition of the unity, implies the ability of the BPs to restore constants exactly. Indeed, it suffices to set $a_k = 1$ for all k in formula (1) below.

For each sequence $a_k \in R$, $k = 0, 1, \ldots, N$ the following function

$$w_N(x) = \sum_{k=0}^{N} a_k \cdot B_k^{(N)}(x) \tag{1}$$

is an N - th order polynomial in x. Let f be a continuous function on X. Then, it is well known that selecting $a_k = f(k/(N+1))$, $k = 0, 1, \ldots, N$ in (1) we obtain $w_N(x) \to f(x)$, uniformly in X, as $N \to \infty$.

The following expression is of importance for a proper scaling of integrals containing the BPs

$$\int_X B_k^{(N)}(x) dx = (N+1)^{-1}, \quad k = 0, 1, \ldots, N. \tag{BP 2}$$

Proposed Descriptors

Let $C^p(X)$, $p = 0, 1, 2\ldots$ denote the space of p-times differentiable functions in X with the convention that $C(X) = C^0(X)$ is the space of all functions that are continuous in X. Define the inner product

$$\forall f, g \in C(X) \quad <f,g> = \int_X f(x)\, g(x)\, dx. \tag{2}$$

As descriptors of function (signal) $f \in C(X)$, denoted further as $d_k(f)$ (or d_k for brevity), we propose to take

$$d_k(f) = (N+1) < f, B_k^{(N)} > =$$ (3)

$$= (N+1) \int_X f(x) B_k^{(N)}(x)\, dx, \quad k = 0, 1, \ldots, N.$$

Note that $d_k(f)$'s depend also on N, but this dependence is not displayed, unless necessary.

It is worth mentioning also the normalized version of these descriptors, denoted further as $\breve{d}_k(f)$, that for $f \in C(X)$ is defined as follows

$$\breve{d}_k(f) = \frac{(N+1) < f, B_k^{(N)} >}{\max_{x \in X} |f(x)|}, \quad k = 0, 1, \ldots, N.$$ (4)

Note that $\breve{d}_k(f)$ is well defined, since for $f \in C(X)$ the maximum in the compact set X is attained. Furthermore,

$$\forall f \in C(X) \quad -1 \le \breve{d}_k(f) \le 1$$ (5)

and $\breve{d}_k(f) = \pm 1$ for $f(x) = \pm 1$, $x \in X$. To prove this fact, it suffices to observe that

$$| < f, B_k^{(N)} > | = | \int_X f(x) B_k^{(N)}(x)\, dx | \le \int_X |f(x)| B_k^{(N)}(x)\, dx \le$$ (6)

$$\le \max_{x \in X} |f(x)| \int_X B_k^{(N)}(x)\, dx = \max_{x \in X} |f(x)|/(N+1),$$

due to (BP1) and (BP2).

Additionally, $\breve{d}_k(f) = 0$, if f is orthogonal to $B_k^{(N)}$. Thus, $\breve{d}_k(f)$'s are descriptors that are well suited for classification problems. One can interpret descriptors $d_k(f)$ and $\breve{d}_k(f)$ as indicators to what extent f is close to (or fits) $B_k^{(N)}$. Note that $d_k(f)$ and $\breve{d}_k(f)$ depend also on N, but this dependence is not displayed, unless necessary.

Sensitivity of Descriptors to Function Shapes

These descriptors are – to some extent – shape sensitive in the sense that is explained below. Our starting point is the following well-known – formula for iterative calculations of the derivative, denoted as D_x, of $B_k^{(N)}(x)$

$$D_x B_k^{(N)}(x) = N \cdot [B_{k-1}^{(N-1)}(x) - B_k^{(N-1)}(x)], \quad k = 0, 1, \ldots, N.$$ (BP 3)

Then, multiplying both sides of (BP 3) by $f \in C^1(X)$, integrating over X with the aid of the integration by parts (for $1 \le k \le (N-1)$ we have $B_k^{(N)}(0) = B_k^{(N)}(1) = 0$) and shifting index k we immediately obtain

$$< D_x f, B_{k+1}^{(N+1)} > = d_{k+1}(f) - d_k(f) \quad k = 0, 1, \ldots, (N-1).$$ (7)

These relationships can be interpreted as follows: if f is strictly increasing (decreasing) in X, then the left-hand side of (7) is positive (negative). Thus also the difference $d_{k+1}(f) - d_k(f)$ retains this property. In other words, if f is strictly increasing (decreasing) in X then also the sequence of $d_k(f)$ is, and this statement holds in a natural way, i.e., without having a priori knowledge or our intervention by imposing constraints. Dividing both sides of (7) by $\max_{x \in X} |f(x)|$ we conclude that this monotonicity preserving property holds also for the normalized descriptors $\check{d}_k(f)$'s.

Assuming that $f \in C^2(X)$ and repeating the similar reasoning for $D_x^2 f(x)$, we come to the conclusion that if $D_x^2 f(x) > 0$, $x \in X$, which implies the convexity of f, then also sequences $d_k(f)$'s and $\check{d}_k(f)$'s are also convex in the sense that their second order differences are positive.

These properties, important for classification of the descriptor sequence, are illustrated in Fig. 1.

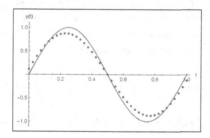

Fig. 1. Descriptors $d_k(f)$'s (dots) for $N = 50$ of function $f(x) = \sin(2 \pi x)$, $x \in X$.

3   Learning Descriptors from Noisy Samples of Functional Data

In practice, the data is not available in functional form $f \in C(X)$, which means that the proposed descriptors cannot be computed directly. Most often we only have samples of f values, observed with noise. We adopt a standard description of this type of sampling, assuming that the samples are taken at equidistant points x_i (e.g., instants of time or spatial variables), with random additive perturbations ϵ_i, $i = 1, 2, \ldots, n$. We assume that these disturbances have zero expected values and finite variances. For simplicity, we assume that these variances are equal, and denote them by $0 < \sigma^2 < \infty$. In summary, the functional data samples y_i, $i = 1, 2, \ldots, n$ are of the form

$$y_i = f(x_i) + \epsilon_i, \quad \mathbb{E}\epsilon_i = 0, \quad \mathbb{E}\epsilon_i^2 = \sigma^2 < \infty, \quad i = 1, 2, \ldots, n, \qquad (8)$$

$\mathbb{E}[\epsilon_i \epsilon_j] = 0$ for $i \neq j$, where $\mathbb{E}$ is the expectation of a random variable.

Problem statement: having observations (x_i, y_i), $i = 1, 2, \ldots, n$ at our disposal, our aim is to propose a learning algorithm for estimating descriptors $d_k(f)$, $k = 0, 1, \ldots, N$. For the sake of simplicity we assume that the original sample points are already transformed to $x_i \in [0, 1]$ and $\Delta_n \overset{def}{=} x_{i+1} - x_i = 1/n$.

In the remainder of this paper, we will denote the descriptor estimates as $\hat{d}_k^{(n)}(\bar{y})$, $k = 0, 1, \ldots, N$, where $\bar{y}$ is a column vector of ordered observations y_i, $i = 1, 2, \ldots, n$ with possible upper indices when several functional elements f are considered.

According to (3), a natural and simple algorithm for $\hat{d}_k^{(n)}(\bar{y})$ is of the form

$$\hat{d}_k^{(n)}(\bar{y}) = \frac{N+1}{n} \sum_{i=1}^{n} y_i \, B_k^{(N)}(x_i), \quad k = 0, 1, \ldots, N. \tag{9}$$

Notice that noisy observations y_i's are directly inserted into (9) without any prefiltering (see [22] for a discussion on the advantages of using pre- or post-filtering). Nevertheless, $\hat{d}_k^{(n)}(\bar{y})$ still have satisfactory statistical properties, as stated below. One can consider more robust estimators of the expectation, e.g., the median or the trimmed mean in (9), but here we confine our attention to the classic mean, since Bernstein polynomials have a hidden ability to mitigate large errors.

Notice that for the bias $\delta_{kn} \overset{def}{=} d_k(f) - \mathbb{E}[\hat{d}_k^{(n)}(\bar{y})]$ we have

$$\delta_{kn} = (N+1) \Delta_n \sum_{i=1}^{n} [f(\tilde{x}_{ki}) \, B_k^{(N)}(\tilde{x}_{ki}) - f(x_i) \, B_k^{(N)}(x_i)], \tag{10}$$

where $\tilde{x}_{ki}$'s are intermediate points in $I_i \overset{def}{=} [x_i - \Delta_n/2, x_i + \Delta_n/2]$ when the mean value theorem is applied to the integrals

$$\int_{I_i} f(x) \, B_k^{(N)}(x) \, dx = \Delta_n \, f(\tilde{x}_{ki}) \, B_k^{(N)}(\tilde{x}_{ki}).$$

Lemma 1. *If f has a continuous derivative in $[0, 1]$, then $|\delta_{kn}| = O(N/n)$ and the learning algorithm $\hat{d}_k^{(n)}(\bar{y})$ is asymptotically unbiased, as $n \to \infty$, for each finite and fixed N, $k = 0, 1, \ldots, N$.*

Indeed, the modulus of each summand in (1) is bounded by Δ_n multiplied by by the maximum over $[0, 1]$ of the modulus of the derivative of $f(x) \, B_k^{(N)}(x)$, which – in turn – is bounded by

$$\max_{x} |f(x)| + N \max_{x} |f'(x)|.$$

due to BP3). This finishes the proof, since this bound is uniform in k.

For the variance of $\hat{d}_k^{(n)}(\bar{y})$ we have for $k = 0, 1, \ldots, N$

$$\mathrm{VAR}[\hat{d}_k^{(n)}(\bar{y})] = \left(\frac{N+1}{n}\right)^2 \mathbb{E}\left[\sum_{i=1}^{n} \epsilon_i B_k^{(N)}(x_i)\right]^2 \leq \sigma^2 \frac{(N+1)^2}{n}, \tag{11}$$

due to the uncorrelatedness of ϵ_i's and the fact that $0 \leq B_k^{(N)}(x) \leq 1$.

Lemma 2. *Under the assumptions of Lemma 1, $\hat{d}_k^{(n)}(\bar{y})$'s are consistent in the mean squared error (MSE) sense as $n \to \infty$, for each finite and fixed N, $k = 0, 1, \ldots, N$.*

Indeed, it is well known that the MSE can be expressed as the sum of the variance and the squared bias. Thus, the result follows directly for Lemma 1 and (11).

Notice that the above results hold also in the case when f is a random element and descriptors $d_k(f)$'s are random variables. To this end, it suffices to consider the expectations as conditional ones, given $d_k(f)$'s.

4 Learning Classifiers Based on Bernstein Descriptors

We assume that random element f is drawn from a (sub-)class of continuously differentiable functions $\mathcal{F} : X \to R$. Two nonempty subsets $\mathcal{F}_I$ and $\mathcal{F}_{II}$ are distinguished in $\mathcal{F}$ and f is drawn from one of them with a priori probabilities $p_I > 0$, $p_{II} > 0$, respectively, and $p_I + p_{II} = 1$. These probabilities are unknown, but their estimation by fractions in the learning sequence is a simple task, unless there is no large imbalance between samples from class I and II in a learning sequence.

f is represented by random vector vector $\bar{d}(f)$ of its descriptors $d_k(f)$, $k = 0, 1, \ldots, N$, assuming fixed $N > 1$. Its choice is discussed later on. Probability distributions of $\bar{d}(f)$ depend on whether f is from class I or II, but they are unknown. Also $\bar{d}(f)$ is not directly available.

The only information that we have at our disposal is contained in a learning sequence, which is of the form:

$$\mathcal{L}_L \stackrel{def}{=} \{(\bar{y}^{(1)}, j_1), (\bar{y}^{(2)}, j_2), \ldots, (\bar{y}^{(L)}, j_L)\}, \tag{12}$$

where $j_k \in \{I, II\}$ are class labels, assumed to be correct, while $\bar{y}^{(k)}$ are vectors of equidistant samples from random elements $f^{(k)}$, drawn either from $\mathcal{F}_I$ or $\mathcal{F}_{II}$. These samples are taken at x_i, $i = 1, 2, \ldots, n$, according to (8), $k = 1, 2, \ldots, L$.

Now, our aim is to present an algorithm of learning, tuning, testing and selecting a classifier that classifies a random element f to classes I or II, based on its random samples $\bar{y}$ and using the estimates of the Bernstein descriptors.

To this end, let us denote by

$$cl.\,parameters = \text{LEARN}[cl.\,name, learning\,seq.]$$

a generic learning procedure that takes a classifier name and a learning sequence as its inputs and provides tuning parameters of the classifier after learning as its outputs.

As $cl.\,name$ one may select, e.g., one of the frequently used classifiers listed in Table 1 or even an ensemble of classifiers. We denote such a class of considered classifiers as $\mathcal{CL}$. The second tool that we need is a testing procedure:

Table 1. Examples of frequently used classifiers.

Acronym	Classifier
LogR	The logistic regression classifier
SVM	The support vector machine
DecT	The decision tree classifier
gbTr	The gradient boosted trees
RFor	The random forests classifier
k-NN	The k nearest neighbors classifier

$$\{accuracy,\ precision, \dots\} = \mathrm{TEST}[cl.\ name,\ parameters,\ testing\ seq.]$$

that takes the classifier name, its parameters and a testing sequence as inputs. Its output is a list of commonly used indicators of classifiers' quality, e.g., the accuracy, precision, recall, specificity, F1 and possibly many others. The TEST runs in a standard way, namely, it the selected classifier (with *parameters* from the learning procedure) on a supplied *testing sequence* and calculates the quality indicators. In a more advanced version, the testing inside TEST is performed many times on randomly selected subsequences of the *testing sequence* and the resulting indicators are averaged. It is further assumed that the TEST procedure is used in this more advanced version.

Selection and Learning a Classifier Based on Bernstein Descriptors

Data acquisition Collect samples of random elements, ask an expert to classify them and form learning sequence $\mathcal{L}_L$

Learning descriptors Select the order N of Bernstein descriptors. For the vector of samples $\bar{y}^{(l)}$ in $\mathcal{L}_L$ estimate the elements of the following list:

$$\bar{d}(\bar{y}^{(l)}) \overset{def}{=} \{\hat{d}_k^{(n)}(\bar{y}^{(l)}),\ k = 0,\ 1, \dots,\ N\}, \tag{13}$$

using (9). To each $\bar{d}(\bar{y}^{(l)})$ attach label j_l that corresponds to $\bar{y}^{(l)}$ in $\mathcal{L}_L$ and form a transformed learning sequence $\mathcal{D}_L$ from pairs $(\bar{d}(\bar{y}^{(l)}), j_l)$, $l = 1, 2, \dots, L$.

Optional step $\mathcal{D}_L$ augmentation. Extend $\mathcal{D}_L$ by copying each of its elements $\eta > 1$ times and replacing $\bar{d}(\bar{y}^{(l)})$ vectors by their randomly perturbed copies with zero mean, but keeping the same class label. Perturbations by additive Gaussian random vectors are the first choice. Slightly abusing the notation, we shall further denote this extended learning sequence again by $\mathcal{D}_L$.

Preparations Select classifier CL_{cur} from $\mathcal{CL}$ and split at random $\mathcal{D}_L$ into two disjoint and covering $\mathcal{D}_L$ sets: tuning set $\mathcal{DL}_{L1}$ and testing set $\mathcal{DT}_{L2}$, $L1 + L2 = L$.

Learning Run CL_{cur} *parameters* $= \mathrm{LEARN}[CL_{cur}, \mathcal{DL}_{L1}]$.

Testing and Validation. Run the testing procedure:

$$\{accuracy, \, precision, ...\} = \text{TEST}[CL_{cur}, \, CL_{cur} \, parameters, \, \mathcal{DT}_{L2}]$$

and decide whether the quality indicators are satisfactory.
IF YES – STOP and provide CL_{cur}, CL_{cur} *parameters* as the final result.
OTHERWISE
IF the admissible number of trials to select a proper classifier is not reached, then GO TO the Preparations step.
OTHERWISE
IF $N < n$ increase N and GO TO the Learning descriptors step.
OTHERWISE
Declare the failure of the learning process and STOP.

If failure occurred, one may consider enlarging the number of observations n and/or extending the set of considered classifiers.

Testing on Samples from Shocks and Vibrations

Operators' cabins of large working machines repetitively undergo shocks and vibrations (see [31] for examples of signals of this kind and [32] for their interpretation). The data in [31] consists of $N = 43$ curves, sampled at $n = 1024$ equidistant points each. An expert assigned label I (light working conditions) or II (heavy working conditions) to each series of signal samples.

An optional step – data augmentation was applied, providing $\mathcal{D}_L$ with $L = 43000$. This was done by adding the Gaussian noise with zero mean and the disperssion 0.05 to the estimates obtained in the learning descriptors step of the algorithm.

The algorithm of learning and selecting good classifiers was run on the augmented data. The list of tested classifiers is presented in Table 1. Only two of them, namely the logistic regression and the SVM provided accuracy larger than 0.95 (for the LogR – 0.98 and for the SVM –0.951 were obtained). Other quality indicators of these classieifers were high: the recall was larger than 0.98 in both cases, the precision attained by the LogR was 0.98 and 0.93, respectively, by the SVM. The Cohen kappa coefficient was equal to 0.96 for the LogR and 0.9 for the SVM.

Conclusions and Possible Extensions. Summarizing, the proposed approach of selecting the descriptors based on Bernstein polynomials and testing an adequate classifier occurred to be successful in classifying functional data from their noisy samples.

These descriptors can also be used for estimating an observed signal by applying the following kernel $\mathcal{K}(x, \, x') \overset{def}{=} (N+1) \sum_{k=0}^{N} B_k^{(N)}(x) \, B_k^{(N)}(x')$, $x, \, x' \in X$. Although kernel $\mathcal{K}$ has different properties than those typically used in nonparametric regression estimation by Parzen kernel methods, it can be applied for change detection problems in a similar way as it was recently proposed in [8,9]. Our descriptors can also be used as a part of generating signature hybrid descriptors in a way similar to the one proposed recently in [35]. Another way of their applications include novelty detection in ways found fruitful in [13] and [24].

References

1. Abdulla, L., Al-Ani, M.: A review study for electrocardiogram signal classification. UHD J. Sci. Technol. **4**(1), 103–117 (2020) https://doi.org/10.21928/uhdjst.v4n1y
2. Ahsan, M.R., Ibrahimy, M.I., Khalifa, O.O., et al.: EMG signal classification for human computer interaction: a review. Eur. J. Sci. Res. **33**(3), 480–501 (2009)
3. Augustyniak, P., Tadeusiewicz, R.: Optimization of ECG procedures chain for reliability and data reduction. In: Ubiquitous Cardiology: Emerging Wireless Telemedical Appl. pp. 202–227. IGI Global (2009)
4. Azlan, W.A., Low, Y.F.: Feature extraction of electroencephalogram (EEG) signal - a review. In: 2014 IEEE Conference on Biomedical Engineering and Sciences (IECBES), pp. 801–806 (2014). https://doi.org/10.1109/IECBES.2014.7047620
5. Chen, W.. Ditzian, Z.: Best polynomial and Durrmeyer approximation in $L_p(S)$ Indagationes Mathematicae **2**, 437–452 (1991)
6. Ciesielski, Z.: Nonparametric polynomial density estimation. Probab. Math. Stat. **9**(1), 1–10 (1988)
7. Derrienic, M.M.: On multivariate approximation by Bernstein-type polynomials. J. Approxim. Theory **45**, 155–166 (1985)
8. Galkowski, T., Krzyzak, A., Filutowicz, Z.: A new approach to detection of changes in multidimensional patterns. J. Artifi. Intell. Soft Comput. Res. **10**(2), 125–136 (2020). https://doi.org/10.2478/jaiscr-2020-0009
9. Galkowski, T., Krzyzak, A., Patora-Wysocka, Z., Filutowicz, Z., Wang, L.: A new approach to detection of changes in multidimensional patterns - Part II. J. Artifi. Intell. Soft Comput. Res. **11**(3), 217–227 (2021). https://doi.org/10.2478/jaiscr-2021-0013
10. Gandhi, T., Panigrahi, B.K., Anand, S.: A comparative study of wavelet families for EEG signal classification. Neurocomputing **74**(17), 3051–3057 (2011)
11. Garrett, D., Peterson, D.A., Anderson, C.W., Thaut, M.H.: Comparison of linear, nonlinear, and feature selection methods for EEG signal classification. IEEE Trans. Neural Syst. Rehabil. Eng. **11**(2), 141–144 (2003). https://doi.org/10.1109/TNSRE.2003.814441
12. Harris, T., Tucker, J.D., Li, B., Shand, L.: Elastic depths for detecting shape anomalies in functional data. Technometrics **63**(4), 466–476 (2020)
13. Homenda, W., Jastrzebska, A., Pedrycz, W., Yu, F.: Combining classifiers for foreign pattern rejection. J. Artifi. Intell. Soft Comput. Res. **10**(2), 75–94 (2020). https://doi.org/10.2478/jaiscr-2020-0006
14. Lorentz, G.G.: Bernstein Polynomials. American Mathematical Soc. (2013)
15. Lotte, F., Congedo, M., Lécuyer, A., Lamarche, F., Arnaldi, B.: A review of classification algorithms for EEG-based brain–computer interfaces. J. Neural Eng. **4**(2), R1–R13 (2007). https://doi.org/10.1088/1741-2560/4/2/r01
16. Marron, J.S., Ramsay, J.O., Sangalli, L.M., Srivastava, A.: Functional data analysis of amplitude and phase variation. Stat. Sci. **30**(4), 468–484 (2015). https://doi.org/10.1214/15-STS524
17. Mironovova, M., Bíla, J.: Fast Fourier transform for feature extraction and neural network for classification of electrocardiogram signals. In: 2015 Fourth International Conference on Future Generation Communication Technology (FGCT), pp. 1–6 (2015) https://doi.org/10.1109/FGCT.2015.7300244
18. Mohammad, A.J., Mohammad, I.J.: Summation-Integral Bernstein Type Of Neural Network Operators. Asian J. Math. Comput. Res. 74–86 (2017)

19. Pepelyshev, A., Rafajłłowicz, E., Steland, A.: Estimation of the quantile function using Bernstein-Durrmeyer polynomials. J. Nonpara. Stat. **26**(1), 1–20 (2014)
20. Rafajlowicz, E.: Nonparametric least squares estimation of a regression function. Statistics **19**(3), 349–358 (1988)
21. Rafajłowicz, E., Skubalska-Rafajłowicz, E.: Nonparametric regression estimation by Bernstein-Durrmeyer polynomials. Tatra Mt. Math. Publ. **17**, 227–239 (1999)
22. Pawlak, M., Rafajlowicz, E., Krzyzak, A.: Postfiltering versus prefiltering for signal recovery from noisy samples. IEEE Trans. Inf. Theory **49**(12), 3195–3212 (2003)
23. Rafajłowicz, W., Rafajłowicz, E.: Learning shape sensitive descriptors for classifying functional data. In: Rutkowski, L., Scherer, R., Korytkowski, M., Pedrycz, W., Tadeusiewicz, R., Zurada, J.M. (eds.) ICAISC 2021. LNCS (LNAI), vol. 12854, pp. 485–495. Springer, Cham (2021). https://doi.org/10.1007/978-3-030-87986-0_43
24. Rafajłowicz, W.: Learning novelty detection outside a class of random curves with application to covid-19 growth. J. Artifi. Intell. Soft Comput. Res. **11**(3), 195–215 (2021)
25. Rafajłowicz, W.: Learning Decision Sequences For Repetitive Processes—Selected Algorithms. SSDC, vol. 401. Springer, Cham (2022). https://doi.org/10.1007/978-3-030-88396-6
26. Rutkowski, L.: A general approach for nonparametric fitting of functions and their derivatives with applications to linear circuits identification. IEEE Trans. Circ. Syst. **33**(8), 812–818 (1986). https://doi.org/10.1109/TCS.1986.1086001
27. Rutkowski, L., Rafajłowicz, E.: On optimal global rate of convergence of some nonparametric identification procedures. IEEE Trans. Autom. Control **AC-34**, 1089–1091 (1989)
28. Skubalska-Rafajłowicz, E., Rafajłowicz, E.: Classifying functional data from orthogonal projections – model, properties and fast implementation. In: Paszynski, M., Kranzlmüller, D., Krzhizhanovskaya, V.V., Dongarra, J.J., Sloot, P.M.A. (eds.) ICCS 2021. LNCS, vol. 12744, pp. 26–39. Springer, Cham (2021). https://doi.org/10.1007/978-3-030-77967-2_3
29. Srivastava, A., Klassen, E., Joshi, S.H., Jermyn, I.H.: Shape analysis of elastic curves in euclidean spaces. IEEE J. Sel. Areas Commun. **10**(2), 391–400 (1992). https://doi.org/10.1109/49.126990
30. Wang, C., Zhang, H., Fan, W., Fan, X.: A new wind power prediction method based on chaotic theory and Bernstein Neural Network. Energy **117**, 259–271 (2016)
31. Więckowski, J.: Data from vibration in SchRs1200, Mendeley Data, V1 (2021). http://dx.doi.org/10.17632/htddgv2p3b.1
32. Więckowski, J., Rafajlowicz, W., Moczko, P., Rafajlowicz, E.: Data from vibration measurement in a bucket wheel excavator operator's cabin with the aim of vibrations damping. In: Data in Brief, p. 106836 (2021)
33. Wszolek, W., Tadeusiewicz, R., Chyla, A.: Recognition of selected helicopter types based on the generated acoustic signal with application of artificial intelligence methods. In: INTER-NOISE and NOISE-CON Congress and Conference Proceedings, vol. 2001(4), pp. 1734–1737. Institute of Noise Control Engineering (2001)
34. Xie, W., Chkrebtii, O., Kurtek, S.: Visualization and outlier detection for multivariate elastic curve data. IEEE Trans. Visual Comput. Graphics **26**(11), 3353–3364 (2020). https://doi.org/10.1109/TVCG.2019.2921541
35. Zalasinski, M., Cpalka, K., Laskowski, L., Wunsch, D., Przybyszewski, K.: An algorithm for the evolutionary-fuzzy generation of on-line signature hybrid descriptors. J. Arti. Intell. Soft Comput. Res. **10**(3), 173–187 (2020). https://doi.org/10.2478/jaiscr-2020-0012

Artificial Intelligence in Modeling and Simulation

Multiscale Multifractal Detrended Analysis of Speculative Attacks Dynamics in Cryptocurrencies

David Alaminos[1]([email]) [iD] and M. Belén Salas[2,3]([email])

[1] Department of Business, Universitat de Barcelona, Barcelona, Spain
alaminos@ub.edu
[2] Department of Finance and Accounting, Universidad de Málaga, Málaga, Spain
belensalas@uma.es
[3] Cátedra de Economía y Finanzas Sostenibles, Universidad de Málaga, Málaga, Spain

Abstract. Cryptocurrencies have drawn the interest of both scholars and professionals due to their decentralised, unique payment system supported by blockchain technology and their autonomy from sovereign governments, centralised organisations, and banking systems. Numerous works have studied, on the one hand, the behavior of cryptocurrencies, and on the other hand, the multifractal model in financial markets. Nevertheless, the limitations of existing models exist, and the literature calls for more research into multifractal analysis techniques applied to finance, as the methodology widely used in previous research is the regression model and machine learning methods. This study introduces a new model for predicting unexpected situations of speculative attacks in the cryptocurrency market, applying the method of Multiscale Multifractal Detrended Fluctuation Analysis, which shows excellent precision results. Our approach has a high impact potential on the forecast of possible speculative actions over the value of cryptocurrencies and against the risks derived from the control of cryptocurrencies by private entities, so the question of the possible effect on the financial system is of great importance.

Keywords: Cryptocurrencies · Speculative attacks · Multiscale Multifractal Detrended Fluctuation Analysis · Expectations · Financial system

JEL Codes: C63 · D84 · F3 · G15

MSC Codes: 28A80 · 91B28

1 Introduction

Following the onset of the financial crisis worldwide and the associated lack of credibility of the current financial sector, cryptocurrencies have attracted a great deal of interest since their emergence in 2009, since they are becoming more and more significant in the international finance market. Cryptocurrencies are virtual currencies based on the

© The Author(s), under exclusive license to Springer Nature Switzerland AG 2023
L. Rutkowski et al. (Eds.): ICAISC 2022, LNAI 13588, pp. 325–339, 2023.
https://doi.org/10.1007/978-3-031-23492-7_28

Internet that employ cryptographic features to process and carry out payment transactions securely [1, 2]. Unlike other financial assets, cryptocurrencies are not subject to any higher authority, they are infinitely divisible and their value is supported by the safety of an algorithm that makes it possible to track all transactions [3]. The key benefits of cryptocurrencies are that they are transparent and accessible 24 h a day. Cryptocurrency transactions are recorded in an open, public accounting ledger known as the blockchain. This decentralized structure provides cryptocurrencies with unprecedented transparency, making them an exception to traditional government policies [4].

In a decentralized framework such as Bitcoin or Ethereum, each member must perform mining activities. These activities take place on a chain of blocks (blockchain), which is an associated open record of every exchange that happens in the system for a particular intention and which is available to everyone. This avoids the need for a central regulator and gives the authorities directly involved in the transaction control [5]. For this reason, the cryptocurrency market is a volatile market that has been characterized in recent years by its strong ups and downs, with price volatility being, therefore, one of the main problems of decentralized cryptocurrencies [6, 7]. It is important to clearly understand the volatility changes in cryptocurrencies because such changes can affect investors' risk, altering their respective cryptocurrency investments. Various investigations have attempted to both interpret and forecast the Bitcoin price, thus [8] shows that it is complicated to justify Bitcoin prices by conventional financial analysis, and [9] concluded that the Bitcoin price is unforeseeable, although its volatility can be forecast by its historical data. This misunderstanding of Bitcoin's price formulation has sparked a discussion about its future role in the Bitcoin market. The research by [10] detected that during the period 2011–2020 the price of bitcoin fluctuated violently due to speculative activities, and concluded that the prices in the Bitcoin market are very volatile since they show nonstationary behavior, in which the distribution of the statistical data differs across time. This is because it is politically sensitive, engaging in speculative behaviour and political risk.

An important characteristic of a cryptocurrency is that it is particularly prone to price speculation, caused by investors who exploit the exchange rate to make a profit [11]. Some authors have investigated price speculation in the market for cryptocurrencies [12–15]. For their part, [12] determined that because the blockchain system has an open nature without permissions, the 51% attack is a typical threat in the cryptocurrency market. They concluded that an attacker performing a 51% attack can perform a double-spending attack and prevent any further transactions from being confirmed. Therefore, early detection of a speculative attack is very relevant to restrict the potential damage it may cause. The greater the investment in cryptocurrencies, the greater the risk that price speculation in the cryptocurrency markets might influence to other financial trainings and ultimately to the world economy [16–18]. Therefore, financial organisations and policymakers need to understand how the cryptocurrency market works to formulate the regulation in this system and estimate potential systemic risks.

Cryptocurrencies have a definite impact when planning investments in the financial markets, given their consideration as investment assets and the speculative attacks that the cryptocurrency market can suffer. Many works have demonstrated that financial markets are dynamic non-linear complex systems with chaotic and fractal patterns [19–21]. To research multifractality in non-linear time series, [22–24], apply the Multifractal Detrended Fluctuations Analysis (MF-DFA), the most often used approach to investigate it. [24] establishes that the multifractal model has been shown to correctly measure the complexity of economic systems since it can describe a financial time series through its multifractal spectrum; Thus, this type of analysis offers the possibility of studying the local regularity of the time series, which is useful in detecting speculative attacks on the markets. Likewise, [23] proposes that the analysis with multifractal analysis techniques could be extended to other markets such as currencies and government bonds, and contribute to the prior and timely detection of financial falls, that is, serve as a risk management tool.

Finally, many researchers have been attracted by the multi-fractal characteristics of financial systems. Previously, methods of fractal analysis, like detrended fluctuation analysis (DFA) [25] and multifractal detrended fluctuation analysis (MF-DFA) [22, 26, 27] have been established to identify long-range autocorrelation and to explore the efficiency of the market. Others models have explored the volatility of the market and described the market's non-linear characteristics [19, 20, 23]. For their side, [28] developed the multiscale multifractal analysis (MMA) method, providing new ways to measure the non-linearity of time series and analysing heart rate variability. So, the MF-DFA and MMA methods are well used for studying the multifractality of time series, as they are a powerful tool for investigating multifractal characteristics in an unstable time series, and have been also established for multiplicity series [22, 28]. Remarkable studies [29, 30] exhibit the multi-fractality of the foreign exchange market, supporting evidence that suggests the adoption of the MF-DFA technique. Finally, [26, 27] state that in future researches should be applied in financial markets, including foreign exchange markets, trade, commodities, and predictability of cryptocurrency time series; since it demonstrates that the level of multifractality for a wide range of foreign exchange markets is linked to the phase of the evolution of the market.

The objective of this study is to apply a speculative attack model to demonstrate that this attack mechanism can also occur in the cryptocurrency market, specifically with Bitcoin and Ethereum. The method used is that of MF-DFA and MMA, verifying that it may be the best methodology to detect and predict this type of unexpected situation. Recently, the MF-DFA has been used extensively in many sophisticated dynamical structures, for example, geophysical [31], traffic control [32], and financial market [33]. Numerous works of multifractal analysis have been applied to economics and finance, such as stock markets, interest, and exchange rates, as they exhibit a multifractal nature, often accompanied by the existence of long-term temporal interactions and robust tailed likelihood distributions [22, 26]. On the other hand, several authors have introduced the method MMA in others fields like traffic signals [20, 34] introduce the MMA to examine the characteristics of short- and long-range financial time series and to offer new tools to monitor the non-linearity of the time series. They state that future research into improving the performance of this method is desirable, as there is a demand for an increasing number of financial time series to be studied worldwide, as financial markets. The contribution to the literature that we are making with our study is the application of two methodologies, MF-DFA and MMA, which complement each other and have not been used together in previous research on cryptocurrency markets, obtaining very precise results. Most of the previous research has applied statistical methods, especially the regression model and machine learning methods, but the literature demands future research into combating speculative attacks on cryptocurrencies with different methodologies, as regardless of the technique chosen, there is a potential need for a method to combat speculative attacks on cryptocurrencies [4, 6]. In addition, the literature calls for techniques that can further refine the blockchain's function, potentially improving cybersecurity in the digital world. This cyber security improvement would greatly decrease cyber attacks [5]. Also, previous studies require the multifractal analysis techniques applied to finance [20]. Finally, most of the research on the behavior of cryptocurrencies has focused on Bitcoin. [35] and [2] propose as future lines of research to extend the analysis of market behavior to other cryptocurrencies. Thus our study has also covered Ethereum, these currencies representing 70% of the total flow in 2018. In addition, each cryptocurrency follows its specific global financial market trend and is unrelated to stock markets, which makes it adequate for incorporation into global investment strategies [5]. This study is structured as described below. Section 2 outlines the methodology used. Section 3 analyzes the results achieved. The article finishes by outlining the conclusions of the study.

2 Methodology

2.1 Speculative Attacks' Model

Since the success of the attack is decided in period 2, we first consider the small players' action from period 1, and then consider the big player's decision whether or not to start an early attack. A possible delayed strike by the big investor would be the rest of his L-credit following any speculation advanced on.

According [36], we will suppose that the small players play an activation strategy where agents assail the coin if the signal drops under a certain critical value x^*. Like in this approach of the equilibrium unique to the model is defined by two crucial variables: x^* and a fundamental minus crucial parameter for early speculation by the big investor, $(\theta - \lambda)$. If $\theta - \lambda \leq (\theta - \lambda)^*$, the coin would collapse.

We first discuss the equilibrium of the given activation strategies, we, therefore, examine the optimal strategies for activation. Certainly, if the approach is activated, a minor agent i would attempt to raid the coin if his signal $x_i \leq x^*$. The likelihood of occurrence depends on the true economic situation, $\theta - \lambda$, as described below

$$prob\big[x_i \leq x^* | \theta - \lambda\big] = prob\big[\theta - \lambda + \sigma \varepsilon_i \leq x^*\big] = prob\left[\varepsilon_i \leq \frac{x^* - (\theta - \lambda)}{\sigma}\right]$$
$$= F\left(\frac{x^* - (\theta - \lambda)}{\sigma}\right) \quad (1)$$

Because there is a small-agent continuum, and their noise conditions are separate, a joint confusion about the conduct of little actors is absent. Therefore, the density of attacking minor agents, ξ, is the same as this likelihood. Since F (.) is tightly rising, the impact of a speculative attack is narrowly declining in θ-λ; the lower the force of the economic fundamentals, the weaker the big investor's early speculation, the further little agents will strike.

A successful speculative attack would occur if the aggregate of minor speculative actors outweighs the power of the economic principles, minus the speculation early on by the big actor, i.e., if

$$F\left(\frac{x^* - (\theta - \lambda)}{\sigma}\right) \geq \theta - \lambda \tag{2}$$

Hence, the crucial variable $(\theta - \lambda)^*$, for the set of minor actors attacking is enough to provoke a devaluation, as follow

$$F\left(\frac{x^* - (\theta - \lambda)^*}{\sigma}\right) = (\theta - \lambda)^* \tag{3}$$

For smaller amounts, in which $\theta - \lambda \leq (\theta - \lambda)^*$, the impact of speculation (the left-handed of (3)) is higher, and the force of the fixed exchange rate (the right-handed of (3)) smaller, which implies that aggression has more success. Consequently, for higher parameters, where $\theta - \lambda > (\theta - \lambda)^*$, the occurrence of speculation is shorter and the fixed exchange rate strength higher, so an assault will not have success.

We, therefore, obtain the activation-optimal approaches of the minor actors. An investor notices a signal x_i and, for this signal, the likelihood of a successful offense is denoted by

$$prob\left[\theta - \lambda \leq (\theta - \lambda)^* | x_i\right] = prob\left[x_i - \sigma \varepsilon_i \leq (\theta - \lambda)^*\right]$$
$$= prob\left[\varepsilon_i \geq \frac{x_i - (\theta - \lambda)^*}{\sigma}\right] = 1 - F\left(\frac{x_i - (\theta - \lambda)^*}{\sigma}\right) = F\left(\frac{(\theta - \lambda)^* - x_i}{\sigma}\right) \tag{4}$$

where the last equation is derived from f(.), $F(v) = 1 - F(-v)$. The reward requested from hitting the coin for agent i, by speculation unit, is therefore

$$(1 - t)F\left(\frac{(\theta - \lambda)^* - x_i}{\sigma}\right) - t\left(1 - F\left(\frac{(\theta - \lambda)^* - x_i}{\sigma}\right)\right) = F\left(\frac{(\theta - \lambda)^* - x_i}{\sigma}\right) - t \tag{5}$$

In an activation optimal strategy, the reward anticipated of the coin attack for the marginal player has to be zero, i.e. the optimal cut x^* in the activation, strategy is provided by

$$F\left(\frac{(\theta - \lambda)^* - x^*}{\sigma}\right) = t \tag{6}$$

To resolve the balance, we redesign (6) to get $(\theta - \lambda)^* = x^* + \sigma F^{-1}(t)$. Replacing into (3), we obtain

$$(\theta - \lambda)^* = F\left(\frac{x^* - (x^* + \sigma F^{-1}(t))}{\sigma}\right), or (\theta - \lambda)^* = F\left(-F^{-1}(t)\right)$$
$$= 1 - F\left(-F^{-1}(t)\right) = 1 - t \tag{7}$$

So, the crucial parameters are

$$(\theta - \lambda)^* = 1 - t, \text{ and} \tag{8a}$$

$$x^* = 1 - t - \sigma F^{-1}(t) \tag{8b}$$

These crucial parameters match with the only novelty being the speculation of the major agent λ.

Next, we consider the big player's decision whether or not to speculate in period 1, and, if so, to what extent. There is no incertitude in the small players' combined behaviour, hence the large player can perfectly predict its speculation, save for the noise of its signal. From (8), a devaluation will occur if the fundamental $\theta \le \theta^* \equiv 1 - t + \lambda$.

The likelihood of aggression being successful can be expressed as

$$prob[\theta \le 1 - t + \lambda | y] = prob[y - \tau\eta \le 1 - t + \lambda | y] = prob\left[\frac{y - \lambda - (1-t)}{\tau} \le \eta | y\right]$$
$$= G\left(\frac{1 - t + \lambda - y}{\tau}\right) \tag{9}$$

where we once again employ the distribution symmetry. If the attack is successful, the major investor also wishes to speculate in period 3, so that the total amount of speculation is L. But we also have the risk, which occurs with likelihood q, that speculation in period 3 is much excessively delayed, hence the big investor benefits only from his speculating early λ. The payoff desired to attack in quantity $\lambda \ge 0$ in an early phase is therefore

$$E\pi = G\left(\frac{1 - t + \lambda - y}{\tau}\right)(L(1 - q) + \lambda q) - t\lambda \tag{10}$$

The first requirement for an indoor solution λ^* is

$$\frac{\partial E\pi}{\partial \lambda} = g\left(\frac{1 - t + \lambda^* - y}{\tau}\right)\frac{1}{\tau}(L(1 - q) + \lambda^* q) + G\left(\frac{1 - t + \lambda^* - y}{\tau}\right)q - t = 0 \tag{11}$$

Since $E\pi$ is a function continuous of λ, which is fixed on the closed interval [0,L], we assume the existence of an early optimum quantity of speculation λ, which maximises the profit expectation. Otherwise, the optimal λ is neither single nor necessarily internal. Indeed, if the charges of early speculation, t, are low enough, the efficient early speculation is the same as the credit restriction L.

Proposition 1. Requirements for speculation at an early stage by major investors.

i.   There is a value critical to the charges of early speculation $\underline{t} > 0$ as if $0 < t < \underline{t}$, for certain values of the other parameters, indicating that the efficient early speculation is the higher restriction, $\lambda = L$.
ii. For some variables of other parameters, there is a crucial value for the charges of early speculation $\bar{t} > 0$ as if $t > \bar{t}$, so the efficient early speculation is zero, $\lambda = 0$.

The second condition if the solution is internal is

$$\frac{\partial^2 E\pi}{\partial^2 \lambda} = g'\left(\frac{1-t+\lambda^*-y}{\tau}\right)\frac{1}{\tau^2}(L(1-q)+\lambda^*q) + 2g\left(\frac{1-t+\lambda^*-y}{\tau}\right)\frac{q}{\tau} < 0$$
(12)

The second term of (12) is a positive term, which implies that the first term should be contradictory, i.e. that $g'(.) < 0$ in an internal solution. To investigate the impact of rising costs of speculation t on efficient early speculation, note that (11) could be denoted by H(λ,t) = 0, which implies the definition of the efficient early speculation λ^* to be a function of costs of speculation. Deferring concerning t, we get $H_1 \frac{d\lambda^*}{dt} + H_2 = 0$, or $d\lambda^*/dt = -H_2/H_1$, where $H_2 \equiv \frac{\partial^2 E\pi}{\partial\lambda\partial t}$ and $H_1 \equiv \frac{\partial^2 E\pi}{\partial^2 \lambda} < 0$ from the second requirement. It deduces, thus, that the sign of $\frac{d\lambda^*}{dt}$ is the same as the sign of $H_2 \equiv \frac{\partial^2 E\pi}{\partial\lambda\partial t}$. We distinguish the first-order condition (11) about t, leading to

$$\frac{\partial^2 E\pi}{\partial\lambda\partial t} = -g'\left(\frac{1-t+\lambda^*-y}{\tau}\right)\frac{1}{\tau^2}(L(1-q)+\lambda^*q) - g\left(\frac{1-t+\lambda^*-y}{\tau}\right)\frac{q}{\tau} - 1$$
(13)

A rise in speculation charges t concerns efficient early speculation through the three terms in (13), with the second and third terms being negative. The second term reflects the reduction in speculation by minor investors due to upper charges of speculation. The third term incorporates the direct impact of higher speculation charges, making speculation more costly, and leading to minus early speculation.

For the very first term, though, we assume that it is affirmative, since $g'(.) < 0$. As greater speculation costs decrease speculation by minor agents, the projected impact of greater early speculation by the major player on the probability of success rises.

The model might be changed to accept N > 1 major traders as described below. First, for tractability reasons, we will disregard information imbalances between the large investors, under the assumption that all of them are observing the identical signal y. Next, we presume that the costs of speculation funds are convex, implying that the charges of early speculation λ_j for agent, j is tc(λ_j), for which c(.) is convex and strongly positive. The estimated profit from early speculation of player j would be $E\pi_j = G\left(\frac{1-t+\lambda-y}{\tau}\right)(L_j(1-q)+\lambda_j q) - tc(\lambda_j)$ where $\lambda = \sum_j \lambda_j$, and L_j is player j's credit limit. However, similar results to Proposition 1 can be deduced, which implies that if costs are not above a critical value, there will be no early speculation. The first requirement for efficient indoor early speculation λ_j^* of agent j is

$$\frac{\partial E\pi}{\partial\lambda_j} = g\left(\frac{1-t+\lambda^*-y}{\tau}\right)\frac{1}{\tau}\left(L_j(1-q)+\lambda_j^*q\right) + G\left(\frac{1-t+\lambda^*-y}{\tau}\right)q$$
$$- tc'\left(\lambda_j^*\right) = 0$$
(14)

In an equilibrium interior in pure strategies, all major traders would soon speculate in the quantity shown by (14).

2.2 Multifractal Detrended Fluctuation Analysis (MF-DFA)

MF-DFA is an efficient number methodology that examines the scaling characteristics of oscillations through the computation of a set of multifractal fluctuation functions $F_q(s)$ [37, 38]. First, we must split the time series of length N into a series of adjacent windows of distance s [the number of adjacent windows is $N_s \equiv$ int(N/s)]. As s is normally not a factor of the longitude of the time series, parts at the end of the time series could be ignored. To evade this, we once more construct a series of adjacent windows of distance s, but from the very end of the time series. Fluctuations are aggregates of squares of the local differences among the time series interlinked through time and a deviation polynomial adjusted to the data inside the window provided.

The MF-DFA method is a generalisation of the DFA one for analysing the characterization of multifractal nonstationary time series [20]. Suppose X(i) (i) = 1,2,...,N is a time series of dimension N and the MF-DFA method is written in the next stages.

Step 1. Calculate the appropriate time series profile X(i) by the equation below

$$X(i) = \log \sum_{k=1}^{i} (x_k - \langle x \rangle), i = 1, 2, \ldots, N, \tag{15}$$

where $\langle x \rangle$ are the average values of all-time series elements X(i).

Step 2. Divide X(i) into $t = N/s$ equal-length non-overlapping divisions s, where the mathematical operator ⊓ obtains the first integrate that is greater than or equaN/s. In most cases, there will be a small section left at the end of the time series because the dimension N of the time series is usually not a multiple of the time scale s. Likewise, the same operation is used again to consider the residual portion at the end of the time series, from the other end. In this way, we can get 2t segments together.

Step 3. Compute the local tendencies for every section of the 2t ones by a q-order adjustment of each series with the coming equation.

$$F^m(s, v) = \frac{1}{s} \sum_{k=1}^{s} (X((v-1)s+k) - X_v(k))^m, k = 1, 2, \ldots, t \tag{16}$$

and

$$F^m(s, v) = \frac{1}{s} \sum_{k=1}^{s} (X((N-(v-t)s+k) - X_v(k))^m, k = 1, 2, \ldots, t \tag{17}$$

whereas $X_v(k)$ is the polynomial adjusted in the vth section and m is the order number of the fitting polynomial of the MF-DFA. MF-DFAs of different order vary in the deviation capacity of the time series, although the value of m has a small effect on the time series of the multifractal. Therefore, in our study, we employ the quadratic MF-DFA, i.e. the *m* is equal to 2.

Step 4. Mean of all sections to derive the alternation of order q.

$$F(s, q) = \left\{ \frac{1}{2t} \sum_{k=1}^{2t} \left[F^2(s, k) \right]^{\frac{q}{2}} \right\}^{\frac{1}{q}}, k = 1, 2, \ldots, 2t, \tag{18}$$

where index parameter q may be any non-zero real number. If the parameter q is 0, we apply the next oscillation function:

$$F(s, q) = exp\left\{\frac{1}{4t} \sum_{k=1}^{2t} ln\left[F^2(s, k)\right]\right\}, \quad k = 1, 2, \ldots, 2t, \tag{19}$$

Step 5. Establish the scaling behaviour of the oscillation function by analysing the logarithmic plots of F(s, q) versus time scale s as below.

$$F(s, q) \sim s^{h(q)}.$$

h(q) is referred to as the generalised Hurst exponent and is the so-called Hurst exponent h while q is 2. Usually, the time series is monofractal whereas the h(q) is independent of q, the time series is multifractal meanwhile the h(q) is dependent of q.

The generalized Hurst exponent h(q) is a function of the magnitude of the oscillations. The interpretation of the values of h is as shown below [37, 38] h $\in (0, 0.5)$ indicates antipersistency of the time series, to say, the long-term anti-correlations, which means that big values are more probable to be traced by little values and vice versa. h = 0.5 denotes non-correlated noise, where no correlations exists. h $\in (0.5, 1)$ represents the persistence of the time series, i.e. the long-term correlation, implying that high values are susceptible to being tracked by big data and the other way around. The greater the h, the more strongly the correlations in the time series are. H = 1 indicates that the time series is 1/f noise. h > 1 denotes that the time series is unsteady. h = 1.5 suggests Brownian action (integrated white noise). h $\geq$ 2 indicates black noise.

2.3 Multiscale Multifractal Analysis (MMA)

MMA method was introduced in [38], and is a method of time series analysis, aimed at describing the scaling characteristics of the fluctuations in the signal under analysis. The central output of the process is obtention of the so-called Hurst surface h (q, s), defined by the local Hurst exponent h (fluctuation scale exponent) the multifractal parameter q [27] and the observation scale s (data window width).

The MMA is a standard generalisation of MF-DFA method [39] and provides a broader analysis of the properties of the fluctuations including more worldwide and robust results. The method, as a generalisation, correlates directly with prior fractal signal analysis methods. The results of the basic MF-DFA method [25] are expressed on the Hurst surface h (q, s) by one (or two) single points belonging to the exponent (or 1 and 2). The MMA obviates the requirement to make initial suppositions about the time scale of the research difficulty. This novel methodology is likely to simultaneously characterise the monofractality or multifractality of time series over a large variety of scales and can easily be implemented to cross-data It can recognise fractal properties of time series correctly for even relatively small scales, and is capable of recognising different fractal characteristics at small and large scales at the same time. The implementation of the method requires mentioning the following considerations. First, the fluctuation function Fq (s) is calculated and its graph is constructed in the log-log plane. From knowing Fq (s), the dependence h (q) of the generalized Hurst exponents in order q is determined, and lastly, the spectrum of singularity f (α) can be estimated with a Legendre transform this last part not required for MMA. The MF-DFA's final results are highly dependent on various user decisions in the initial step [34]. The first decision to consider is the mean reduction in calculating the profile of the data, as it has been pointed out that this step could be redundant with the trend reduction procedure. The second decision is the selection of the order of the trend polynomial implemented in every data window to derive Fq (s). The third decision is the proper range of degrees s, where the family of curves Fq (s) should be computed, too small degrees make the trend action run on a set of only a few points and scales too large could skew the calculation due to the small number of values in Fq (s) for large s. So, to estimate correctly the scales and describe the real information on the fractal properties for each time scale associated with the signal, a multifractal frequency spectrum with a varying scale range is calculated using a moving window of adjustment, sweeping the entire range of the scales along with the graph Fq (s).

3 Results

We use daily USD exchange rates for Bitcoin (BTC) and Ethereum (ETH) from Kraken Cryptocurrency exchange. Daily data on cryptocurrencies were gathered from the Coinmarketcap.com and CryproCompare.com websites. The observation data used in this study were collected during February 1st 2011 to December 1st 2020.

The oscillating parts of various frequency scales can be defined as central mode functions (IMFs), that is, nonlinear stationary signals, showing the amplitudes of the time-varying oscillations of different characteristic scales. We extracted 10 IMFs (IMF1-IMF10) and a r residue, and subsequently generated the Hurst analysis for every IMF and the residue. Table 1 reports the Hurst exponent, corresponding to the log-log plot slope of (R/S) and the fractional size in D = 2-H of each IMF and the r-residue. Based on it, we notice the components' various characteristics. The IMF1-IMF10 dual linear association responds to the two Hurst exponents shown in Table 1. H1, derived from the fine-scale ln(n), is higher than 0.5, showing far-reaching correspondence and consistent behaviour. In contrast, H2, which is obtained from the large-scale slope ln(n), is smaller than 0.5, indicating dynamic antipersistence. Despite the similar dual fractal behaviour of IMF1-IMF10, they have different breakpoints concerning their fractal condition and different long-term period. The Hurst exponents of IMF10 and the residual r are both close to 1, revealing their consistent behaviour and their long-range characteristics of correlation.

The Hurst exponents and fractal dimensions are recomputed and shown in Table 2, which represent their various long-term properties of correspondence. The Hurst exponent of Large Players (Bitcoin-USD Dollar) and Small Players (Ethereum-USD Dollar) is lower than 0.5, suggesting the existence of anti-persistence correlations. In the case of Large Players (Ethereum-USD Dollar) and Small Players (Bitcoin-USD Dollar) are higher than 0.5, stating that the two sub-sequences correspondence are significant and assiduous. The other statistical indices of the four sub-sequences are also given in Table 2. Large Players (Bitcoin-USD Dollar) have the highest variance contribution rate 61.52%,

Table 1. Hurst exponents and fractal dimensions of the elements

	IMF1	IMF2	IMF3	IMF4	IMF5	IMF6	IMF7	IMF8	IMF9	IMF10	r
H1	0.6342	0.6450	0.6684	0.7073	0.7823	0.8631	0.8835	0.9084	0.9229	0.9793	0.9862
H2	0.0573	0.2123	0.2736	0.1545	0.2416	0.2721	0.2644	0.2082	0.0702	0.3203	0.1431
D1	3.8264	7.6101	7.8280	8.6292	7.5387	4.8890	5.7437	6.8900	2.1579	5.4947	8.9180
D2	5.8539	8.0401	11.5514	8.7597	7.9766	8.2374	7.1293	9.2422	5.9321	9.4161	10.6411

indicating that it is the predominant driver of the fluctuation and dynamic behaviour of the cryptocurrency market.

Table 2. Hurst exponents, fractal dimension, and other statistical indices of four subsequences.

	H	D	Mean	Variance	Variance contribution rate
Large players (Bitcoin-USD Dollar)	0.45278149	14.529	0.058789179	0.74034347	61.52%
Large players (Ethereum-USD Dollar)	0.79258158	14.639	0.04180298	0.45419066	37.75%
Small players (Bitcoin-USD Dollar)	0.58127372	18.652	0.075472619	0.95044125	0.7898%
Small players (Ethereum-USD Dollar)	0.42134992	13.474	0.054519784	0.68657815	57.06%

In addition, Table 3 is given the parameter ΔS and $\Delta \alpha$ of the multifractal features of the four sub-sequences of scale. The value ΔS was determined by R-L to measure numerically the amount of shift of the singularity spectrum, where $R = \alpha_{max} - \alpha_0$, $L = \alpha_0 - \alpha_{min}$; remark that α_0 is the maximum value concerning to the multifractal spectrum. The ΔS of the Small Players (Ethereum-USD Dollar) and the Small Players (Bitcoin-USD Dollar) are 0.05678804 and 0.086605 respectively, whereas the ΔS of the Large Players (Ethereum-USD Dollar) is 0.17512579, indicating that the spectrum of the Large Players (Ethereum-USD Dollar) has more asymmetry than the Small Players (Ethereum-USD Dollar) and the Small Players (Bitcoin-USD Dollar), and this means that the Large Players (Ethereum-USD Dollar) is more unaffected by large magnitude fluctuations [40]. The index $\Delta \alpha$ stands for strength of the multifractal characteristic. The $\Delta \alpha$ of the Large Players (Ethereum-USD Dollar) is the highest of the four ones, meaning that it has the highest multifractality strength. The Small Players (Ethereum-USD Dollar) have the smallest $\Delta \alpha$, which implies that the multifractality of this is the mildest among the four subsequences.

Table 3. Indices of the multifractal characteristics of the four scale subsequences.

	$\Delta\alpha$	R	L	ΔS
Large players (Bitcoin-USD Dollar)	0.39508023	0.16484096	0.230239262	−0.0653983
Large players (Ethereum-USD Dollar)	1.05795926	0.44141674	0.386303262	0.17512579
Small players (Bitcoin-USD Dollar)	0.52319289	0.21829394	0.304898947	0.086605
Small players (Ethereum-USD Dollar)	0.343064	0.14313821	0.125266559	0.05678804

4 Conclusions

This study has developed a speculative attack model of the cryptocurrency market, Bitcoin and Ethereum. The selected period has been from February 1st 2011 to December 1st 2020. The method used was MF-DFA and MMA, verifying that it may be the best methodology for detecting and predicting such unexpected situations of speculative attacks. Specifically, the goal has been to demonstrate that this attack mechanism can also occur in the cryptocurrency market by improving accuracy levels with the Multiscale multifractal analysis methodology.

The results achieved show that the various frequency-time components IMFs of the collected magnitude series exhibit anti-persistent and persistent behavior. From there, four different scaling sub-sequences are identified and overlaid: Large Players (Bitcoin-USD Dollar), Large Players (Ethereum-USD Dollar), Small Players (Bitcoin-USD Dollar), and Small Players (Ethereum-USD Dollar). The Large Players (Bitcoin-USD Dollar) subsequence, which has the greatest variance contribution rate, is the key factor pushing the dynamic fluctuation and behaviour of the cryptocurrency market, therefore, major investors in Bitcoin dominate the market. However, the MF-DFA method shows that the four subsequences are marked by varying multifractality, which suggests that they have diverse heterogeneity: the Large Players (Ethereum-USD Dollar) possesses the strongest multifractality, the Small Players (Bitcoin-USD Dollar) and the Large Players (Bitcoin-USD Dollar) are the following two and the Small Players (Ethereum-USD Dollar) has the mildest multifractality. This indicates that the Large Players (Ethereum-USD Dollar) subsequence is more heterogeneous than the others.

In comparison to prior investigations, this research has developed a speculative attack model applying the methodology Multiscale MF-DFA, which complement each other and have not been used together in previous studies of cryptocurrencies. Most previous works have applied statistical and deep learning methods.

This research offers an important contribution to the international financial markets field, as the findings have significant implications for investors, market participants, and policymakers, who seek to derive economic and financial benefits from cryptocurrencies and to understand the properties of this market. Our research analyzes the behavior of cryptocurrencies under uncertainty and helps to improve market volatility forecasting in the face of potential speculative attacks by cryptocurrencies, potentially enabling investors to improve the risk-adjusted performance of their portfolios. In addition, interested market players will gain valuable insights into whether cryptocurrencies could be used to guide monetary policy and portfolio construction in a global environment. Finally, some future directions of research may be oriented to the study of different strategies of speculative attacks on cryptocurrencies in situations of severe liquidity drops and the possible contagions that can influence other cryptocurrencies or currencies.

Acknowledgement. This research was funded by Universitat de Barcelona (Convocatòria d'Àrees Emergents, Project Code: AS017634).

References

1. Paule-Vianez, J., Prado-Román, C., Gómez-Martínez, R.: Economic policy uncertainty and Bitcoin. Is Bitcoin a safe-haven asset?. Europ. J. Manage. Bus. Econ. **29**(3), 347–363 (2020). https://doi.org/10.1108/EJMBE-07-2019-0116
2. Yang, B., Sun, Y., Wang, S.: A novel two-stage approach for cryptocurrency analysis. Int. Rev. Financ. Anal. (2020).https://doi.org/10.1016/j.irfa.2020.101567
3. Abakah, E.J.A., Gil-Alana, L.A., Madigu, G., Romero-Rojo, F.: Volatility persistence in cryptocurrency markets under structural breaks. Int. Rev. Econ. Finance **69**, 680–691 (2020). ISSN 1059–0560. https://doi.org/10.1016/j.iref.2020.06.035
4. Grobys, K., Sapkota, N.: Predicting cryptocurrency defaults. Appl. Econ. **52**(46), 5060–5076 (2020). https://doi.org/10.1080/00036846.2020.1752903
5. Poongodi, M.: Prediction of the price of Ethereum blockchain cryptocurrency in an industrial finance system. Comput. Electr. Eng. **81**, 106527 (2020). ISSN 0045–7906. https://doi.org/10.1016/j.compeleceng.2019.106527
6. Mudassir, M., Bennbaia, S., Unal, D., Hammoudeh, M.: Time-series forecasting of Bitocoin prices using high-dimensional features: a machine learning approach. Neural Comput. Appl. 1–15 (2020). https://doi.org/10.1007/s00521-020-05129-6
7. Van Hijfte, S.: Decoding Blockchain for Business. 1st ed. New York: Apress. ISBN-13 (pbk): 978–1–4842–6136–1 ISBN-13 (electronic): 978–1–4842–6137–8 (2020). https://doi.org/10.1007/978-1-4842-6137-8
8. Kristoufek, L.: What are the main drivers of the Bitcoin price? Evidence from wavelet coherence analysis. PLoS ONE **10**(4), e0123923 (2015)
9. Aalborg, H.A., Molnar, P., de Vries, J.E.: What can explain the price, volatility and trading volume of Bitcoin? Financ. Res. Lett. **29**, 255–265 (2019)
10. Li, Y., Wang, Z., Wang, H., Wu, M., Xie, L.: Identifying price bubble periods in the Bitcoin market-based on GSADF model. Qual. Quant. **55**(5), 1829–1844 (2021). https://doi.org/10.1007/s11135-020-01077-4
11. Wei,Y.M., Dukes, A.: Cryptocurrency Adoption with Speculative Price Bubbles. Marketing Science Published online in Articles in Advance 08 Oct 2020 (2020). https://doi.org/10.1287/mksc.2020.1247
12. Di Pietro, R., Raponi, S., Caprolu, M., Cresci, S.: New dimensions of information warfare. In: New Dimensions of Information Warfare. AIS, vol. 84, pp. 1–4. Springer, Cham (2021). https://doi.org/10.1007/978-3-030-60618-3_1
13. Chaim, P., Laurini, M.P.: Is Bitcoin a bubble? Phys. A **517**, 222–232 (2019). https://doi.org/10.1016/j.physa.2018.11.031

14. Li, Z.-Z., Tao, R., Su, C.-W., Lobonţ, O.-R.: Does Bitcoin bubble burst? Qual. Quant. **53**(1), 91–105 (2018). https://doi.org/10.1007/s11135-018-0728-3
15. Cheah, E.T., Fry, J.: Speculative bubbles in bitcoin markets? an empirical investigation into the fundamental value of bitcoin. Econ. Lett. **130**, 32–36 (2015). https://doi.org/10.1016/j.econlet.2015.02.029
16. Lambrecht, M., Sofianos, A., Xu, Y.: Does mining fuel bubbles? an experimental study on cryptocurrency markets. AWI Discussion Paper Series No. 703. University of Heidelberg, Department of Economics, Heidelberg (2021). https://doi.org/10.11588/heidok.00030059
17. Manaa, M., et al.: Crypto-Assets: Implications for financial stability, monetary policy, and payments and market infrastructures. ECB Occasional Paper, No. 223 (2019)
18. Guo, F., Chen, C.R. Huang, Y.S: Markets contagion during financial crisis: a regime-switching approach. Int. Rev. Econ. Finance **20**, 95–109 (2011)
19. Wang, H.Y., Wang, T.T.: Multifractal analysis of the Chinese stock, bond and fund markets. Phys. A **512**, 280–292 (2018). https://doi.org/10.1016/j.physa.2018.08.067
20. Yujun, Y., Jianping, L., Yimei, Y.: Multiscale multifractal multiproperty analysis of financial time series based on Rényi entropy. Int. J. Mod. Phys. C **28**(2), 1750028 (2017). https://doi.org/10.1142/S0129183117500280
21. Zeng, Y., Wang, J., Xu, K.: Complexity and multifractal behaviors of multiscale-continuum percolation financial system for Chinese stock markets. Phys. A **471**, 364–376 (2017)
22. Fernandes, L.H.S., De Araújo, F.H.A., Silva, I.E.M.: The (in) efficiency of NYMEX energy futures: a multifractal analysis. Phys. **556**, 124783 (2020). https://doi.org/10.1016/j.physa.2020.124783
23. Shahzad, S.J.H., Nor, S.M., Mensi, W., Kumar, R.R.: Examining the efficiency and interdependence of US credit and stock markets through MF-DFA and MF-DXA approaches. Phys. A **417**, 351–363 (2017)
24. Rendón, S.: Stock crack detection using multifractal analysis (local and pointwise Hölder exponents): Stock Index of Mexico IPC and FX USD/MXN. MPRA (Munich Personal RePEc Archive) Paper No. 47699 (2013). https://mpra.ub.uni-muenchen.de/47699/
25. Peng, C.K., Buldyrev, S.V., Havlin, S., Simon, M., Stanley, H.E., Goldberger, A.L.: Mosaic organization of DNA nucleotides. Phys. Rev **E49**, 1685–1689 (1994)
26. Figliola A., Rosenblatt M., Serrano, E.P.: Local regularity analysis of market index for the 2008 economical crisis. Revista de Matemática: Teoría y Aplicaciones,**19** (1), 65–78 (2012). (ISSN 1409–2433)
27. Kantelhardt, J.W., Zschiegner, S.A., Koscielny-Bunde, E., Havlin, S., Bunde, A., Stanley, H.E.: Multifractal detrended fluctuation analysis of nonstationary time series. Phys. A **316**, 87–114 (2002). https://doi.org/10.1016/S0378-4371(02)01383-3
28. Aijing, L., Hui, M., Pengjian, S.: The scaling properties of stock markets based on modified multifractal detrended fluctuation analysis. Phys. A **436**, 525 (2015)
29. Wang, Y., Liu, L., Gu, R.: Analysis of efficiency for Shenzhen stock market based on multifractal detrended fluctuation analysis. Int. Rev. Financ. Anal. **18**, 271–276 (2009)
30. Yuan, Y., Zhuang, X.T., Jin, X.: Measuring multifractality of stock price fluctuation using multifractal detrended fluctuation analysis. Phys. A **388**, 2189–2197 (2009)
31. Subhakar, D., Chandrasekhar, E.: Reservoir characterization using multifractal detrended fluctuation analysis of geophysical well-log data. Phys. A **445**, 57–65 (2016). https://doi.org/10.1016/j.physa.2015.10.103
32. Xu, Y., Feng, H.: Revisiting multifractality of TCP traffic using multifractal detrended fluctuation analysis. J. Stat. Mech. Theory Exp. **2014**(2), P02007 (2014). https://doi.org/10.1088/1742-5468/2014/02/P02007
33. Tiwari, A.K., Albulescu, C.T., Yoon, S.M.: A multifractal detrended fluctuation analysis of financial market efficiency: comparison using dow jones sector ETF indices. Phys. A **483**, 182–192 (2017). https://doi.org/10.1016/j.physa.2017.05.007

34. Wang, J., Shang, P., Cui, X.: Multiscale multifractal analysis of traffic signals to uncover richer structures Phys. Rev. E **89**, 032916 (2014)
35. Scharnowski, S.: Understanding bitcoin liquidity. Finance Res. Lett. **38**, 101477 (2021). ISSN 1544–6123. https://doi.org/10.1016/j.frl.2020.101477
36. Corsetti, G., Dasgupta, A., Morris, S., Shin, H.S.: Does one Soros make a difference? a theory of currency crises with large and small traders. Rev. Econ. Stud. **71**(1), 87–114 (2004)
37. Liu, H., Zhang, X., Zhang, X.: Multiscale multifractal analyisis on air traffic flow time series: a single airport departure flight case. Phys. A (2019). https://doi.org/10.1016/j.physa.2019.123585
38. Gierałtowski, J., Zebrowski, J.J., Baranowski, R.: Multiscale multifractal analysis of heart rate variability recordings with a large number of occurrences of arrhythmia. Phys. Rev. E **85**, 021915 (2012). https://doi.org/10.1103/PhysRevE.85.021915
39. Alaminos, D., Aguilar-Vijande, F., Sánchez-Serrano, J.R.: Neural networks for estimating speculative attacks models. Entropy **23**(1), 106 (2021)
40. Goldberger, A., et al.: PhysioBank, physiotoolkit, and physionet: components of a new research resource for complex physiologic signals. Circulation **101**(23), e215–e220 (2000). https://doi.org/10.1161/01.CIR.101.23.e215

A Heuristic Repair Algorithm for the Hospitals/Residents Problem with Ties

Son Thanh Cao, Le Quoc Anh, and Hoang Huu Viet[✉]

School of Engineering and Technology, Vinh University, Vinh City, Vietnam
{sonct,anhlq,viethh}@vinhuni.edu.vn

Abstract. The Hospitals/Residents problem with Ties is a many-to-one stable matching problem and it has several practical applications. In this paper, we present a heuristic repair algorithm to find a stable matching with maximal size for this problem. Our approach is to apply a random-restart algorithm used commonly to deal with constraint satisfaction problems. At each iteration, our algorithm finds and removes the conflicted pairs in terms of preference ranks between hospitals and residents to improve rapidly the stability of the matching. Experimental results show that our approach is efficient in terms of execution time and solution quality for the problem of large sizes.

Keywords: Hospitals/residents with ties · Heuristic repair · Undominated blocking pair · Weakly stable matching

1 Introduction

In 1962, Gale and Shapley introduced the Hospitals/Residents problem (HR) under the name *"College Admissions Problem"* [3]. An instance of the HR involves a set of residents and a set of hospitals, in which each of them ranks a subset of the other set in a strict order of preference and each hospital has a *capacity* to indicate the maximum number of residents that can be assigned to it. Solving such a problem is to find a *matching* of residents and hospitals, in which each resident is assigned to at most one hospital and each hospital does not exceed its capacity. Moreover, the matching must be *stable* or it admits no *blocking pair*, where a blocking pair (r, h) for the matching is a resident r and a hospital h such that (i) r and h rank each other; (ii) r either is unassigned or prefers h to the hospital assigned to it; and (iii) h either is under-subscribed or prefers r to the worst resident assigned to it. HR can be found in applications such as the National Resident Matching Program (NRMP) in the US [18], the Scottish Pre-registration house officer Allocations (SPA) matching scheme [7], or the Canadian Resident Matching Service (CARMS) in Canada [1].

Recently, there are several variations of HR have been proposed by researchers [2,8,13,15]. The most popular one is a natural generalization of HR

© The Author(s), under exclusive license to Springer Nature Switzerland AG 2023
L. Rutkowski et al. (Eds.): ICAISC 2022, LNAI 13588, pp. 340–352, 2023.
https://doi.org/10.1007/978-3-031-23492-7_29

known as the Hospitals/Residents problem with Ties (HRT) [8,13], where both residents and hospitals can rank a subset of the other set with ties. Accordingly, there are three criteria of stable matchings consisting of *weak stability*, *strong stability*, and *super-stability* [8]. Among these criteria, the problem of finding weakly stable matchings has been an active field of researchers for several years since its practical applications. Irving et al. [8] showed that an instance of HR may have more than one stable matching and every stable matching is the same size, while an instance of HRT may have more than one weakly stable matching with different sizes. The problem of finding a weakly stable matching with the maximum number of residents assigned to hospitals is known as MAX-HRT and shown to be NP-hard [8].

In the last few years, several algorithms to solve MAX-HRT were introduced in the literature. Manlove et al. [14] proved that the size of the largest stable matching was at most twice the size of the smallest one for any HRT instance. Kwanashie et al. [12] presented an integer programming approach to find a stable matching. Munera et al. [16] proposed an adaptive search algorithm for the stable matching with ties and incomplete lists (SMTI) [10,14] and its extension to deal with MAX-HRT. Kir'aly [11] described ingenious approximation algorithms for MAX-HRT. However, all the algorithms mentioned above are inefficient to solve MAX-HRT of large sizes.

In this paper, we propose a heuristic repair algorithm to solve MAX-HRT. For brevity, hereinafter, we refer to a weakly stable matching as a stable matching and MAX-HRT as HRT. Our idea is to improve the stability of a randomly generated matching. At each iteration, our algorithm finds a set of undominated blocking pairs of a matching from the residents' point of view, then it removes the best blocking pair for each hospital such that it does not only remove as many blocking pairs from the residents' point of view as possible but also removes as many blocking pairs as possible from the hospitals' point of view. Experimental results show that our algorithm is efficient in solving HRT of large sizes.

The remainder of this paper is structured as follows. Section 2 reminds the main definitions for HRT, Sect. 3 presents our proposed algorithm, Sect. 4 discusses our experimental results, and Sect. 5 concludes our work.

2   Background

In this section, we remind the background for HRT [4,8]. An instance I of HRT involves a set of residents, denoted by $\mathcal{R} = \{r_1, r_2, \cdots, r_n\}$, and a set of hospitals, denoted by $\mathcal{H} = \{h_1, h_2, \cdots, h_m\}$, in which each $r_i \in \mathcal{R}$ ranks a subset of $\mathcal{H}$ in its preference list and each $h_j \in \mathcal{H}$ ranks a subset of $\mathcal{R}$ in its preference list. Moreover, each h_j has a *capacity* $c_j \in \mathbb{Z}^+$ to indicate the maximum number of residents that can be assigned to it. We denote a set of *acceptable* pairs by $\mathcal{A} = \{(r_i, h_j) \in \mathcal{R} \times \mathcal{H}\}$, where r_i and h_j must rank each other.

An *assignment* M is a subset of $\mathcal{A}$. If $(r_i, h_j) \in M$, we say that r_i is assigned to h_j and h_j is assigned r_i, and we denote $M(h_j)$ by the set of residents assigned to h_j and $M(r_i) = h_j$, respectively. If r_i is unassigned in M, then we denote by

$M(r_i) = \varnothing$. A hospital $h_j \in \mathcal{H}$ is called *under-subscribed, full*, or *over-subscribed* if $|M(h_j)| < c_j$, $|M(h_j)| = c_j$, or $|M(h_j)| > c_j$, respectively.

Definition 1 (matching). *A matching is an assignment M such that $|M(r_i)| \leq 1$ for each $r_i \in \mathcal{R}$, and $|M(h_j)| \leq c_j$ for each $h_j \in \mathcal{H}$, meaning that each resident is assigned to at most one hospital, and no hospital is over-subscribed.*

Given a matching M and a pair $(r_i, h_j) \in \mathcal{A}$, if r_i strictly prefers h_j to $M(r_i)$, then we denote by $h_j \prec_{r_i} M(r_i)$; if h_j strictly prefers r_i to the worst resident in $M(h_j)$, then we denote by $r_i \prec_{h_j} M(h_j)$.

Definition 2 (blocking pair). *A pair $(r_i, h_j) \in \mathcal{R} \times \mathcal{H}$ is a blocking pair for a matching M if (i) $(r_i, h_j) \in \mathcal{A}$; (ii) $M(r_i) = \varnothing$ or $h_j \prec_{r_i} M(r_i)$; and (iii) $|M(h_j)| < c_j$ or $r_i \prec_{h_j} M(h_j)$.*

Definition 3 (stable matching). *A matching M is called stable if it admits no blocking pairs, otherwise, it is called unstable.*

Definition 4 (matching size). *The size of a stable matching M, denoted by $|M|$, is the number of residents assigned to hospitals in M. If $|M| = n$, then M is called perfect. Otherwise, M is called non-perfect.*

Definition 5 (dominated blocking pair). *A blocking pair $(r_i, h_j) \in \mathcal{R} \times \mathcal{H}$ dominates a blocking pair $(r_i, h_k) \in \mathcal{R} \times \mathcal{H}$ from the residents' point of view if r_i prefers h_j to h_k.*

Definition 6 (undominated blocking pair). *A blocking pair $(r_i, h_j) \in \mathcal{R} \times \mathcal{H}$ is called an undominated blocking pair (UBP) if there exists no other blocking pair that dominates it from the residents' point of view.*

The concepts of the dominated and undominated blocking pairs were given in [4] and then they were applied to solve efficiently the SMTI problem [5,17]. In this paper, we apply these concepts to solve HRT. Given a matching M and a blocking pair $(r_i, h_j) \in \mathcal{R} \times \mathcal{H}$ for M, we call an operation of removing (r_i, h_j) for M means that r_i is assigned to h_j, or $M(r_i) = h_j$. We assume that there exist two blocking pairs, denoted by $(r_i, h_j) \in \mathcal{R} \times \mathcal{H}$ and $(r_i, h_k) \in \mathcal{R} \times \mathcal{H}$, for M, where (r_i, h_j) dominates (r_i, h_k) from the residents' point of view. If we remove (r_i, h_j) for M to obtain a matching M' from M, i.e. $M'(r_i) = h_j$, and the other pairs of M' are the same as those of M, except if $M'(h_j) > c_j$, then the worst resident in $M'(h_j)$ becomes unassigned. As a result, the blocking pair (r_i, h_k) is removed for M'. Otherwise, if we remove (r_i, h_k) for M to obtain a matching M' from M, then the blocking pair (r_i, h_j) still remains for M'. This follows that if we remove an UBP (r_i, h_j) for a matching M, then all the blocking pairs formed by r_i from the residents' point of view will be removed for M. We have equivalent concepts of the dominated and undominated blocking pairs from the hospitals' point of view. Accordingly, if we remove an UBP (r_i, h_j) from the hospitals' point of view, then all the blocking pairs formed by h_j will be removed for M.

Table 1. An instance of HRT of eight residents and four hospitals

Residents	Preference lists	Hospitals	Preference lists	Capacities
r_1	h_1 $(h_2$ $h_3)$ h_4	h_1	r_8 r_2 r_7 r_1 r_6 r_5 r_3 r_4	$c_1 = 3$
r_2	h_4 h_1 h_2 h_3	h_2	r_6 r_2 r_1 r_4 r_3 r_7	$c_2 = 6$
r_3	h_1 h_3 h_4 h_2	h_3	r_6 r_2 r_1 r_4 r_5 r_8 r_7 r_3	$c_3 = 3$
r_4	$(h_1$ $h_4)$ h_2 h_3	h_4	r_2 r_5 r_4 $(r_7$ $r_8)$ r_1 r_3	$c_4 = 4$
r_5	h_3 h_1 h_4			
r_6	h_2 h_1 h_3			
r_7	h_2 h_4 h_1 h_3			
r_8	h_1 h_3 h_4			

We consider an HRT instance consisting of 8 residents and 4 hospitals shown in Table 1. In residents' preference lists, for example, the notation r_1: h_1 $(h_2$ $h_3)$ h_4 means r_1 strictly prefers h_1 to h_2 and h_3, which are equally preferred. We have similar notations in the hospitals' preference lists. The matching $M = \{(r_1, \varnothing),$ $(r_2, \varnothing), (r_3, h_1), (r_4, h_1), (r_5, h_3), (r_6, h_1), (r_7, h_3), (r_8, \varnothing)\}$ is unstable because there exist blocking pairs such as $(r_1, h_1), (r_1, h_2), (r_1, h_3), (r_1, h_4), (r_2, h_1)$ for M. The blocking pair (r_1, h_1) dominates the blocking pair (r_1, h_4) from the residents' point of view and the blocking pair (r_1, h_1) is undominated since there exists no blocking pairs dominating it from the residents' point of view. If we remove (r_1, h_1) for M to obtain a matching M', i.e. $M' = \{(r_1, h_1), (r_2, \varnothing),$ $(r_3, h_1), (r_4, \varnothing), (r_5, h_3), (r_6, h_1), (r_7, h_3), (r_8, \varnothing)\}$, then all the UBPs formed by r_1 from the residents' point of view are removed for M'. The matching $M = \{(r_1, h_1), (r_2, h_4), (r_3, h_1), (r_4, h_4), (r_5, h_3), (r_6, h_2), (r_7, h_2), (r_8, h_1)\}$ is perfect since M is stable and $|M| = 8$.

3   Algorithm for HRT

In this section, we propose an algorithm of repairing undominated blocking pairs, called heuristic repair algorithm, to solve MAX-HRT. Given an arbitrary matching M of an instance I of HRT, we assume that there exists a set $X = \{(r_i, h_j)|(r_i, h_j) \in \mathcal{R} \times \mathcal{H}\}$ of UBPs from the residents' point of view for M. As we mentioned above, if we remove only an UBP $(r_i, h_j) \in X$ for M (i.e. $M(r_i) = h_j$), then all the blocking pairs formed by r_i will be removed for M. If so, we were wasted time in finding the remaining pairs in X. Obviously, we cannot remove every pair $(r_i, h_j) \in X$, since if there exist two pairs $(r_i, h_j) \in X$ and $(r_k, h_j) \in X$, then we remove (r_i, h_j) or (r_k, h_j) for M (i.e. $M(r_i) = h_j$ or $M(r_k) = h_j$)? Our question is that which pairs $(r_i, h_j) \in X$ should be removed in M such that we can rapidly obtain the stability of M. To answer this question, we first analyze the instance of HRT given in Table 1. We assume that given an unstable matching $M = \{(r_1, \varnothing), (r_2, \varnothing), (r_3, h_1), (r_4, h_1), (r_5, h_3), (r_6, h_1),$ $(r_7, h_3), (r_8, \varnothing)\}$, then the set of UBPs from the residents' point of view for M

is $X = \{(r_1, h_1), (r_2, h_4), (r_6, h_2), (r_7, h_2), (r_8, h_1)\}$. Since X is a set of UBPs from the residents' point of view, each $r_i \in X$ belongs to only one element of X, while each $h_j \in X$ is not so. This means that we can partition $X = X_1 \cup X_2 \cup X_3$, where $X_1 = \{(r_1, h_1), (r_8, h_1)\}$, $X_2 = \{(r_2, h_4)\}$, and $X_3 = \{(r_6, h_2), (r_7, h_2)\}$. If we remove a pair $(r_i, h_j) \in X_t (t = 1, 2, 3)$ that (r_i, h_j) dominates all the other $(r_k, h_j) \in X_t$ from the hospitals' point of view, then all (r_k, h_j) formed by h_j from the hospitals' point of view are removed for M.

As with the analysis above, our idea to solve HRT is that at each iteration of our algorithm, we do the following: (i) finding a set X of UBPs for an unstable matching M from the residents' point of view; (ii) partitioning $X = X_1 \cup X_2 \cup \cdots \cup X_l$ such that each $X_t (t = 1, 2, \cdots, l)$ consists of blocking pairs $(r_i, h_j) \in X$ formed by a unique $h_j \in X$; and (iii) removing a pair $(r_i, h_j) \in X_t (t = 1, 2, \cdots, l)$ that (r_i, h_j) dominates all the other $(r_k, h_j) \in X_t$ from the hospitals' point of view. By doing so, our idea is not to remove all the blocking pairs formed by r_i from the residents' point of view but also reject as many blocking pairs formed by h_j from the hospitals' point of view as possible to obtain a stable matching of an HRT instance as quickly as possible.

Our algorithm is shown in Algorithm 1. To avoid getting stuck in local maxima, we use the mechanism of the random-restart hill climbing algorithm [19]. Specifically, our algorithm finds a maximum stable matching, denoted by M_{best}, from a randomly generated matching M. At each iteration, our algorithm runs as follows. First, the algorithm finds a set X of UBPs for M from the residents' point of view (line 4). Second, the algorithm checks if X is empty, then if M_{best} is worse than M in terms of the matching size, M is assigned to M_{best} (lines 6-8). Next, the algorithm checks if M_{best} is perfect, then it returns M_{best} (lines 9-11), otherwise, it restarts at a randomly generated matching M and continues the next iteration (lines 12-13). Third, the algorithm checks if a small probability of p is accepted, it chooses a random pair $(r_i, h_j) \in X$ and removes it for M (lines 15-22). Otherwise, it iterates for each $h_j \in X$ to select a pair $(r_i, h_j) \in X$ that h_j prefers r_i to r_k for all $(r_k, h_j) \in X$ and removes (r_i, h_j) for M (lines 24-25). When the algorithm removes a blocking pair (r_i, h_j) for M, i.e. $M(r_i) = h_j$, and if h_j is over-subscribed, then it removes the pair $(r_z, h_j) \in M$ such that h_j is full, where r_z is the worst resident assigned to h_j in M (lines 26-29). Finally, the algorithm repeats until either M_{best} is a perfect matching or a maximum number of iterations is reached. In the latter case, the algorithm returns either a maximum stable matching found so far or an unstable matching. We note that to find an UBP $(r_i, h_j) \in X$ from the residents' point of view for M, the algorithm runs an iteration for each hospital h_j in ascending order of ranks in r_i's preference list and returns the first blocking pair encountered, then (r_i, h_j) is an undominated blocking pair.

An execution of our algorithm for the HRT instance shown in Table 1 is illustrated as in Table 2. We assume that the probability to choose a random pair in X is $p = 0$ and the algorithm starts from a random matching $M_0 = \{(r_1, \varnothing), (r_2, \varnothing), (r_3, h_1), (r_4, h_1), (r_5, h_3), (r_6, h_1), (r_7, h_3), (r_8, \varnothing)\}$. At the first iteration, the algorithm finds a set $X_0 = \{(r_1, h_1), (r_2, h_4), (r_6, h_2), (r_7, h_2), (r_8, h_1)\}$

Algorithm 1: Heuristic Repair Algorithm

Input: - An HRT instance I of size $n \times m$
- A small probability p.
- The maximum iterations max_iters.

Output: A matching M_{best}.

1. $M :=$ a randomly generated matching;
2. $M_{best} := M$;
3. **for** $iter := 1$ **to** max_iters **do**
4. $X :=$ a set of undominated blocking pairs for M;
5.   **if** $(X = \emptyset)$ **then**
6.   **if** $(|M_{best}| < |M|)$ **then**
7. $M_{best} := M$;
8.   **end**
9.   **if** $(|M_{best}| = n)$ **then**
10.   break;
11.   **end**
12. $M :=$ a randomly generated matching;
13.   continue;
14.   **end**
15.   **if** *(a small probability of p)* **then**
16.   take a random pair $(r_i, h_j) \in X$;
17. $M(r_i) := h_j$;
18.   **if** $(h_j$ *is over-subscribed* $)$ **then**
19. $r_z :=$ worst resident in $M(h_j)$;
20. $M(r_z) := \varnothing$;
21.   **end**
22.   **else**
23.   **for** *(each* $h_j \in X)$ **do**
24.   select $(r_i, h_j) \in X$ such that h_j prefers r_j to r_k, $\forall(r_k, h_j) \in X$;
25. $M(r_i) := h_j$;
26.   **if** $(h_j$ *is over-subscribed* $)$ **then**
27. $r_z :=$ worst resident in $M(h_j)$;
28. $M(r_z) := \varnothing$;
29.   **end**
30.   **end**
31.   **end**
32. **end**
33. **return** M_{best};

---

of UBPs from the residents' point of view for M_0. Since (r_8, h_1) dominates (r_1, h_1) from the hospitals' point of view (i.e. h_1 prefers r_8 to r_1) and (r_6, h_2) dominates (r_7, h_2) from the hospitals' point of view (i.e. h_2 prefers r_6 to r_7), the algorithm removes (r_2, h_4), (r_6, h_2) and (r_8, h_1) to obtain a matching $M_1 = \{(r_1, \varnothing), (r_2, h_4), (r_3, h_1), (r_4, \varnothing), (r_5, h_3), (r_6, h_2), (r_7, h_3), (r_8, h_1)\}$. At the second iteration, the algorithm finds a set $X_1 = \{(r_1, h_1), (r_4, h_1), (r_7, h_2)\}$ of UBPs from the residents' point of view for M_1. It should be noted that at the first iteration, (r_8, h_1) dominated (r_1, h_1) and we removed (r_8, h_1) for M_0, but there exists $(r_1, h_1) \in X_1$ for M_1, since $(r_1, h_1) \in X_1$ is an UBP found from the residents' point of view. It is explained similarly for $(r_7, h_2) \in X_1$. Since (r_1, h_1) dominates (r_4, h_1) from the hospitals' point of view, the algorithm removes (r_1, h_1) and (r_7, h_2) to obtain a matching M_2. The algorithm repeats until the fourth iteration, where $X_3 = \{\emptyset\}$, and it returns a perfect matching M_3.

Table 2. An execution of the algorithm for HRT in Table 1

Iter.	Input	UBPs	Remove	Output
1	M_0	$X_0 = \{(r_1, h_1),$ $(r_2, h_4), (r_6, h_2),$ $(r_7, h_2), (r_8, h_1)\}$	$\{(r_2, h_4),$ $(r_6, h_2),$ $(r_8, h_1)\}$	$M_1 = \{(r_1, \varnothing), (r_2, h_4), (r_3, h_1),$ $(r_4, \varnothing), (r_5, h_3), (r_6, h_2), (r_7, h_3),$ $(r_8, h_1)\}$
2	M_1	$X_1 = \{(r_1, h_1),$ $(r_4, h_1), (r_7, h_2)\}$	$\{(r_1, h_1),$ $(r_7, h_2)\}$	$M_2 = \{(r_1, h_1), (r_2, h_4), (r_3, h_1),$ $(r_4, \varnothing), (r_5, h_3), (r_6, h_2), (r_7, h_2),$ $(r_8, h_1)\}$
3	M_2	$X_2 = \{(r_4, h_4)\}$	$\{(r_4, h_4)\}$	$M_3 = \{(r_1, h_1), (r_2, h_4), (r_3, h_1),$ $(r_4, h_4), (r_5, h_3), (r_6, h_2), (r_7, h_2),$ $(r_8, h_1)\}$
4	M_3	$X_3 = \{\emptyset\}$		

4   Experiments

In this section, we evaluate the performance of our heuristic repair algorithm, namely HR, for HRT. To do this, we applied the SMTI generator [6] to generate HRT instances with parameters (n, m, p_1, p_2), where n is the number of residents, m is the number of hospitals, p_1 is the probability of incompleteness, and p_2 is the probability of ties. Without loss of generality, we assume that in each generated instance, the preference lists of residents and hospitals consist of acceptance pairs. Otherwise, we run a preprocessing procedure to remove unacceptance pairs in HRT instances. We implemented all experiments by Matlab 2019a on a personal computer with a Core i7-8550U CPU 1.8GHz and 16 GB memory.

4.1   Comparison with Local Search

In this section, we present an experiment to compare the execution time and solution quality found by HR with those found by Local Search (LS) [4]. We set the probability $p = 0.03$ and the maximum number of iterations to 500 in both HR and LS algorithms.

Experiment 1. We chose $n = 100$, $m = 10$, $p_1 \in [0.1, 0.8]$ with step 0.1, and $p_2 \in [0.0, 1.0]$ with step 0.1. For each combination of parameters (n, m, p_1, p_2), we randomly generated 100 HRT instances, in which the capacity c_j of each hospital $h_j \in \mathcal{H}$ is generated randomly and $c_j \in [1, q]$, where q is the total number of residents ranked by hospital $h_j \in \mathcal{H}$. Then, we ran HR, LS and averaged results. Figure 1(a) shows the percentage of perfect matchings found by HR and LS. When p_1 varies from 0.1 to 0.4, both HR and LS always find 100% of perfect matchings (therefore, they are not depicted in Fig. 1(a)), while p_1 varies from 0.5 to 0.8, the percentage of perfect matchings found by HR is slightly higher than that found by LS. Figure 1(b) shows the average execution time of HR and LS. The experimental results show that HR runs about 100 times faster than LS for any p_1 and p_2. On average, the execution time of HR increases from about 0.008(s) to 0.02(s), while that of LS increases from about 0.5(s) to 43.5(s) for any value of p_2. In contrast, when p_2 varies from 0.0 to 1.0, the execution time of both HR and LS decreases slightly for any value of p_1. This can be explained as follows. Although LS considers only UBPs, the number of such UBPs is very large, i.e. the number of neighbor matchings is very large because a neighbor is generated by removing a blocking pair in the set of UBPs. This increases significantly the execution time of LS. However, HR finds the set of UBPs and removes many blocking pairs in the set of UBPs to generate a new matching for the next iteration without evaluating the cost of matchings as in LS and therefore, HR runs much faster than LS.

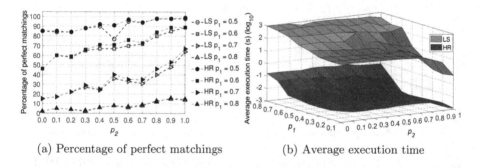

(a) Percentage of perfect matchings   (b) Average execution time

Fig. 1. Comparing solution quality and execution time of HR and LS algorithms

4.2   Experiments for HRT of Large Sizes

In this section, we present experimental results for HRT instances of large sizes to consider the behavior of our algorithm. We set $p = 0.03$ and $max\_iters = 1000$ in HR.

Experiment 2. We chose $n = 1000$, $m = 50$, $p_1 \in [0.1, 0.8]$ with step 0.1, and $p_2 \in [0.0, 1.0]$ with step 0.1. For each combination of parameters (n, m, p_1, p_2), we randomly generated 100 HRT instances, in which c_j of each hospital $h_j \in \mathcal{H}$ is generated randomly and $c_j \in [1, q]$, where q is the total number of residents ranked by hospital $h_j \in \mathcal{H}$. Our experimental results show that when $p_1 = 0.8$, HR finds 98% of perfect matchings for $p_2 \in \{0.0, 0.2, 0.3\}$ and 99% of perfect matchings for $p_2 \in \{0.4, 0.8\}$. For the remaining values of p_1 and p_2, HR finds 100% of perfect matchings. Figure 2(a) shows the average capacity in generated instances. For each $p_1 \in [0.1, 0.8]$, the average capacity of hospitals is about $0.5n(1-p_1)$ residents (i.e. from 450 residents to 100 residents). When p_2 increases from 0.0 to 1.0, the average capacity of hospitals remains unchanged. When $p_1 = 0.8$, meaning that h_j has the smallest capacity c_j, and therefore some instances may have no perfect matchings and HR cannot find perfect matchings for these instances. Figure 2(b) shows the average number of iterations used by HR. When p_1 increases from 0.1 to 0.8, the number of iterations used by HR slightly decreases. When p_2 increases from 0.0 to 0.9, the number of iterations used by HR increases. However, when $p_2 = 1.0$, the number of iterations used by HR decreases rapidly because the probability of ties is 100%, meaning that the ranks of hospitals in residents' preference lists are the same. Therefore, HR only considers the first accepted hospital instead of all hospitals in order to find an UBP from the resident's point of view. We can see that although the generated instances have large sizes, HR used a small number of iterations, about 40 to 100, to find perfect matchings.

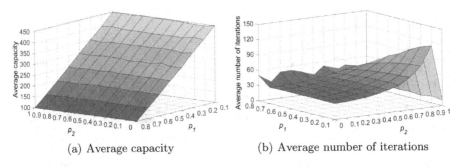

(a) Average capacity   (b) Average number of iterations

Fig. 2. Average capacity of instances and average number of iterations used by HR for $n = 1000$ and $m = 50$

Experiment 3. In this experiment, we chose $n \in [100, 1000]$ with step 100, $m \in [10, 50]$ with step 5, $p_1 = 0.5$, and $p_2 = 0.5$. For each combination of parameters (n, m, p_1, p_2), we randomly generated 100 HRT instances, in which the capacity of each hospital is chosen as in Experiment 2. Figure 3(a) shows the percentage of perfect matchings found by HR. We see that when $m \in [20, 50]$, HR always finds 100% of perfect matchings. When $m = 10$ and n increases from 100 to 1000, HR finds about from 85% down to 47% of perfect matchings, respectively, and the number of unassigned residents in stable matchings is about from 1 to 2 unassigned residents as shown in Fig. 3(b). When $m = 15$, HR finds about 98% of perfect matchings for all values of $n \in [100, 1000]$.

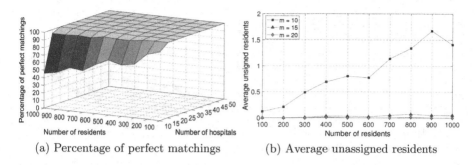

(a) Percentage of perfect matchings   (b) Average unassigned residents

Fig. 3. Percentage of perfect matchings and average unassigned residents for $n \in [100, 1000]$ and $m \in [10, 50]$

Experiment 4. In this last experiment, we evaluated the effect of capacities of hospitals on perfect matchings found by HR. To do this, we chose the values of n, m, p_1 and p_2 as in Experiment 3. We changed the capacity of each hospital as follows.

First, we considered a popular case, where $c_j = n/m$, meaning that the total capacity of hospitals is equal to the number of residents. The experimental results, depicted in Fig. 4(a), show that HR finds 90% of perfect matchings for $n \in [100, 1000]$ and $m \in [20, 50]$. When $m = 10$, HR finds about from 85% (at $n = 100$) down to 1% (at $n = 1000$) of perfect matchings. Figure 4(b) shows the average execution time found by HR. When m increases from 20 to 50 and n increases from 100 to 1000, the execution time found by HR increases about from 0.01(s) to 1.5(s). However, when $m = 10$ and n increases from 100 to 1000, the execution time found by HR increases about from 0.02(s) to 4.5(s), since the percentage of perfect matchings found by HR decreases, meaning that HR used many iterations to find perfect matchings for generated instances.

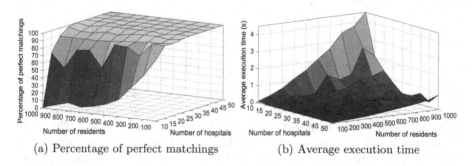

(a) Percentage of perfect matchings   (b) Average execution time

Fig. 4. Percentage of perfect matchings and average execution time found by HR, where $c_j = n/m$

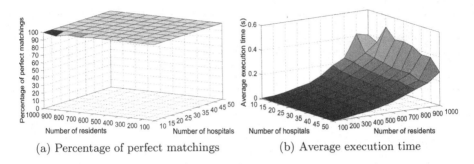

(a) Percentage of perfect matchings   (b) Average execution time

Fig. 5. Percentage of perfect matchings and average execution time found by HR, where $c_j = [0.2q, 0.6q]$

Second, we randomly generated $c_j \in [0.2q, 0.6q]$, where q is the total number of residents ranked by hospital $h_j \in \mathcal{H}$. This means that each hospital ranks about 50% of residents (since $p_1 = 0.5$), but selects only about from 10% to 30% of the total of ranked residents. Figure 5(a) shows that when $(n, m) = (700, 10)$, HR finds 99% of perfect matchings, and when $(n, m) = (900, 10)$, HR finds 97% of perfect matchings. For the remaining values of n and m, HR finds 100% of perfect matchings. In this case, the percentage of perfect matchings found by HR is higher than that when $c_j = n/m$, meaning that the capacity for each hospital strongly affects the solution quality of HRT. Figure 5(b) shows the average execution time found by HR. When n increases from 100 to 1000, the average execution time of HR increases only about from 0.01(s) to 0.4(s). We see that when $n = 1000$ and $m \in [10, 50]$, the execution time of HR is very small, about 0.4(s), meaning that HR is efficient for solving HRT instances of large sizes.

5   Conclusions

In this paper, we proposed a heuristic repair algorithm to solve HRT. The algorithm starts to search a solution of the problem from a random matching. At each iteration, the algorithm finds a set of undominated blocking pairs from the residents' point of view for the matching. Then, the algorithm removes the best undominated blocking pair for each hospital such that it does not only remove many blocking pairs from the residents' of view as possible but also removes as many blocking pairs as possible from the hospitals' point of view. The algorithm repeats until it finds a perfect matching or reaches a maximum number of iterations. Experiments showed that our algorithm is efficient in terms of execution time and solution quality for HRT of large sizes. In the future, we plan to extend this approach to find strongly stable matchings or super-stable matchings for HRT [8,9].

References

1. Canadian resident matching service (CaRMS). http://www.carms.ca/
2. Biró, P., Manlove, D.F., McBride, I.: The hospitals/residents problem with couples: Complexity and integer programming models. In: Proceeding of SEA 2014: 13th International Symposium on Experimental Algorithms, Copenhagen, Denmark, pp. 10–21 (June 2014)
3. Gale, D., Shapley, L.S.: College admissions and the stability of marriage. Am. Math. Mon. **9**(1), 9–15 (1962)
4. Gelain, M., Pini, M.S., Rossi, F., Venable, K.B., Walsh, T.: Local search for stable marriage problems with ties and incomplete lists. In: Proceedings of 11th Pacific Rim International Conference on Artificial Intelligence, Daegu, Korea, pp. 64–75 (August 2010)
5. Gelain, M., Pini, M.S., Rossi, F., Venable, K.B., Walsh, T.: Local search approaches in stable matching problems. Algorithms **6**(4), 591–617 (2013)
6. Gent, I.P., Prosser, P.: An empirical study of the stable marriage problem with ties and incomplete lists. In: Proceedings of the 15th European Conference on Artificial Intelligence, Lyon, France, pp. 141–145 (July 2002)
7. Irving, R.W.: Matching medical students to pairs of hospitals: A new variation on a well-known theme. In: Proceedings of ESA 1998: the 6th Annual European Symposium, Venice, Italy, pp. 381–392 (August 1998)
8. Irving, R.W., Manlove, D.F., Scott, S.: The hospitals/residents problem with ties. In: Proceedings of the 7th Scandinavian Workshop on Algorithm Theory, Bergen, Norway, pp. 259–271 (July 2000)
9. Irving, R.W., Manlove, D.F., Scott, S.: Strong stability in the hospitals/residents problem. In: Alt, H., Habib, M. (eds.) STACS 2003. LNCS, vol. 2607, pp. 439–450. Springer, Heidelberg (2003). https://doi.org/10.1007/3-540-36494-3_39
10. Iwama, K., Miyazaki, S., Morita, Y., Manlove, D.: Stable marriage with incomplete lists and ties. In: Proceedings of International Colloquium on Automata, Languages, and Programming, Prague, Czech Republic, pp. 443–452 (July 1999)
11. Király, Z.: Linear time local approximation algorithm for maximum stable marriage. Algorithms **6**(1), 471–484 (2013)
12. Kwanashie, A., Manlove, D.F.: An integer programming approach to the hospitals/residents problem with ties. In: Proceedings of the International Conference on Operations Research, pp. 263–269. Erasmus University Rotterdam (September 2013)
13. Manlove, D.: Algorithmics of Matching Under Preferences, vol. 2. World Scientific (2013)
14. Manlove, D.F., Irving, R.W., Iwama, K., Miyazaki, S., Morita, Y.: Hard variants of stable marriage. Theoret. Comput. Sci. **276**(1–2), 261–279 (2002)
15. Manlove, D.F., McBride, I., Trimble, J.: "Almost-stable" matchings in the hospitals / residents problem with couples. Constraints **22**(1), 50–72 (2017)
16. Munera, D., Diaz, D., Abreu, S., Rossi, F., Saraswat, V., Codognet, P.: A local search algorithm for SMTI and its extension to HRT problems. In: Proceedings of the 3rd International Workshop on Matching Under Preferences, pp. 66–77. University of Glasgow, UK (April 2015)
17. Munera, D., Diaz, D., Abreu, S., Rossi, F., Saraswat, V., Codognet, P.: Solving hard stable matching problems via local search and cooperative parallelization. In: Proceedings of the Twenty-Ninth AAAI Conference on Artificial Intelligence, Austin, Texas, pp. 1212–1218 (January 2015)

18. Roth, A.E.: The evolution of the labor market for medical interns and residents: A case study in game theory. J. Polit. Econ. **92**(6), 991–1016 (1984)
19. Russel, S., Norvig, P.: Artificial Intelligence: A Modern Approach, 3rd edn. Prentice Hall Press, Upper Saddle River (2009)

The Impact of Data Preprocessing on Prediction Effectiveness

Adam Kiersztyn[1]($^{(\boxtimes)}$) and Krystyna Kiersztyn[2]

[1] Department of Computer Science, Lublin University of Technology, Lublin, Poland
a.kiersztyn@pollub.pl, adam.kiersztyn.pl@gmail.com
[2] Department of Mathematical Modelling, The John Paul II Catholic University of Lublin, Lublin, Poland
krystyna.kiersztyn@gmail.com

Abstract. This study considers a very important issue, which is the impact of preprocessing on model performance. On the example of data describing taxicab trips in New York City, a model predicting the average speed of a trip was built. The effectiveness of the obtained model was examined using relative error. The results were compared with the models obtained after prior data cleaning from the records containing missing data. Additionally, the effect of removing outliers on model quality was examined. An integral part of the paper is the description of a new method of anomaly detection. The author's method involves fuzzy classification of the declared distance into three classes. As an indicator to allow for classification, the percentage of redundant distance with respect to Manhattan distance was selected. The results of a wide range of numerical experiments confirm the necessity of preprocessing. Comparison of a number of competing anomaly detection and prediction model building methods allows for reasonable generalization of the obtained conclusions. Additionally, the skillful use of fuzzy sets for anomaly detection allowed the development of a method that can be generalized to other transportation issues.

Keywords: Preprocessing · Prediction model · Outlier detection · Anomaly detection · Filling gaps

1 Introduction

A key component of the prediction process, as well as other advanced data analysis methods, is proper preprocessing [2,9,25,27]. The most important steps in data preparation include removing/filling data gaps [15,19,20,28], and detecting outliers and anomalies [10–12,14,17,26].

Most methods in the prediction process build models [3,23] that have coincide to varying degrees with empirical values. A number of model building methods and techniques are considered [4,6,8,21,22]. The performance differences

The work was co-financed by the Lublin University of Technology Scientific Fund: FD-20/IT-3/002.

between the methods are comparable, and in many cases, the quality of the model is determined by the training set [16].

In this study, a novel approach for detecting anomalies in transportation data is presented. Using operations on fuzzy sets, a method allowing the detection of anomalies in taxicab trip distances is proposed. The method enables the identification of courses that are, with a high probability, an attempt to cheat the customers or the taxicab corporation. In the process of anomaly detection the author's fuzzy modification of the three sigma rule is used [13,14].

Additionally, the effect of preprocessing on the effectiveness of taxicab travel speed prediction is presented. A series of numerical experiments were conducted demonstrating the impact of data preprocessing on the effectiveness of the travel time prediction model.

The work is organised as follows. Section 2 provides a theoretical description of the proposed method. In the next Sect. 3, the results of numerical experiments are presented. Finally, Sect. 4 contains conclusions and future work directions.

2   Prerequisites

For travel in an urban area, the route length is usually longer than a straight line connecting the beginning and end of the trip. When considering trip distance, it is important to take into account the nature of the city and the arrangement of the streets with each other. In the case of newly developed U.S. cities, we are usually dealing with a grid of streets and blocks. In this case, it is advisable to use a dedicated taxicab metric (1), also called the Manhattan distance.

$$d_M\left((x_1,y_1),(x_2,y_2)\right) = |x_1 - x_2| + |y_1 - y_2| \tag{1}$$

If the declared (empirical) trip distance d_e significantly deviates from the value determined by the Manhattan metric d_M, we can assume that there is an anomaly. As an indicator to determine the level of redundant distance can be one given by Eq. (2).

$$\delta = \frac{d_e - d_M}{d_e} \tag{2}$$

The proposed novel indicator differs from the classical relative error in that it omits an absolute value. This is a deliberate action dictated by the fact that in some cases the declared trip distance is smaller than the value obtained using the Manhattan metric. In order to determine the direction and degree of deviation from the norm, fuzzy descriptors describing the three phenomena were proposed. The membership functions for each descriptor are given by the formulas (3), (4), (5).

$$\mu_{low}(\delta) = \begin{cases} 1, \delta \in (-\infty; 0) \\ \frac{0.05-\delta}{0.05}, \delta \in [0; 0.05] \\ 0, \delta \in (0.05; \infty) \end{cases} \tag{3}$$

$$\mu_{normal}(\delta) = \begin{cases} 0, \delta \in (-\infty; 0) \vee (1; \infty) \\ \frac{\delta}{0.05}, \delta \in [0; 0.05] \\ 1, \delta \in (0.05; 0.5] \\ \frac{1-\delta}{0.5}, \delta \in (0.5; 1] \end{cases} \tag{4}$$

$$\mu_{hight}(\delta) = \begin{cases} 0, \delta \in (-\infty; 0.5) \\ \frac{\delta - 0.5}{0.5}, \delta \in [0.5; 1] \\ 1, \delta \in (1; \infty) \end{cases} \tag{5}$$

The membership degree to particular descriptors determines the anomaly level of a given trip. The membership degree to particular descriptors determines the anomaly level of a given trip. If the degree of membership in the "high" descriptor is dominant, then it is most probably an attempt to defraud the client and deliberately extend the trip distance by the driver. On the other hand, if the membership degree of the "low" descriptor is prevalent, it can be assumed that it is an attempt to defraud the cab company or the tax authorities.

3 Numerical Experiments

The numerical experiments were limited to the analysis of a subset of data describing single taxicab trips in New York City [7]. For the clarity of the presented results, the analysis was limited to 100,000 randomly selected trips. Even a brief analysis of the considered data allows us to conclude that the data are not free from outliers. This is clearly evidenced by the visualization of points representing the beginnings of individual trips (cf. Fig. 1).

In many cases, the beginning and the end coordinates of the trip were incorrectly recorded as points with coordinates (0,0). In the analyzed dataset, there were 1953 such entries. Additionally, in a number of cases, there were missing data in trip distance and time or it was incorrectly indicated that these values were equal to zero. 592 such items were found. A novel anomaly detection method presented in the theoretical section was used to detect more sophisticated anomalies involving unreasonably declared trip distances (Fig. 2).

It is easy to observe that the coordinates of the start of a trip do not predefine whether a trip is suspected of being anomalous. It turns out that a significant proportion of the anomaly elements are those that fall into the "low" category with the highest degree. Moreover, it appears that more than one-third of the trips fall into this category to a degree that exceeds the membership degree in the "normal" category.

For the purpose of testing the effect of preprocessing on model quality, it was proposed to consider three well-known prediction methods, namely Gradient Boosted Tree (GBT), Tree Ensamble (TE) and Random Forest (RF). Based on each of these methods, an average travel speed model is created.

As part of the numerical experiments, the available dataset was randomly divided into a training and a testing set. Different number of elements (in percentage) were assigned to the training set. The quality of the model was examined on the elements of the training set by calculating the relative error. For each

Fig. 1. The location of the beginning of the trips

percentage of elements in the training set, the entire experiment was independently repeated 10 times. The results of basic relative error statistics for different methods and different sizes of the training set are shown in Table 1.

By analyzing the results presented in Table 1, it can be observed that the smallest error of just over 2% is obtained using the GBT method. However, it should be noted that this method has a tendency to overtraining. In the case of too large training set (at the level of 80 and 90% of elements) we obtain worse values of statistics than for the model built on the basis of 70% of available elements. For other methods, the value of median and quartiles decreases as the number of elements in the training set increases. This is a correct relation which proves stability of a given method.

After applying a simple preprocessing based on rejecting elements with incorrect start and end trip coordinates, we obtain slightly different error statistics for each method (cf. Table 2).

It can be clearly observed that after discarding the invalid data, the TE method has gained the most. By comparing the median values, a slight improvement of the model can be noted. Additionally, if records with zero values for

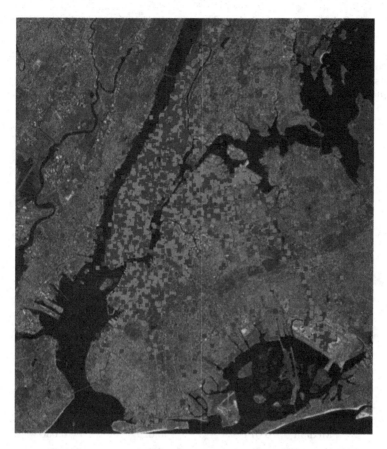

Fig. 2. Visualization of normal (green) and abnormal (red) records (Color figure online)

travel time and travel distance are removed (see Table 3), the effectiveness of the model increases when using 80% of the elements in the training set. At the same time, at 90%, one can observe an overtraining of the analysed methods.

Simple preprocessing methods consisting of removing records with incomplete data and records with obvious incorrect values can slightly improve the quality of the model. Additionally, if outliers are taken into account in the analysis, it is possible to significantly improve individual models.

Four popular outlier detection methods, namely Elliptical Envelope (EE) [24], Gaussian Mixture (GM) [1], Isolation Forest (IF) [18], and Local Outlier Factor (LOF) [5], were proposed for outlier detection. Based on the results obtained from each method, 1% of the elements with the highest degree of anomaly were rejected. For the remaining elements, a split was made between the training and testing set. Table 4 presents the basic relative error statistics with 90% of the elements in the training set.

Table 1. Relative error statistics when considering all records in the dataset

The percentage of elements in the training set	Statistic	GBT relative error	TE relative error	RF relative error
90	Average	0.024	0.062	0.063
	Median	0.014	0.024	0.025
	Quartile 1	0.007	0.010	0.011
	Quartile 3	0.025	0.051	0.053
80	Average	0.024	0.063	0.061
	Median	0.013	0.027	0.026
	Quartile 1	0.006	0.011	0.012
	Quartile 3	0.025	0.057	0.056
70	Average	0.024	0.067	0.064
	Median	0.013	0.028	0.027
	Quartile 1	0.006	0.012	0.011
	Quartile 3	0.024	0.060	0.057
60	Average	0.025	0.067	0.067
	Median	0.014	0.029	0.028
	Quartile 1	0.007	0.012	0.012
	Quartile 3	0.026	0.060	0.061
50	Average	0.025	0.071	0.074
	Median	0.014	0.029	0.030
	Quartile 1	0.006	0.013	0.013
	Quartile 3	0.025	0.062	0.064

The analysis of the values of the basic statistics in Table 4 leads to very interesting conclusions. Undoubtedly, removing outliers significantly increases the efficiency of the model. The differences between the individual ones are noticeable especially in the case of the mean, which we know to be very sensitive to outliers. The median values for individual outlier detection methods strongly confirm that proper preprocessing is an essential part of model preparation.

Applying the author's method of detecting anomalies in trip distance also improves the quality of the model (cf. Table 5). Skillful detection of anomalies using fuzzy techniques allowed to increase the stability of the prediction. It should be noted that 3/4 of the modeled values have a relative error smaller than 0.00051, which is practically negligible.

Table 2. Relative error statistics for dropping records with zero start or end trip coordinates

The percentage of elements in the training set	Statistic	GBT relative error	TE relative error	RF relative error
90	Average	0.025	0.061	0.061
	Median	0.013	0.024	0.024
	Quartile 1	0.006	0.010	0.010
	Quartile 3	0.025	0.051	0.053
80	Average	0.025	0.065	0.063
	Median	0.014	0.027	0.026
	Quartile 1	0.007	0.012	0.011
	Quartile 3	0.026	0.057	0.054

Table 3. Relative error statistics when considering only records with both positive time and trip distance values

The percentage of elements in the training set	Statistic	GBT relative error	TE relative error	RF relative error
90	Average	0.029	0.064	0.063
	Median	0.014	0.025	0.024
	Quartile 1	0.006	0.011	0.011
	Quartile 3	0.026	0.053	0.051
80	Average	0.026	0.063	0.062
	Median	0.014	0.026	0.025
	Minimum	0	0	0
	Quartile 1	0.006	0.011	0.011
	Quartile 3	0.025	0.055	0.055

Table 4. Relative error statistics after removing the top 1% outliers within each outlier detection method

Method	Statistic	GBT relative error	TE relative error	RF relative error
EE	Average	0.00162	0.00590	0.00640
	Median	0.00024	0.00071	0.00074
	Quartile 1	0.00006	0.00013	0.00015
	Quartile 3	0.00075	0.00293	0.00305
GM	Average	0.00492	0.00196	0.00225
	Median	0.00018	0.00009	0.00008
	Quartile 1	0.00004	0.00001	0.00001
	Quartile 3	0.00062	0.00048	0.00047
IF	Average	0.00680	0.00875	0.00882
	Median	0.00013	0.00083	0.00081
	Quartile 1	0.00003	0.00016	0.00016
	Quartile 3	0.00042	0.00338	0.00348
LOF	Average	0.03399	0.02754	0.02477
	Median	0.00016	0.00117	0.00114
	Quartile 1	0.00004	0.00022	0.00021
	Quartile 3	0.00055	0.00488	0.00483

Table 5. Relative error statistics of models for records where the membership degree of the "normal" class exceeds 0.5

The percentage of elements in the training set	Statistic	GBT relative error	TE relative error	RF relative error
90	Average	0.00338	0.00973	0.00967
	Median	0.00015	0.00063	0.00065
	Quartile 1	0.00003	0.00013	0.00013
	Quartile 3	0.00051	0.00272	0.00295
80	Average	0.01180	0.01011	0.00920
	Median	0.00017	0.00063	0.00061
	Quartile 1	0.00004	0.00012	0.00011
	Quartile 3	0.00053	0.00273	0.00261

4 Conclusions and Future Work

In a series of numerical experiments, it was shown that proper preprocessing is an essential part of the process of building predictive models. Comparison of selected prediction methods confirms the fact that detection and removing outliers improves model quality. Additionally, appropriate reaction to obviously incorrect values and missing data allows to increase model efficiency.

The proposed innovative method of detecting anomalies in the declared trip distance makes it possible to catch unusual dependencies. These relationships cannot be discovered by typical outlier detection methods. The use of fuzzy descriptors describing selected categories of relationships between the declared trip distance and the distance on the map allows to significantly improve the efficiency of the model estimating the average trip speed.

It is planned to conduct analyses using more descriptors describing a larger number of features. Moreover, it is intended to verify the conclusions on a larger number of databases and to transfer the obtained results to other related issues.

References

1. Aitkin, M., Wilson, G.T.: Mixture models, outliers, and the EM algorithm. Technometrics **22**(3), 325–331 (1980)
2. Alasadi, S.A., Bhaya, W.S.: Review of data preprocessing techniques in data mining. J. Eng. Appl. Sci. **12**(16), 4102–4107 (2017)
3. Arabameri, A., Pradhan, B., Rezaei, K., Sohrabi, M., Kalantari, Z.: Gis-based landslide susceptibility mapping using numerical risk factor bivariate model and its ensemble with linear multivariate regression and boosted regression tree algorithms. J. Mt. Sci. **16**(3), 595–618 (2019)
4. Berthold, M.R.: Mixed fuzzy rule formation. Int. J. Approx. Reason. **32**(2–3), 67–84 (2003)
5. Breunig, M.M., Kriegel, H.P., Ng, R.T., Sander, J.: LOF: identifying density-based local outliers. In: Proceedings of the 2000 ACM SIGMOD International Conference on Management of Data, pp. 93–104 (2000). https://doi.org/10.1145/342009.335388
6. Coppersmith, D., Hong, S.J., Hosking, J.R.: Partitioning nominal attributes in decision trees. Data Min. Knowl. Discov. **3**(2), 197–217 (1999)
7. Donovan, B., Work, D.: New York city taxi trip data (2010–2013) (2014). https://doi.org/10.13012/J8PN93H8
8. Friedman, J.H.: Stochastic gradient boosting. Comput. Stat. Data Anal. **38**(4), 367–378 (2002)
9. Kamiran, F., Calders, T.: Data preprocessing techniques for classification without discrimination. Knowl. Inf. Syst. **33**(1), 1–33 (2012)
10. Karczmarek, P., Kiersztyn, A., Pedrycz, W.: Fuzzy set-based isolation forest. In: 2020 IEEE International Conference on Fuzzy Systems (FUZZ-IEEE), pp. 1–6. IEEE (2020)
11. Karczmarek, P., Kiersztyn, A., Pedrycz, W., Al, E.: K-means-based isolation forest. Knowl.-Based Syst. **195**, 105659 (2020)
12. Karczmarek, P., Kiersztyn, A., Pedrycz, W., Czerwiński, D.: Fuzzy c-means-based isolation forest. Appl. Soft Comput. **106**, 107354 (2021)
13. Kiersztyn, A., Karczmarek, P., Kiersztyn, K., Pedrycz, W.: The concept of detecting and classifying anomalies in large data sets on a basis of information granules. In: 2020 IEEE International Conference on Fuzzy Systems (FUZZ-IEEE), pp. 1–7. IEEE (2020)

14. Kiersztyn, A., Karczmarek, P., Kiersztyn, K., Pedrycz, W.: Detection and classification of anomalies in large data sets on the basis of information granules. IEEE Trans. Fuzzy Syst. **30**(8), 2850–2860 (2021)
15. Kiersztyn, A., et al.: Data imputation in related time series using fuzzy set-based techniques. In: 2020 IEEE International Conference on Fuzzy Systems (FUZZ-IEEE), pp. 1–8. IEEE (2020)
16. Kiersztyn, A., et al.: A comprehensive analysis of the impact of selecting the training set elements on the correctness of classification for highly variable ecological data. In: 2021 IEEE International Conference on Fuzzy Systems (FUZZ-IEEE), pp. 1–6. IEEE (2021)
17. Kiersztyn, K.: Intuitively adaptable outlier detector. Stat. Anal. Data Min.: ASA-Data Sci. J. **15**(4), 463–479 (2021)
18. Liu, F.T., Ting, K.M., Zhou, Z.H.: Isolation-based anomaly detection. ACM Trans. Knowl. Discov. Data **6**(1) (2012). https://doi.org/10.1145/2133360.2133363
19. Łopucki, R., Kiersztyn, A., Pitucha, G., Kitowski, I.: Handling missing data in ecological studies: ignoring gaps in the dataset can distort the inference. Ecol. Modell. **468**, 109964 (2022)
20. Osman, M.S., Abu-Mahfouz, A.M., Page, P.R.: A survey on data imputation techniques: water distribution system as a use case. IEEE Access **6**, 63279–63291 (2018)
21. Piironen, J., Vehtari, A.: Comparison of Bayesian predictive methods for model selection. Stat. Comput. **27**(3), 711–735 (2017)
22. Priyanka, K.D.: Decision tree classifier: a detailed survey. Int. J. Inf. Decis. Sci. **12**(3), 246–269 (2020)
23. Raval, K.M.: Data mining techniques. Int. J. Adv. Res. Comput. Sci. Softw. Eng. **2**(10) (2012)
24. Rousseeuw, P.J., Driessen, K.V.: A fast algorithm for the minimum covariance determinant estimator. Technometrics **41**(3), 212–223 (1999). https://doi.org/10.1080/00401706.1999.10485670
25. Vijayarani, S., Ilamathi, M.J., Nithya, M., et al.: Preprocessing techniques for text mining-an overview. Int. J. Comput. Sci. Commun. Netw. **5**(1), 7–16 (2015)
26. Wang, H., Bah, M.J., Hammad, M.: Progress in outlier detection techniques: a survey. IEEE Access **7**, 107964–108000 (2019)
27. Wu, C., Chau, K.W., Fan, C.: Prediction of rainfall time series using modular artificial neural networks coupled with data-preprocessing techniques. J. Hydrol. **389**(1–2), 146–167 (2010)
28. Zhang, Z.: Missing data imputation: focusing on single imputation. Ann. Transl. Med. **4**(1) (2016)

Convergence of RBF Networks Regression Function Estimates and Classifiers

Adam Krzyżak[1,2], Tomasz Gałkowski[3(✉)], and Marian Partyka[4]

[1] Department of Computer Science and Software Engineering, Concordia University,
1455 de Maisonneuve Blvd. West, Montreal H3G 1M8, Canada
krzyzak@cs.concordia.ca
[2] Department of Electrical Engineering, Westpomeranian University of Technology,
ul. Sikorskiego 37, 70-313 Szczecin, Poland
[3] Institute of Computational Intelligence, Częstochowa University of Technology,
Częstochowa, al. Armii Krajowej 36, 42-200 Częstochowa, Poland
tomasz.galkowski@pcz.pl
[4] Department of Knowledge Engineering, Faculty of Production Engineering and
Logistics, Opole University of Technology, ul. Ozimska 75, 45-370 Opole, Poland
m.partyka@po.opole.pl

Abstract. In the paper convergence of the RBF network regression estimates and classifiers with so-called regular radial kernels is investigated. The parameters of the network are trained by minimizing the empirical risk on the training data. We analyze MISE convergence by utilizing the machine learning theory techniques such as VC dimension and covering numbers and the error bounds involving them. The performance of the normalized RBF network regression estimates is also tested in simulations.

Keywords: Nonlinear regression · Classification · RBF networks · Convergence · VC dimension · Covering numbers · Simulations

1 Introduction

Artificial neural network have been applied in machine learning from early days of the field starting with two layer perceptrons of Rosenblatt perceptrons, followed by multilayer perceptrons Duda, Hart and Stork [16] and finally entering

A. Krzyżak—Research of the first author was supported by the Alexander von Humboldt Foundation and the Natural Sciences and Engineering Research Council of Canada under Grant RGPIN-2015-06412. He carried out this research at the Westpomeranian University of Technology during his sabbatical leave from Concordia University.

T. Gałkowski—Research of the second author was supported by the project financed with the program of the Polish Minister of Science and Higher Education under the name "Regional Initiative of Excellence" in the years 2019 - 2022 project number 020/RID/2018/19 the amount of financing 12,000,000 PLN.

L. Rutkowski et al. (Eds.): ICAISC 2022, LNAI 13588, pp. 363–376, 2023.
https://doi.org/10.1007/978-3-031-23492-7_31

into the domain of deep learning [20]. In the literature on classical neural networks several types of feed-forward neural networks have been discussed. They include: multilayer perceptrons (MLP), radial basis function (RBF) networks, normalized radial basis function (NRBF) networks and deep networks. These neural network models have been applied in different application areas including interpolation, classification, data smoothing and regression. Convergence analysis of MLP has been studied by Cybenko [9], White [64], Hornik et al. [29], Barron [2], Anthony and Bartlett [1], Devroye et al. [11], Györfi et al. [25], Ripley [53], Haykin [27], Hastie et al. [26]. Deep networks have been thoroughly surveyed in Bengio et al. [20] and their convergence properties were recently investigated by Kohler and Krzyżak [31] and Bauer and Kohler [3]. These two latter papers are one of the first papers to analyze convergence properties of deep multilayer networks. RBF networks have been introduced by Moody and Darken [45] and their properties investigated by Park and Sandberg [48,49], Girosi and Anzellotti [18], Girosi et al. [19], Xu et al. [66], Krzyżak et al. [34], Krzyżak and Linder [35], Krzyżak and Niemann [36], Györfi et al. [25], Krzyżak and Schäfer [41] and Krzyżak and Partyka [37,39].

In this paper we consider the radial basis function (RBF) networks with one hidden layer consisting of k hidden nodes with a fixed kernel $\phi : \mathcal{R}_+ \to \mathcal{R}$:

$$f_k(x) = \sum_{i=1}^{k} w_i \phi \left(\|x - c_i\|_{A_i} \right) \tag{1}$$

where

$$\|x - c_i\|_{A_i}^2 = [x - c_i]^T A_i [x - c_i]$$

These networks form a of functions satisfying the following conditions:

(i) **radial basis function condition:** $\phi : \mathcal{R}_0^+ \to \mathcal{R}^+$ is a left-continuous, monotone decreasing function, the so-called *kernel*.

(ii) **centre condition:** $c_1, ..., c_k \in \mathcal{R}^d$ are the so-called *centre vectors* with $\|c_i\| \leq R$ for all $i = 1, ..., k$.

(iii) **receptive field condition:** $A_1, ..., A_k$ are symmetric, positive definite, real $d \times d$-matrices each of which satisfies the eigenvalue inequalities $\ell \leq \lambda_{min}(A_i) \leq \lambda_{max}(A_i) \leq L$. Here, $\lambda_{min}(A_i)$ and $\lambda_{max}(A_i)$ are the minimal and the maximal eigenvalue of A_i, respectively. A_i specifies the *receptive field* about the centre c_i.

(iv) **weight condition:** $w_1, ..., w_k \in \mathcal{R}$ are the *weights* satisfying $\sum_{i=1}^{k} |w_i| \leq b$ for all $i = 1, ..., k$.

Throughout the paper we use the convention $0/0 = 0$. Popular kernels satisfying (i) are:

– **Window-type kernels.** These are kernels for which some $\delta > 0$ exists such that $\phi(t) \notin (0, \delta)$ for all $t \in \mathcal{R}_0^+$. The classical naive kernel $\phi(t) = \mathbf{1}_{[0,1]}(t)$ is a member of this class.

- **Non-window-type kernels with bounded support.** These comprise all kernels with support of the form $[0, s]$ which are right-continuous in s. For example, for $\phi(t) = \max\{1 - t, 0\}$, $\phi(x^T x)$ is the Epanechnikov kernel.
- **Regular radial kernels.** These kernels are nonnegative, monotonically decreasing, left continuous, $\int_{\mathcal{R}^d} \phi(\|x\|)dx \neq 0$, and $\int_{\mathcal{R}^d} \phi(\|x\|)dx < \infty$, where $\|\cdot\|$ is the Euclidean norm on $\mathcal{R}^d$. Regular kernels include naive kernels, Epanechnikov kernels, exponential kernels and the Gaussian kernels. Note that the regular kernels are bounded.

Let us denote the parameter vector $(w_0, \ldots, w_k, c_1, \ldots, c_k, A_1, \ldots, A_k)$ by θ. It is assumed that the kernel is fixed, while network parameters $w_i, c_i, A_i, i = 1, \ldots, k$ are learned from the data. The most popular radial functions ϕ are:

- $\phi(x) = e^{-x^2}$ (Gaussian kernel)
- $\phi(x) = e^{-x}$ (exponential kernel)
- $\phi(x) = (1 - x^2)_+$ (truncated parabolic or Epanechnikov kernel)
- $\phi(x) = \frac{1}{\sqrt{x^2+c^2}}$ (inverse multiquadratic)

All these kernels are nonincreasing. In the literature on approximation by means of radial basis functions the following monotonically increasing kernels were considered

- $\phi(x) = \sqrt{x^2 + c^2}$ (multiquadratic)
- $\phi(x) = x^{2n} \log x$ (thin plate spline)

They play important role in interpolation and approximation with radial functions [19], but are not considered in this paper.

Standard RBF networks have been introduced by Broomhead and Lowe [8] and Moody and Darken [45]. Their approximation error was studied by Park and Sandberg [48,49]. These result have been generalized by Krzyżak, Linder and Lugosi [34], who also showed weak and strong universal consistency of RBF networks for a large class of radial kernels in the least squares estimation problem and classification. The rate of approximation of RBF networks was investigated by Girosi and Anzellotti [18]. The rates of convergence of RBF networks trained by complexity regularization have been investigated in regression estimation problem by Krzyżak and Linder [35].

Normalized RBF networks are generalizations of standard RBF networks and are defined by

$$f_k(x) = \frac{\sum_{i=1}^k w_i \phi(\|x - c_i\|_{A_i})}{\sum_{i=1}^k \phi(\|x - c_i\|_{A_i})}. \tag{2}$$

Normalized RBF networks (2) have been originally investigated by Moody and Darken [45] and Specht [59]. Further results were obtained by Shorten and Murray-Smith [58]. Normalized RBF networks (NRBF) are related to the classical nonparametric kernel regression estimate also called the Nadaraya-Watson estimate (3):

$$r_n(x) = \frac{\sum_{i=1}^n Y_i K(\frac{x-X_i}{h_n})}{\sum_{i=1}^n K(\frac{x-X_i}{h_n})} \tag{3}$$

where $K : \mathcal{R}^d \to \mathcal{R}$ is a kernel and h_n is a smoothing sequence (bandwidth) of positive real numbers. The estimate has been introduced by Nadaraya [46] and Watson [63] and studied by Devroye and Wagner [14], Krzyżak [32], Krzyżak and Pawlak [40] and Györfi et al. [25]. Its recursive versions were investigated in [24,25]. Kernel methods have been applied in regression estimation and classification [15,25,50,54,55] and also in change detection [17]. For parametric methods for novelty detection in COVID-19 prognosis curves refer to [52]. Other nonparametric regression estimation techniques include nearest-neighbor estimate [10,12,25], partitioning estimate [4,25], orthogonal series estimate [23,25], tree estimate [7,26] and Breiman random forest [6,30,57]. Alternative classification techniques have been discussed in [28,44,47].

In the analysis of the NRBF nets (2) presented in [66] and in [37,38] the authors analyzed convergence of the normalized RBF by exploiting the relationship between their mean integrated square error (MISE) and MISE of the kernel regression estimate, however these results were valid only on the training data, i.e., no generalization was shown. Generalization ability of NRBF networks and their convergence was investigated in [41].

This paper investigates generalization ability and weak convergence of the RBF network (1) with parameters trained by the empirical risk minimization with applications in nonlinear function learning and classification. In this paper we will use specialized tools from computational learning theory such as VC dimension and covering numbers to analyze generalization ability of RBF networks with so-called regular kernels. The paper is organized as follows. In Sect. 2 the algorithm for nonlinear function learning is presented. In Sect. 3 the RBF network classifier is discussed. In Sect. 4 convergence properties of the classical RBF net regression estimates and classification rules are discussed. Some simulation results are presented in Sect. 5 and conclusions are given in Sect. 6.

2   Nonlinear Regression Estimation

Let (X,Y), (X_1,Y_1), (X_2,Y_2), ..., (X_n,Y_n) be independent, identically distributed, $\mathcal{R}^d \times \mathcal{R}$–valued random variables with $\mathbf{E}Y^2 < \infty$, and let $m(x) = \mathbf{E}(Y|X = x)$ be the corresponding nonlinear regression function. Let μ denote the distribution of X. It is well-known that regression function R minimizes L_2 error:

$$\mathbf{E}|m(X) - Y|^2 = \min_{f:\mathcal{R}^d \to \mathcal{R}} \mathbf{E}|f(X) - Y|^2.$$

Our aim is to estimate m from the i.i.d. observations of random vector (X,Y)

$$D_n = \{(X_1,Y_1),\ldots,(X_n,Y_n)\}$$

using RBF network (1). We train the network using so-called empirical risk minimization by choosing its parameters that minimize the empirical L_2 risk

$$\frac{1}{n}\sum_{j=1}^{n}|f(X_j) - Y_j|^2 \tag{4}$$

on the training data D_n, that is we choose RBF network m_n in the class

$$\mathcal{F}_n = \{f_k = f_\theta : \theta \in \Theta_n\} = \left\{ \sum_{i=1}^{k} w_i \phi \left(\|x - c_i\|_{A_i}\right) : \sum_{i=0}^{k_n} |w_i| \leq b_n \right\} \quad (5)$$

where

$$\Theta_n = \{\theta = (w_1, \ldots, w_{k_n}, c_1, \ldots, c_{k_n}, A_1, \ldots, A_{k_n})\}.$$

so that

$$\frac{1}{n} \sum_{j=1}^{n} |m_n(X_j) - Y_j|^2 = \min_{f \in \mathcal{F}_n} \frac{1}{n} \sum_{j=1}^{n} |f_\theta(X_j) - Y_j|^2. \quad (6)$$

We measure the performance of the RBF network estimates by the mean squared error

$$\mathbf{E}\left[|m_n(X) - m(X)|^2\right] = \mathbf{E}\left[\int |m_n(x) - m(x)|^2 \mu(dx)\right].$$

This approach has been investigated among others by Zeger and Lugosi [43] and by Györfi et al. [25].

Initial analysis of convergence of m_n was carried out in [34] using Vapnik-Chervonenkis dimension concept introduced by Vapnik and Chervonenkis [60, 61] and covering numbers which are basic tools of computational learning theory (CLT) and of machine learning. They were applied in nonparametric regression learning by many researchers (for in-depth survey of the main results in CLT and their applications in nonparametric regression refer to [25]). In this paper we use machine learning tools of CLT to analyze generalization ability and convergence of the RBF networks with regular kernels. In our analysis we are motivated by the results of presented in [34, 41] and [25].

3   RBF Classification Rules

Let (Y, X) be a pair of random variables taking values in the set $\{1, ..., M\}$, whose elements are called classes, and in R^d, respectively. The problem is to classify X, i.e. to decide on Y. Let us define *a posteriori* class probabilities

$$p_i(x) = P\{Y = i | X = x\}, i = 1, \cdots, M, x \in R^d.$$

The Bayes classification rule

$$\Psi^*(X) = i \text{ if } p_i(X) > p_j(X), j < i, \text{ and } p_i(X) > p_j(X), j > i$$

minimizes the probability of error. The Bayes risk L^* is defined by

$$P\{\Psi^*(X) \neq Y\} = \inf_{\Psi: R^d \to \{1, ..., M\}} P\{\Psi(X) \neq Y\}.$$

The local Bayes risk is equal to $P\{\Psi^*(X) \neq Y \mid X = x\}$. Observe that $p_i(x) = E\{I_{\{Y=i\}} \mid X = x\}$ may be viewed as a regression function of the indicator of the event $\{Y = i\}$. Given the learning sequence $V_n = \{(Y_1, X_1), ..., (Y_n, X_n)\}$ of independent observations of the pair (Y, X), we may learn $p_i(x)$ using RBF nets mimicking (6), i.e.,

$$\frac{1}{n} \sum_{j=1}^{n} |\hat{p}_{in}(X_j) - I_{\{Y_j = i\}}|^2 = \min_{f \in \mathcal{F}_n} \frac{1}{n} \sum_{j=1}^{n} |f_Y(X_j) - I_{\{Y_j = i\}}|^2. \qquad (7)$$

We propose plug-in RBF classifier with parameters learned by (7) resulting in the classification rule Ψ_n which classifies every $x \in R^d$ to any class maximizing $\hat{p}_{in}(x)$. The global performance of Ψ_n is measured by $L_n = P\{\Psi_n(X) \neq \theta \mid V_n\}$ and the local performance by $L_n(x) = P\{\Psi_n(x) \neq \theta \mid V_n\}$. A rule is said to be weakly, strongly, or completely Bayes risk consistent (BRC) if $L_n \to L^*$, in probability, almost surely, or completely, respectively, as $n \to \infty$, see, e.g., Wolverton and Wagner [65] and Greblicki [21].

In the next section we discuss convergence of the RBF regression estimate m_n as well as plug-in classification rule induced by it.

4   Convergence

In this section we present convergence results for the RBF regression estimates and resulting plug-in classification rules.

4.1   Convergence Results

We have the following convergence results for the RBF network m_n and classification rule Ψ_n with regular radial kernels.

Theorem 1. *Let $|Y| \leq L < \infty$ a.s.. Consider a family $\mathcal{F}_n$ of RBF networks defined by (5), with $k_n \geq 1$, and let K be a regular radial kernel. If*

$$k_n, b_n \to \infty$$

and

$$k_n b_n^4 \log(k_n b_n^2)/n \to 0$$

as $n \to \infty$, then the RBF network m_n minimizing the empirical L_2 risk over $\mathcal{F}_n = \{f_\theta : \theta \in \Theta_n\}$ is is consistent, i.e.,

$$\mathbf{E}\left[|m_n(X) - m(X)|^2\right] \to 0 \ \ as \ n \to \infty \qquad (8)$$

and consequently

$$\mathbf{E}\left[|L_n(X) - L^*(X)|^2\right] \to 0 \ \ as \ n \to \infty \qquad (9)$$

for all distributions of (X, Y) with $|Y| \leq L < \infty$.

Theorem 1 provides conditions for mean square convergence of the RBF regression estimates m_n and classifiers Ψ_n for all distributions of the data with bounded Y. The latter condition is naturally satisfied in classification.

4.2   Survey of Main Tools Used in Convergence Analysis

We will first introduce basic tools from CLT required in the analysis of convergence of algorithms m_n and Ψ_n discussed in this paper, see [25].

We will start with the definition of the $\epsilon - cover$ and the covering numbers.

Definition 1. *Let $\epsilon > 0$ and let $\mathcal{G}$ be a set of functions $\mathcal{R}^d \to \mathcal{R}$. Every finite collection of functions $g_1, \ldots, g_N : \mathcal{R}^d \to \mathcal{R}$ with the property that for every $g \in \mathcal{G}$ there is a $j = j(g) \in \{1, \ldots, N\}$ such that*

$$\|g - g_j\|_\infty := \sup_z |g(z) - g_j(z)| < \epsilon$$

*is called an ϵ-**cover** of $\mathcal{G}$ with respect to $\| \cdot \|_\infty$.*

Definition 2. *Let $\epsilon > 0$ and let $\mathcal{G}$ be a set of functions $\mathcal{R}^d \to \mathcal{R}$. Let $\mathcal{N}(\epsilon, \mathcal{G}, \| \cdot \|_\infty)$ be the size of the smallest ϵ-cover of $\mathcal{G}$ w.r.t. $\| \cdot \|_\infty$. Take $\mathcal{N}(\epsilon, \mathcal{G}, \| \cdot \|_\infty) = \infty$ if no finite ϵ-cover exists. Then $\mathcal{N}(\epsilon, \mathcal{G}, \| \cdot \|_\infty)$ is called an ϵ-**covering number** of $\mathcal{G}$ w.r.t. $\| \cdot \|_\infty$ and will be abbreviated to $\mathcal{N}_\infty(\epsilon, \mathcal{G})$.*

Next we define the VC dimension. We begin with the shatter coefficient.

Definition 3. *Let $\mathcal{A}$ be a class of subsets of $\mathcal{R}^d$ and let $n \in \mathcal{N}$.*
(a) For $z_1, \ldots, z_n \in \mathcal{R}^d$ define

$$s(\mathcal{A}, \{z_1, \ldots, z_n\}) = |\{A \cap \{z_1, \ldots, z_n\} : A \in \mathcal{A}\}|,$$

that is, $s(\mathcal{A}, \{z_1, \ldots, z_n\})$ is the number of different subsets of $\{z_1, \ldots, z_n\}$ of the form $A \cap \{z_1, \ldots, z_n\}$, $A \in \mathcal{A}$.
*(b) Let G be a subset of $\mathcal{R}^d$ of size n. One says that $\mathcal{A}$ **shatters** G if $s(\mathcal{A}, G) = 2^n$, i.e., if each subset of G can be represented in the form $A \cap G$ for some $A \in \mathcal{A}$.*
*(c) The nth **shatter coefficient** of $\mathcal{A}$ is*

$$S(\mathcal{A}, n) = \max_{\{z_1, \ldots, z_n\} \subseteq \mathcal{R}^d} s(\mathcal{A}, \{z_1, \ldots, z_n\}).$$

That is, the shatter coefficient is the maximal number of different subsets of n points that can be picked out by sets from $\mathcal{A}$.

We can now define the VC dimension.

Definition 4. *Let $\mathcal{A}$ be a class of subsets of $\mathcal{R}^d$ with $\mathcal{A} \neq \emptyset$. The **VC dimension** (or Vapnik–Chervonenkis dimension) $V_\mathcal{A}$ of $\mathcal{A}$ is defined by*

$$V_\mathcal{A} = \sup \{n \in \mathcal{N} : S(\mathcal{A}, n) = 2^n\},$$

i.e., the VC dimension $V_\mathcal{A}$ is the largest integer n such that there exists a set of n points in $\mathcal{R}^d$ which can be shattered by $\mathcal{A}$.

Convergence of m_n also implies convergence of Ψ_n thanks to plug-in scheme and therefore we will only discuss convergence of m_n. To show convergence of (8) it is sufficient to show for bounded Y that

$$\inf_{f \in \mathcal{F}_n} \int |f(x) - m(x)|^2 \mu(dx) \to 0 \quad (n \to \infty) \tag{10}$$

and

$$\mathbf{E}\left\{\sup_{f \in \mathcal{F}_n} \left| \frac{1}{n} \sum_{i=1}^{n} |f(X_i) - Y_i|^2 - \mathbf{E}\{|f(X) - Y|^2\} \right| \right\} \to 0 \quad (n \to \infty). \tag{11}$$

Approximation error consistency (10) follows from the Lemma 1 below (stated without proof), which implies that $\bigcup_{k=1}^{\infty} \mathcal{F}_k$ is dense in $L_2(\mu)$ for any probability measure μ on $\mathcal{R}^d$ and for RBF networks with regular radial kernels [34]. It is sufficient to restrict the class RBF nets to a subset of the family $\mathcal{F}_n$ of RBF networks by constraining the receptive field matrices A_i to be diagonal with the equal elements, i.e., $A_i = h_i^{-2}I$. Consequently $\mathcal{F}_n$ becomes

$$f_\theta(x) = \sum_{i=1}^{k} w_i K\left(\left\| \frac{x - c_i}{h_i} \right\|^2 \right) + w_0, \tag{12}$$

where $\theta = (w_0, \ldots, w_k, c_1, \ldots, c_k, h_1, \ldots, h_k)$ is the vector of parameters, $w_0, \ldots, w_k \in \mathcal{R}$, $h_1, \ldots, h_k \in \mathcal{R}$, and $c_1, \ldots, c_k \in \mathcal{R}^d$.

Lemma 1. *Assume that K is a regular radial kernel. Let μ be an arbitrary probability measure on $\mathcal{R}^d$. Then the RBF networks given by (12) are dense in $L_2(\mu)$. In particular, if $m \in L_2(\mu)$, then, for any $\epsilon > 0$, there exist parameters $\theta = (w_0, \ldots, w_k, c_1, \ldots, c_k, h_1, \ldots, h_k)$ such that*

$$\int_{\mathcal{R}^d} |f_\theta(x) - m(x)|^2 \mu(dx) < \epsilon. \tag{13}$$

In the next lemma we consider convergence of estimation error (11).

Lemma 2. *Assume $|Y| \leq L < \infty$ a.s. Consider a family of RBF networks defined by (5), with $k = k_n \geq 1$. Assume that K is a regular radial kernel. If*

$$k_n, b_n \to \infty$$

and

$$k_n b_n^4 \log(k_n b_n^2)/n \to 0$$

as $n \to \infty$, then

$$\mathbf{E}\left\{\sup_{f \in \mathcal{F}_n} \left| \frac{1}{n} \sum_{i=1}^{n} |f(X_i) - Y_i|^2 - \mathbf{E}\{|f(X) - Y|^2\} \right| \right\} \to 0 \quad (n \to \infty)$$

for all distributions of (X, Y) with Y bounded.

Proof Outline. The outline of proof of Theorem 1 is presented in [39] and is omitted.

5   Simulation Results

We applied the Nadaraya-Watson-type RBF function estimate given by (3) to estimate the regression function given by:

$$R(x) = 0.3 + 0.3 * x + \exp(-2 * x + 0.5) * \sin(4 * (x + 2.0)) * \cos(12 * (x - 1.0)) * \log(x + 1.1)$$

(black line in the Figures). We generated 400 testing pairs $(X_i, Y_i), i = 1, ..., 400$ in the interval $[0, 1]$. Additive random noise was generated from normal (Gaussian) distribution with zero mean and variance 1 (red points in the Figures). The regression function estimates produced by the Nadaraya-Watson regression estimate (3) are displayed as blue pluses. The Epanechnikov (parabolic) kernel was applied (Figs 1, 2, 3 and 4).

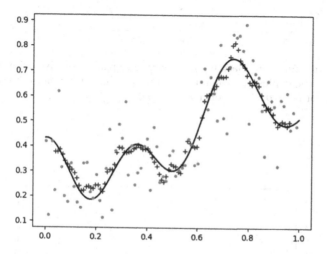

Fig. 1. Estimated function (black line), measurement pairs - additive noise (red points) and estimates (blue pluses). Main parameters of simulation: noise variance $= 0.1$; smoothing factor $h_n = 0.02$. (Color figure online)

In our simulations the bandwidth in Watson-Nadaraya regression estimate was selected arbitrarily. For techniques for automatic bandwidth selection we refer the reader to [22, 42, 56].

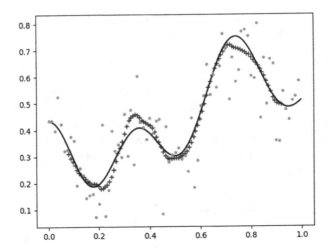

Fig. 2. Estimated function (black line), measurement pairs - additive noise (red points) and estimates (blue pluses). Main parameters of simulation: noise variance $= 0.1$; smoothing factor $h_n = 0.04$. (Color figure online)

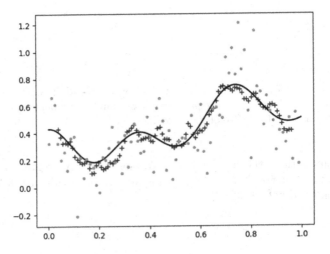

Fig. 3. Estimated function (black line), measurement pairs - additive noise (red points) and estimates (blue pluses). Main parameters of simulation: noise variance $= 0.2$; smoothing factor $h_n = 0.02$. (Color figure online)

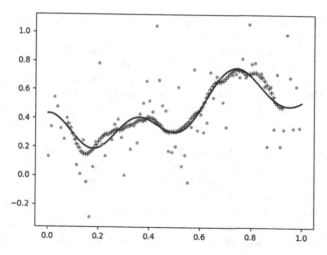

Fig. 4. Estimated function (black line), measurement pairs - additive noise (red points) and estimates (blue pluses). Main parameters of simulation: noise variance $= 0.2$; smoothing factor $h_n = 0.04$. (Color figure online)

6   Conclusions

In the paper we discussed convergence of classical and normalized RBF function regression estimates and classifiers. Simulation results for regression estimation by the normalized RBF networks are also presented. Further experimental studies for both classical and normalized RBF regression estimates and classification rules using data-dependent techniques for parameter selection will be presented in the future work.

References

1. Anthony, M., Bartlett, P.L.: Neural Network Learning: Theoretical Foundations. Cambridge University Press, Cambridge (1999)
2. Barron, A.R.: Universal approximation bounds for superpositions of a sigmoidal function. IEEE Trans. Inf. Theory **39**, 930–945 (1993)
3. Bauer, B., Kohler, M.: On deep learning as a remedy for the curse of dimensionality in nonparametric regression. Ann. Stat. **47**(4), 2261–2285 (2019)
4. Beirlant, J., Györfi, L.: On the asymptotic L_2-error in partitioning regression estimation. J. Stat. Plan. Inference **71**, 93–107 (1998)
5. Bologna, G., Hayashi, Y.: Characterization of symbolic rules embedded in deep DIMLP networks: a challenge to transparency of deep learning. J. Artifi. Intell. Soft Comput. **7**(4), 265–286 (2017)
6. Breiman, L.: Random forests. Mach. Learn. **45**, 5–32 (2001)
7. Breiman, L., Friedman, J.H., Olshen, R.A., Stone, C.J.: Classification and Regression Trees, Wadsworth Advanced Books and Software, CA, Belmont (1984)
8. Broomhead, D.S., Lowe, D.: Multivariable functional interpolation and adaptive networks. Complex Syst. **2**, 321–323 (1988)

9. Cybenko, G.: Approximations by superpositions of sigmoidal functions. Math. Control Signals Syst. **2**, 303–314 (1989)
10. Biau, G., Devroye, L.: Lectures on the Nearest Neighbor Method. SSDS, Springer, Cham (2015). https://doi.org/10.1007/978-3-319-25388-6
11. Devroye, L., Györfi, L., Lugosi, G.: Probabilistic Theory of Pattern Recognition. Springer-Verlag, New York (1996). https://doi.org/10.1007/978-1-4612-0711-5
12. Devroye, L., Györfi, L., Krzyżak, A., Lugosi, G.: On the strong universal consistency of nearest neighbor regression function estimates. Ann. Stat. **22**, 1371–1385 (1994)
13. Devroye, L., Krzyżak, A.: An equivalence theorem for L_1 convergence of the kernel regression estimate. J. Stat. Plan. Inference **23**, 71–82 (1989)
14. Devroye, L.P., Wagner, T.J.: Distribution-free consistency results in nonparametric discrimination and regression function estimation. Ann. Stat. **8**, 231–239 (1980)
15. Duda, P., Rutkowski, L., Jaworski, M., Rutkowska, D.: On the Parzen kernel-based probability density function learning procedures over time-varying streaming data with applications to pattern classification. IEEE Trans. Cybern. **50**(4), 1683–1696 (2020)
16. Duda, R., Hart, P., Stork, D.: Pattern Classification, Wiley, 2nd ed. (2001)
17. Gałkowski, T., Krzyżak, A., Patora-Wysocka, Z., Filutowicz, Z., Wang, L.: A new approach to detection of changes in multidimensional patterns - Part II. J. Artifi. Intell. Soft Comput. Res. **11**(3), 217–227 (2021)
18. Girosi, F., Anzellotti, G.: Rates of convergence for radial basis functions and neural networks. In: Mammone, R.J. (ed.) Artificial Neural Networks for Speech and Vision, pp. 97–113. Chapman and Hall, London (1993)
19. Girosi, F., Jones, M., Poggio, T.: Regularization theory and neural network architectures. Neural Comput. **7**, 219–267 (1995)
20. Goodfellow, I., Bengio, Y., Courville, A.: Deep Learning. The MIT Press (2016)
21. Greblicki, W.: Asymptotically Optimal Probabilistic Algorithms for Pattern Recognition and Identification. Monografie No. 3. Prace Naukowe Instytutu Cybernetyki Technicznej Politechniki Wroclawskiej, Nr. 18. Wroclaw, Poland (1974)
22. Gramacki, A., Gramacki, J.: FFT-based fast bandwidth selector for multivariate kernel density estimation. Comput. Stat. Data Anal. **106**, 27–45 (2017)
23. Greblicki, W., Pawlak, M.: Fourier and Hermite series estimates of regression functions. Ann. Inst. Stat. Math. **37**, 443–454 (1985)
24. Greblicki, W., Pawlak, M.: Necessary and sufficient conditions for Bayes risk consistency of a recursive kernel classification rule. IEEE Trans. Inf. Theory **IT-33**, 408–412 (1987)
25. Györfi, L., Kohler, M., Krzyżak, A., Walk, H.: A Distribution-Free Theory of Nonparametric Regression. Springer Verlag, New York (2002). https://doi.org/10.1007/b97848
26. Hastie, T., Tibshirani, R., Friedman, J.: The Elements of Statistical Learning. SSS, Springer, New York (2009). https://doi.org/10.1007/978-0-387-84858-7
27. Haykin, S.O.: Neural Networks and Learning Machines, 3rd edn. Prentice-Hall, New York (2008)
28. Homenda, W., Jastrzębska, A., Pedrycz, W., Yu, F.: Combining classifiers for foreign pattern rejection. J. Artif. Intell. Soft Comput. Res. **10**(2), 75–94 (2020)
29. Hornik, K., Stinchcombe, S., White, H.: Multilayer feed-forward networks are universal approximators. Neural Netw. **2**, 359–366 (1989)
30. Jordanov, I., Petrov, N., Petrozziello, A.: Classifiers accuracy improvement based on missing data imputation. J. Artifi. Intell. Soft Comput. **8**(8), 31–48 (2018)

31. Kohler, M., Krzyżak, A.: Nonparametric regression based on hierarchical interaction models. IEEE Trans. Inf. Theory **63**, 1620–1630 (2017)
32. Krzyżak, A.: The rates of convergence of kernel regression estimates and classification rules. IEEE Trans. Inf. Theory **IT-32**, 668–679 (1986)
33. Krzyżak, A.: Global convergence of recursive kernel regression estimates with applications in classification and nonlinear system estimation. IEEE Trans. Inf. Theory **IT-38**, 1323–1338 (1992)
34. Krzyżak, A., Linder, T., Lugosi, G.: Nonparametric estimation and classification using radial basis function nets and empirical risk minimization. IEEE Trans. Neural Netw. **7**(2), 475–487 (1996)
35. Krzyżak, A., Linder, T.: Radial basis function networks and complexity regularization in function learning. IEEE Trans. Neural Netw. **9**(2), 247–256 (1998)
36. Krzyżak, A., Niemann, H.: Convergence and rates of convergence of radial basis functions networks in function learning. Nonlinear Anal. **47**, 281–292 (2001)
37. Krzyżak, A., Partyka, M.: Convergence and rates of convergence of normalized recursive radial basis functions networks in function learning and classification. In: Proceedings of 16th International Conference on Artificial Intelligence and Soft Computing, Zakopane, Poland, 11–15 June 2017. Lecture Notes on Artificial Intelligence and Soft Computing, Part I, LNAI, vol. 10245, pp. 107–117. Springer-Verlag, (2017)
38. Krzyżak, A., Partyka, M.: Learning and convergence of the normalized radial basis functions networks. In: Rutkowski, L., Scherer, R., Korytkowski, M., Pedrycz, W., Tadeusiewicz, R., Zurada, J.M. (eds.) ICAISC 2018. LNCS (LNAI), vol. 10841, pp. 118–129. Springer, Cham (2018). https://doi.org/10.1007/978-3-319-91253-0_12
39. Krzyżak, A., Partyka, M.: On Learning and convergence of rbf networks in regression estimation and classification. In: Rutkowski, L., Scherer, R., Korytkowski, M., Pedrycz, W., Tadeusiewicz, R., Zurada, J.M. (eds.) ICAISC 2019. LNCS (LNAI), vol. 11508, pp. 131–142. Springer, Cham (2019). https://doi.org/10.1007/978-3-030-20912-4_13
40. Krzyżak, A., Pawlak, M.: Distribution-free consistency of a nonparametric kernel regression estimate and classification. IEEE Trans. Inf. Theory **IT-30**, 78–81 (1984)
41. Krzyżak, A., Schäfer, D.: Nonparametric regression estimation by normalized radial basis function networks. IEEE Trans. Inf. Theory **51**, 1003–1010 (2005)
42. Lv, J., Pawlak, M.: Bandwidth selection for kernel generalized regression neural networks in identification of Hammerstein systems. J. Artifi. Intell. Soft Comput. Res. **11**(3), 181–194 (2021)
43. Lugosi, G., Zeger, K.: Nonparametric estimation via empirical risk minimization. IEEE Trans. Inf. Theory **41**, 677–687 (1995)
44. Mikołajczyk, A., Grochowski, M., Kwasigroch, A.: Towards explainable classifiers using the counterfactual approach - global explanations for discovering bias in data. J. Artif. Intell. Soft Comput. Res. **11**(1), 51–67 (2021)
45. Moody, J., Darken, J.: Fast learning in networks of locally-tuned processing units. Neural Comput. **1**, 281–294 (1989)
46. Nadaraya, E.A.: On estimating regression. Theory Probability Appli. **9**, 141–142 (1964)
47. Nowicki, R.K., Grzanek, K., Hayashi, Y.: Rough support vector machine for classification with interval and incomplete data. J. Artif. Intell. Soft Comput. Res. **10**(1), 47–56 (2020)
48. Park, J., Sandberg, I.W.: Universal approximation using Radial-Basis-Function networks. Neural Comput. **3**, 246–257 (1991)

49. Park, J., Sandberg, I.W.: Approximation and radial-basis-function networks. Neural Comput. **5**, 305–316 (1993)
50. Pietruczuk, L., Rutkowski, L., Jaworski, M., Duda, P.: The Parzen kernel approach to learning in non-stationary environment. In: Proceedings 2014 International Joint Conference on Neural Networks (IJCNN), pp. 3319–3323 (2014)
51. Pollard, D.: Convergence of Stochastic Processes. Springer Verlag, New York (1984). https://doi.org/10.1007/978-1-4612-5254-2
52. Rafajłowicz, W.: Learning novelty detection outside a class of random curves with application to COVID-19 growth. J. Artif. Intell. Soft Comput. Res. **11**(3), 195–215 (2021)
53. Ripley, B.D.: Pattern Recognition and Neural Networks. Cambridge University Press, Cambridge (2008)
54. Rutkowski, L.: Identification of MISO nonlinear regressions in the presence of a wide class of disturbances. IEEE Trans. Inf. Theory **37**(1), 214–216 (1991)
55. Rutkowski, L.: Adaptive probabilistic neural networks for pattern classification in time-varying environment. IEEE Trans. Neural Networks **15**(4), 811–827 (2004)
56. Samworth, R.J., Wand, M.P.: Asymptotics and optimal bandwidth selection for highest density region estimation. Ann. Stat. **38**(3), 1767–1792 (2010)
57. Scornet, E., Biau, G., Vert, J.-P.: Consistency of random forest. Ann. Stat. **43**(4), 1716–1741 (2015)
58. Shorten, R., Murray-Smith, R.: Side effects of normalising radial basis function networks. Int. J. Neural Syst. **7**, 167–179 (1996)
59. Specht, D.F.: Probabilistic neural networks. Neural Netw. **3**, 109–118 (1990)
60. Vapnik, V.N., Chervonenkis, A.Y.: On the uniform convergence of relative frequencies of events to their probabilities. Theory Probab. Appli. **16**, 264–280 (1971)
61. Vapnik, V.: Estimation of Dependences Based on Empirical Data. ISS, Springer, New York (2006). https://doi.org/10.1007/0-387-34239-7
62. van de Geer, S.: Empirical Processes in M-Estimation. Cambridge University Press, New York (2000)
63. Watson, G.S.: Smooth regression analysis, Sankhya Series A, vol. 26, pp. 359–372
64. White, H.: Connectionist nonparametric regression: multilayer feedforward networks that can learn arbitrary mappings. Neural Netw. **3**, 535–549 (1990)
65. Wolverton, C.T., Wagner, T.J.: Asymptotically optimal discriminant functions for pattern classification. IEEE Trans. Inf. Theory **IT-15** 258–265 (1969)
66. Xu, L., Krzyżak, A., Yuille, A.L.: On radial basis function nets and kernel regression: approximation ability, convergence rate and receptive field size. Neural Netw. **7**, 609–628 (1994)

A Comparative Study of Genetic Programming Variants

Cry Kuranga[✉][iD] and Nelishia Pillay[iD]

Department of Computer Science, University of Pretoria, Lynnwood Road, Hillcrest,
Pretoria 0002, South Africa
kurangacry@gmail.com, nelishia.pillay@up.ac.za

Abstract. Genetic programming tends to optimize complicated structures producing human-competitive results; therefore, it is applied to a wide range of problems such as classification and regression. This work experimentally performs a comparative study of Genetic programming variants, namely gene expression, grammatical evolution, Cartesian, multi-expression programming, and stacked-based as general regression and classification solvers. The analyses will help to understand the strengths of each variant and identify the relative performance of variants that stand relative to each other for the given problem domains. To determine the performance difference between selected GP variants, hyperparameter tuning was performed on each GP variant for each dataset to minimize the performance difference due to implementation. A total of 11 datasets were used in the experiments, seven from the regression benchmark suite, and four from the classification. The obtained results indicate that the choice of Genetic programming variant has an impact on the performance of regression and classification problems. Multi-expression programming exhibits outstanding performance as a regression and classification solver which scales graciously with problem size and complexity whereas other variants were problem-dependent. Future work could consider implementing a multi-expression paradigm with other Genetic programming variants such as grammatical evolution and gene expression programming.

Keywords: Genetic programming · Prediction · Classification

1 Introduction

Genetic programming (GP) is a population-based evolutionary algorithm that breeds computer programs intended to evolve stochastically to produce better programs as the algorithm iterates [1]. GP is based on the concept of 'survival

This work is based on the research supported wholly/in part by the National Research Foundation of South Africa (Grant Numbers 138150). Opinions expressed and conclusions arrived at, are those of the author and are not necessarily to be attributed to the NRF.

of the fittest' and uses evolutionary operators such as selection, mutation, and crossover to evolve programs. Computer programs are expressed as symbolic structures, usual as trees, tailored to achieve a favorable goal, governed by the execution of the program's instruction for given input data [2].

GP tends to optimize complicated structures producing human-competitive results; therefore, it is applied to a wide range of problems such as classification and regression [3]. Also, the development of GP is on the rise which is catalyzed by emerging of new ideas and continual advancement of the existing applications [4]. However, several variants of GP are still yet to be extensively explored as general regression and classification problem solvers.

A canonical GP (tree-based) exhibits several limitations such as domain knowledge being incorporated only through evolutionary operators such as crossover and mutation though knowledge-based issues are widely acknowledged in evolutionary computation [5]. Furthermore, GP is considered computational expensive due to the bloat phenomenon (growth of non-coding branches in a program), even if the search space is properly constrained [6]. Usually, a canonical GP predicts a single variable at a time, hence it is multi-input single-output in nature. The GP variants which include Cartesian and Grammatical evolution among many others were developed to address some of the drawbacks of the canonical GP.

This work experimentally performs a comparative study of GP variants, namely Gene expression, Grammatical evolution, Cartesian, Multi-expression programming, and Stacked-based GP as general regression and classification solvers. The scalability of the GP variants is analyzed on datasets of varying sizes and complexity. Thus, the relationship between the number of iterations, population size, and success rate are analyzed for the given benchmark problems. A canonical GP is used to benchmark the performance of each variant.

This paper is structured as follows: Sect. 2 provides an overview of the GP variants; Sect. 3 discusses the experimental set-up and Sect. 4 presents the results. Section 5 concludes the work and suggests future work.

2   Genetic Programming

This section discusses GP variants used in this work, namely canonical GP, Cartesian GP, Grammatical Evolution, Gene Expression Programming, Stacked GP, and Multi-Expression Program. The GP variants used in this work were selected based on the literature in which they were used as either predictors or classifiers: GEP [7,8], GE [9,10], CGP [11,12], MEP [13,14], SGP [15,16].

The function set used in each variant consists of $\{-, +, *, \text{protected div}\}$ and the terminal set consists of attributes from input data. Each variant uses a fitness function: *mean square error* for regression problems or *precision* for classification problems. A general stopping criterion, termination after convergence is used for each variant. As such, each algorithm runs until it converges, thus, no further improvement in fitness.

2.1 Canonical Genetic Programming

The canonical GP implemented in this work uses a parse tree to represent the solution in which the internal nodes are program primitives whereas terminal nodes are the data inputs. An arithmetic tree classifier is used for classification. Individuals are created using the ramp half-and-half technique. A one-point crossover and uniform point mutation are used. The parsimony pressure of 0.1 per node is applied. Selection is performed using the tournament selection technique.

2.2 Cartesian Genetic Programming

Individuals in Cartesian GP (CGP) are created by randomly selecting nodes to fill out the grid to have individuals with 20 layers of width 3 [4]. The crossover operation uses a uniform crossover whereas the mutation operation replaces nodes randomly at a set rate. Selection is performed using the tournament selection technique.

2.3 Grammatical Evolution

Individuals are created iteratively using a random search baseline model which is based on the tree-grow method in which the best individual found in all trials is recorded. The parsimony pressure of 0.1 per expansion is implemented as well as a max size of 50 expansions. A one-point crossover and a generational replacement are used. The mutation uses the *integer flip per codon* technique. Selection is performed using the tournament selection technique. The following grammar template is used:

$$< e > ::= (< e > + < e >) \,|\, (< e > - < e >) \,|\, (< e > * < e >) \,|$$
$$\text{pdiv}(< e >, < e >) \,|$$
$$x[:, < varidx >] \,|< c >$$
$$< varidx > ::= \text{GE\_RANGE} : \text{dataset\_n\_vars}$$
$$< c > ::= < d > . < d > \,|- < d > . < d >$$
$$< d > ::= \text{GE\_RANGE} : n$$

2.4 Gene Expression Programming

Individuals in Gene Expression Programming (GEP) are created using the ramped half-and-half technique. The user-defined parameters, the length of the head, and the number of the genes are tuned for all datasets. Function symbol: addition is used as gene linking operator. A uniform point mutation and the following crossover operators are used: one-point, two-point, and a gene-crossover. A gene-crossover operator exchanges the entire gene between two chromosomes. Transposition is realized through the use of the insertion sequence in which a segment across the chromosome is chosen randomly and inserted at another position (except the start position) in the head of the gene. Selection is performed using the tournament selection technique.

2.5   Stacked-Based Genetic Programming

Individuals in Stacked-based GP (SGP) consist of genomes that are linear Push programs. The mutation operator replaces a random subsection of the program with another one of a similar size to the original using a Gaussian random offset. The crossover operator selects a random subsection of a donor and places it randomly into the parent, overwriting anything already using that section and increasing the program length if necessary. Selection is performed using the lexicase selection technique, and a size-neutral variation is used.

2.6   Multi-expression Program

The representation in the Multi-expression program (MEP) is linear. The length of the chromosome is constant and consists of a given number of genes. Genes are sub-strings of variable length in which a gene encodes either a function or a terminal. The first symbol in the chromosome is supposed to be a terminal. A uniform crossover is used. In mutation, there is no restriction in symbols changing for other genes - the terminal symbol can be changed to either a function or another terminal symbol and likewise for a function symbol. Selection is performed using the binary tournament selection technique.

3   Experimental Set-Up

It may be very challenging to achieve a numerical comparison of GP variants due to several reasons such as fairness since the same parameters are expected to be used among the variants. Also, the quality of the results can be affected by minor parameters such as the probability of applying various operators. As such, a feasible way is to come up with the optimal settings for each variant to make the comparison fair though promoting the uneven qualities of implementation.

Eleven datasets are used in the experiments, three are from the symbolic regression benchmark suite (S_R), four from predictions (R), and four from classifications (C) [17]. The selected datasets are of varying sizes and complexity. Real-world datasets are standardized. Table 1 presents a summarized description of the datasets.

The mean square error (MSE) is used to ascertain the prediction performance for regression problems. The precision metric is used to ascertain the classification performance which is computed as (TP/TP+FP) in which TP is a true positive and FP is a false positive. Lower values of MSE are favorable while higher values of precision are favorable. The runtime for each GP variant is measured as the time taken between the start and end of the evolutionary process. Hyper-parameter tuning is performed on each GP variant for each dataset to minimize the performance difference due to implementation. As such, the intrinsic capability of each variant is measured. For each variant, the following metrics were measured: the model's accuracy, computation time, the number of generations, and population size. Each variant is executed until it converges.

Table 1. Regression and classification datasets.

Name	Type	Data source	Vars	- Training set - Testing set
Vladislavleva-4	S_R	$y = \frac{10}{5+\sum_{i=1}^{5}(x_i-3)^2}$	5	U[0.05, 6.05, 1024] U[−0.25, 6.35, 5000]
Pagie-1	S_R	$y = \frac{1}{1+x_1^{-4}} + \frac{1}{1+x_2^{-4}}$	2	E[−5, 5, 0.4] E[−5,5,0.1]
Keijzer-6	S_R	$y = \sum_i^x \frac{1}{i}$	1	E[1, 50, 1] E[1, 120, 1]
Dow	R	Dow Chemical	57	596 150
Housing	R	Housing Values	13	354 152
Tower	R	Gas Chromatography data	5	4721 278
PowerPlant	R	Combined cycle power plant	5	7654 1914
Bank Notes	C	Identification of Banknotes	5	2080 520
SpamBase	C	Classification of Spam in Emails	58	3680 920
Credit Card	C	Credit Card Defaulters	25	320 80
Wine Quality	C	Red and white variants of wine	13	5198 1300

The statistical analysis is performed on six populations (GP variants) at $\alpha = 0.05$ significance level. The Kruskal-Wallis nonparametric test was used to perform statistical tests owing to the data characteristics and varying initial population for each variant [16]. The Mann-Whitney posthoc test with Bonferroni correction was used to perform the pairwise comparisons given that there exist statistically significant differences.

The experiments are executed in a python environment. The hyper-parameter tuning is done using GridSearchCV in sklearn's model_selection package which uses cross-validation on the dataset [18]. The experiments are carried out in two mainstream research areas: regression and classification summarized in Table 1.

4 Results and Discussion

Table 2 presents the obtained average results on generalization for the evaluation of six GP variants discussed in Sect. 2 using MSE for regression and precision (prec) for classification problems.

The optimal values of population size (pop), number of iterations (gen), and runtime (in seconds) when the algorithm converged were also reported in Table 2. Each variant was run 30 times for each dataset and the obtained results

C. Kuranga and N. Pillay

Table 2. Results for GP variants on regression and classification problems.

Dataset	Measure	Cano GP	GE	GEP	CGP	MEP	SGP
Vlas	MSE	1.6846	0.1729	2.8893	0.0410	**0.034**	0.0467
	Time	0.6561	0.3864	**0.021**	0.6397	30.9	7908.1
	Pop	100	200.00	**5**	10	200	50
	Gen	300	500	**10**	80	500	100
Keij	MSE	0.0099	0.0045	0.2177	0.0170	**0.0034**	7.3902
	Time	0.1685	0.1586	**0.021**	6.5344	1.8	201.7
	Pop	50	200.00	**5**	10	200	100
	Gen	50	500	**10**	80	500	100
Paige	MSE	9.2292	3.0001	8.6807	0.2296	**0.1091**	0.5683
	Time	2.2426	0.5936	**0.023**	0.6620	2.8	52.1
	Pop	100	200.00	**5**	10	100	50
	Gen	300	500	**10**	100	100	100
Dow	MSE	0.3478	0.1042	**0.0793**	1.3396	0.1401	0.1183
	Time	0.3871	**0.3006**	1210.9	424.6	2.4	6032.17
	Pop	300	255.00	**100**	100	200	**100**
	Gen	350	350	**120**	200	200	400
Housing	MSE	0.1106	0.0992	**0.0056**	0.0163	0.0080	0.0117
	Time	0.9718	**0.8591**	46.6	108.3	1.3	102.8
	Pop	300	255.00	**80**	100	200	100
	Gen	380	400	**100**	100	100	200
Tower	MSE	3548.96	**0.0863**	988.7	5043.8	32.63	8753.8
	Time	1997.2	**9.238**	2197.3	483.5	40.6	51846.6
	Pop	300	255.00	**80**	100	100	100
	Gen	300	350	180	200	200	**100**
P_Plant	MSE	214.4	**0.7375**	18.24	3864.3	29.03	0.1183
	Time	8.3928	**6.2831**	2192.8	462.3	63.8	6032.17
	Pop	350	500.00	**90**	200	200	100
	Gen	1000	1000	500	**100**	600	500
B_Notes	Prec	0.6437	0.9025	0.8077	0.5699	**0.9993**	0.3776
	Time	2.2963	**1.5494**	83.63	12.08	38.2	1831.9
	Pop	300	255.00	**70**	100	100	100
	Gen	1000	1000	250	**100**	**100**	**100**
SpamBase	Prec	0.9103	0.7884	0.8974	0.4179	**0.9247**	0.8293
	Time	3.1837	**2.9387**	162.7	14.41	79.0	27854.1
	Pop	350	255.00	**80**	100	100	100
	Gen	500	500	100	100	100	**50**
C_Card	Prec	**0.9317**	0.7358	0.8007	0.2298	0.8160	0.8031
	Time	**8.7**	10.206	266.7	258.6	83.6	9992.9
	Pop	300	255.00	80	100	100	**20**
	Gen	700	1000	**100**	**100**	**100**	**100**
W_Quality	Prec	0.8473	**0.8674**	0.8496	0.5394	0.9989	0.8377
	Time	**4.0**	4.3	99.67	106.4	42.3	4455.9
	Pop	350	255.00	**80**	100	100	100
	Gen	500	500	300	**100**	**100**	**100**

for the best-found individual for each run were averaged. Statistical analysis of the obtained results through the use of the Kruskal-Wallis non-parametric test and Mann-Whitney post-hoc test indicates that there was a statistically significant difference between the pairwise comparisons of GP variants for all datasets presented in Table 1. As such, the choice of GP variant has an impact on the performance of regression and classification problems.

From the results presented in Table 2, MEP exhibits the best MSE values on symbolic regression problems. The lowest population size and number of iterations were obtained by GEP. For prediction problems, GEP exhibits outstanding performance in terms of MSE on Housing and Dow datasets, GE exhibits the best MSE on Tower whereas SGP on the PowerPlant datasets. However, GE had the highest population size and number of iterations. MEP exhibits the best MSE values on classification problems with very few exceptions. Canonical GP and GE exhibit the best runtime whereas GEP had the lowest population size.

Figure 1 depicts the performance of the GP variants on varying problem sizes and levels of complexity of datasets. As illustrated in Fig. 1, Keijzer 6 dataset was the simplest to model for all the techniques except SGP. The performance of GEP and Canonical GP deteriorated exponentially when the level of complexity increased whereas MEP and CGP scaled gracious. For prediction problems, Housing and Dow datasets were predicted with ease by all techniques.

However, there was a sharp spike in performance deterioration for the PowerPlant dataset, especially for SGP and GEP. Conversely, SGP scaled gracious as the problem size and level of complexity increased and GE tends to be scalable for prediction problems whereas CGP exhibits the worst performance. The performance of canonical GP improves to yield outstanding performance for credit data.

The experimental results revealed the following: MEP exhibits the best overall performance in terms of MSE and precision, ranked the first in two problem domains. GEP has the least iterations and population sizes for all symbolic regression problems though the algorithm may have been trapped in local minima in which the algorithm converged in the first 10 iterations.

The outstanding performance of MEP could have been attributed to its ability to represent multi-expression giving it a higher chance of finding a solution and also, its code-reuse ability. However, the major drawback of MEP is on computational time. An outstanding performance of GEP in prediction problems could have been attributed to the concept of separating the individual into a head and tail which could have provided an effective way of encoding syntactically correct computer programs. It can be concluded that GP variants, MEP and GEP, that use chromosomes of variable length perform better compared to GP variants that use fixed lengths.

The poor performance of SGP could have been attributed to the issue of the low locality that could have possibly destroyed the good sub-trees created by the crossover. The worst performance of CGP on both regression and classification problems could have been attributed to the wasted evaluation and the effect the genetic operator had on traversing the search space.

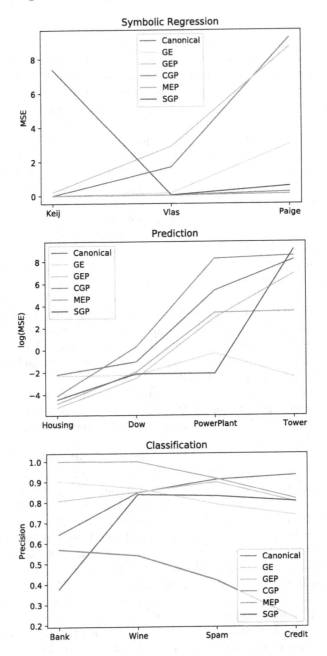

Fig. 1. Scalability performance of GP variants.

5 Conclusion and Future Work

The obtained results suggest that the choice of GP variant has an impact on the performance of regression and classification problems. Generally, the number of iterations, runtime, and population size are directly proportional to the success rate. MEP exhibits an outstanding performance as a regressor and classifier solver while CGP is the worst. GE exhibits exceptional performance on real-world regression problems which are characterized by numerous input variables.

Future work could consider hybridizing GP variants with one another to improve both performance and scalability. Implementing a multi-expression paradigm with other GP variants such as GE and GEP, and algorithm-specific features to other algorithms like having variable chromosome length can also be considered.

References

1. Koza, J.R.: Genetic Programming II: Automatic Discovery of Reusable Subprog. MIT Press, Cambridge (1994)
2. Oltean, M., Grosan, C.: A comparison of several linear genetic programming techniques. Complex Syst. **14**(4), 285–314 (2003)
3. Sette, S., Boullart, L.: Genetic programming: principles and applications. Eng. Appl. Artif. Intell. **14**(6), 727–736 (2001)
4. Ahvanooey, M.T., Li, Q., Wu, M., Wang, S.: A survey of genetic programming and its applications. KSII Trans. Internet Inf. Syst. (TIIS) **13**(6), 1765–1794 (2019)
5. Aguilar-Ruiz, J.S., Riquelme, J.C., Toro, M.: Evolutionary learning of hierarchical decision rules. IEEE Trans. Syst. Man Cybern. Part B (Cybern.) **33**(2), 324–331 (2003)
6. Langdon, W.B., Poli, R.: Fitness causes bloat. In: Chawdhry, P.K., Roy, R., Pant, R.K. (eds.) Soft Computing in Engineering Design and Manufacturing, pp. 13–22. Springer, London (1998). https://doi.org/10.1007/978-1-4471-0427-8_2
7. Sadrossadat, E., Basarir, H., Karrech, A., Durham, R., Fourie, A., Bin, H.: The optimization of cemented hydraulic backfill mixture design parameters for different strength conditions using artificial intelligence algorithms. In: Topal, E. (ed.) MPES 2019. SSGG, pp. 219–227. Springer, Cham (2020). https://doi.org/10.1007/978-3-030-33954-8_28
8. Gholampour, A., Gandomi, A.H., Ozbakkaloglu, T.: New formulations for mechanical properties of recycled aggregate concrete using gene expression programming. Constr. Build. Mater. **130**, 122–145 (2017)
9. Yilmaz, S., Sen, S.: Early detection of botnet activities using grammatical evolution. In: Kaufmann, P., Castillo, P.A. (eds.) EvoApplications 2019. LNCS, vol. 11454, pp. 395–404. Springer, Cham (2019). https://doi.org/10.1007/978-3-030-16692-2_26
10. Contreras, I., Bertachi, A., Biagi, L., Vehí, J., Oviedo, S.: Using grammatical evolution to generate short-term blood glucose prediction models. In: KHD IJCAI, pp. 91–96 (2018)
11. Vasicek, Z.: Bridging the gap between evolvable hardware and industry using cartesian genetic programming. In: Stepney, S., Adamatzky, A. (eds.) Inspired by Nature. ECC, vol. 28, pp. 39–55. Springer, Cham (2018). https://doi.org/10.1007/978-3-319-67997-6_2

12. Elola, A., Del Ser, J., Bilbao, M.N., Perfecto, C., Alexandre, E., Salcedo-Sanz, S.: Hybridizing Cartesian genetic programming and harmony search for adaptive feature construction in supervised learning problems. Appl. Soft Comput. **52**, 760–770 (2017)

13. Wang, H.L., Yin, Z.Y.: High performance prediction of soil compaction parameters using multi expression programming. Eng. Geol. **276** (2020)

14. Fallahpour, A., Wong, K.Y., Rajoo, S., Tian, G.: An evolutionary-based predictive soft computing model for the prediction of electricity consumption using multi expression programming. J. Clean. Prod. **283** (2021)

15. Helmuth, T., Spector, L.: General program synthesis benchmark suite. In: Proceedings of the 2015 Annual Conference on Genetic and Evolutionary Computation, pp. 1039–1046 (2015)

16. Chitty, D.M.: Faster GPU-based genetic programming using a two-dimensional stack. Soft. Comput. **21**(14), 3859–3878 (2017)

17. UCI ML Repository dataset. http://archive.ics.uci.edu/ml/datasets. Accessed 25 Jan 2022

18. Pedregosa, F., et al.: Scikit-learn: machine learning in Python. J. Mach. Learn. Res. **12**, 2825–2830 (2011)

Optimization of Parameterized Behavior Trees in RTS Games

Tomasz Machalewski⬭, Mariusz Marek(✉)⬭, and Adrian Ochmann

Institute of Computer Science, University of Opole,
ul. Oleska 48, 45-052 Opole, Poland
{tmachalewski,mmarek,aochmann}@uni.opole.pl

Abstract. Introduction of Behavior Trees (BTs) impacted the field of
Artificial Intelligence (AI) in games, by providing flexible and natural
representation of non-player characters (NPCs) logic, manageable by
game-designers. Recent trends in the field focused on automatic creation
of AI-agents: from deep- and reinforcement-learning techniques to combinatorial (constrained) optimization and evolution of BTs. In this paper,
we present a novel approach to semi-automatic construction of AI-agents,
that mimic and generalize given human gameplays by adapting and tuning of expert-created BT under a developed similarity metric between
source and BT gameplays. To this end, we formulated mixed discrete-
continuous optimization problem, in which topological and functional
changes of the BT are reflected in numerical variables, and constructed a
dedicated hybrid-metaheuristic. The performance of presented approach
was verified experimentally in a prototype real-time strategy game. Carried out experiments confirmed efficiency and perspectives of presented
approach, which is going to be applied in a commercial game.

Keywords: Real-time strategy · Behavior Tree · Multivariate time
series · Optimization · Metaheuristic

1 Introduction

Artificial Intelligence (AI) in computer games is attributed with great importance and responsibility - at the same time it can breathe life into an otherwise procedural, predictable and recurrent game-world, but can also make it
unplayable and unnatural. Therefore, a vast body of research has been carried
out to model behaviors of Non-Player Characters (NPCs) in games, e.g., [2,14],
providing many representations and algorithms. One of them, Behavior Trees
(BTs), impacted the field by providing flexible and natural representation of
NPCs logic, manageable by game-designers [3,24]. Their success in commercial
games made them implemented either as a part of game engines (CryEngine,
Unreal Engine) or as plugins (Unity) [19]. Nevertheless, increased pressure on
ever better NPCs AI-agents forced complexity of hand-crafted BTs to became
barely-tractable and error-prone, if not created by experienced AI-engineers.

L. Rutkowski et al. (Eds.): ICAISC 2022, LNAI 13588, pp. 387–398, 2023.
https://doi.org/10.1007/978-3-031-23492-7_33

On the other hand, while many just-launched on-line games suffer from player-shortage, the existence of AI with a broad-range of experience and capabilities could increase players retention [4]. Therefore, to handle above challenges, recent trends in the field focused on automatic creation of AI-agents: from deep- and reinforcement-learning techniques [6,9,13,16,21] to combinatorial (constrained) optimization and evolution of BTs [11,18,22].

Although, obtained results are impressive, they still leave room for further development and improvements. While a milestone has been reached with AlphaStar [23], achieving a grandmaster level in StarCraft II and beating over 99% of players, obtaining such results demand excessive resources involved in development, i.e., a lot of effort and training data.

The contribution of this paper is twofold. On the expository side we present a novel approach to semi-automatic construction of AI-agents, that mimic and generalize given human gameplays by adapting and tuning an expert-created BT, comprising a predefined options of topological and functional changes as parameterized nodes. To this end, we formulated mixed discrete-continuous optimization problem, in which parameters of the BT are reflected in numerical variables, and constructed a dedicated hybrid-metaheuristic, guided by developed similarity metric, comparing source and BT gameplays. The performance of presented approach was confirmed experimentally in a prototype Real-Time Strategy (RTS) game.

The proposed approach is well-suited for small gamedev teams, which, with moderate effort of AI-engineers, are enabled to generate different AI-agent instances without resources needed by other methods, and, not less importantly, which are easily interpretable - by nature. In the case of considered game studio, example gameplays, defining playstyle to imitate, were directed by game-designers with use of a special tool, aiding a careful design of gameplays - with play, stop, rewind and post in-game action functionalities.

The rest of the paper is organized as follows. The next section describes the rules of considered real-time strategy (RTS) game. Details of the presented approach are given in Sect. 3, whereas in Sect. 4 settings and results of numerical experiment are presented. The last section concludes the paper.

2   The Game

The prototype game, provided by BAAD Games Studio, is an RTS game, in which two players (red and green) manage their resources to find such a balance between battle and development, that either the other player is destroyed or the player acquired more resources at the end of a 15 min gameplay.

The game is played on a two-dimensional grid of hexagonal cells. Each cell has its *position* on a map, and is either enabled or disabled for the game.

Players manage their game entities under limited *gold* resource. Each game entity has an unique identifier (*id*), *state* describing its current activity (eg., *idle*, *moving*, etc.), *health points* and a *class*. There are two classes of game entities - *units* and *buildings*. A unit represents a *quantity* of movable forces of the same *type*, where *type* (peasant, knight, archer) determines its combat characteristics. Buildings are able to produce resources or entities (at the expense of gold). Each building has its *type*: castle (can settle other buildings), farm (continuously delivers gold unless under attack), barracks (continuously trains assigned forces, up to its capacity limit, unless under attack), and tower (defends cells in the predefined radius by performing distanced attack on enemy units). Quantity, speed and available subset of produced entities depend on a *type* and a *level* of a building.

During the game, all entities are controlled by issuing *actions* with proper parameters. On the engine side, the game consists in *rounds* of 1/10 s. During each round, actions issued by each player are scheduled to be performed at the end of a round (in the order of red-green player).

A *move* (*id, position, proportion*) action commands a *proportion* of the unit *id* to move to the destination *position*, where *proportion* $\in \{0.25, 0.5, 0.75, 1.0\}$ (a new splitted unit is generated if *proportion* < 1.0). The unit performs its movement along the path, given by the pathfinding module, precomputed with avoidance of cells disabled or occupied by enemy entities. In the case when destination position is occupied by an unit of the same player, units are merged if both are of the same type, or the unit stops at the last feasible cell on a path, otherwise.

In each round, every unit executes tasks, ordered by their priority: discovery of enemy unit in the attack range (same cell in the case of peasant and knight, or adjacent cell in the case of archer) and performing an attack; discovery and attack of enemy's building (while encountered on the same cell); continuation of movement along a given path; response to an attack with own attack, resulting in an uninterruptible *battle*.

A *spawn unit* (*id, type, quantity*) action purchases a unit of *quantity* and *type* from already trained in the building *id*. A *settle building* (*id, type, position*) action commands the castle *id* to settle *type* building at *position* on the map. *Upgrade* (*id*) and *repair* (*id*) actions raise the *level* or repair of building *id*, respectively.

The game engine provides full determinism given the same initial *seed* value. Therefore, a gameplay can be recorded and later replayed as an *Action Time-List* (ATL) for each player - an ordered collection of actions with their parameters for each round. Based on this property of the game, there are three types of players:

- Human-players - issuing their actions through user interface - only feasible actions are triggered in ths case.
- ATL-players - precisely replaying a given ATL. In this case, triggering an infeasible action results in *failure* status of the game.
- BHT-players - AI-based players, performing actions according to a logic encoded in a Behavior Tree. A special *query* action provides a data structure describing a current state of the game-world.

The prototype game is targeted to mobile market and has the complete set of features, but is limited in a variety of entity types. Note, that even such defined game poses a real challenge for AI developers to construct an algorithm guiding and managing game entities throughout the whole game.

3   The Methodology

Let $\mathcal{G}(a, b)$ denote a gameplay of red (a) and green (b) players, where a and b are either ATLs or BTs, and let A and B be ATLs of a context (source) gameplay between red and green players, respectively. The goal is to construct a BT T, that mimics and generalizes a playstyle of the red player, i.e., while T could be autonomous AI-player, gameplays $\mathcal{G}(A, B)$ and $\mathcal{G}(T, B)$ should be similar (according to some metric).

Apart from similarity metric, a BT T is said to be *infeasible*, if its corresponding BHT-player disrupts a game reproduced by its ATL opponent, i.e., during the $\mathcal{G}(T, B)$ there is an action issued by B that breaks the game with *failure* status; T is *feasible* otherwise.

3.1   Behavior Trees

Classical Behavior Tree is a hierarchical structure of nodes, where each node is associated either with a task in game-world (leaf node) or performs a control-flow logic (internal node), executed in a depth-first search fashion. Execution of a node returns either *success* or *failure* status to its parent, depending whether its goal was achieved. This status is then used by the parent to execute or prune its remaining children.

Leaf nodes (called *actions*) interact with a game-world by issuing game-actions, described in the previous section. We assume that leaves issue only a valid game-actions, and therefore always return *success* status.

Internal tree nodes control execution flow of their children. A *selector* node sequentially executes its children; if any child returns with *success*, the node stops execution and returns with *success*, otherwise it returns *failure*. A *sequence* node also sequentially executes its children; the node returns with *success* if all its children succeeded, and returns *failure* as soon as any of them fails.

In classical BT leaf nodes called *conditions* are used to check whether a given condition is satisfied in the game-world. As issuing many queries into the game may be inefficient, we developed a caching technique. Let *BlackBoard* (BB) be a set of $(key, value)$ pairs, accessible to all the nodes of a BT, used as a persistent shared memory. Then, a special action node (*GameQuery*) issues the *query* game-action and fills BB with the current state of the game-world, to be read by the other nodes as needed.

3.2   Adaptive Behavior Trees

In the design of the methodology, we abandoned methods constructing BT from scratch, by iteratively evolving its topology, as they were not able to produce, in general, a reasonable BT. Note, while such a solution space can be used in searching for "the best" BT, it is still unsuitable for the considered problem, with almost all solutions evaluated as infeasible. Therefore, we adopt the approach of [20], in which a BT is pre-created by expert AI-designer in such a way, that functional changes to the BT are controlled by node parameters, i.e., nodes logic can be parameterized by discrete or continuous values. Such created BT forms then a general domain to be adaptively tuned, either algorithmically or by a game-designer, to meet the expectations.

To this end, we developed BT nodes that reflect topological BT changes as theirs parameters:

- time-dependent selector - executes one of its children $c_1, \ldots, c_j$ (phases) according to the current *game time* and lengths of their time intervals $l_1, \ldots, l_{j-1}$, respectively.
- switching selector - executes one of its children $c_1, \ldots, c_j$ according to the value $v \in \{1, \ldots, j\}$ of its parameter.
- leaf-nodes with parameterized logic.

Let Adaptive Behavior Tree (ABT) $T(p)$ be a BT with parameterized nodes, where $p = \{p_1, \ldots, p_k\}$ is a vector of BT parameters, $P = P_1 \times \ldots \times P_k$ is a domain of $T(p)$ $(p \in P)$ and P_i is a domain of p_i $(p_i \in P_i)$, $i \in \{1, \ldots, k\}$. Note, all $T(p)$, $p \in P$, are feasible, i.e., each of them can be fully-functional AI-agent.

3.3 Strategy

In RTS games, the problem of developing advanced AI is particularly complicated, because of the necessity to observe a large area and to react to occurring events [17]. The AI must also simultaneously manage many units of different specifications. In addition, all decisions cannot be a mere consequence of a map situation, but must be result of a long-term strategy, dynamically updated during a gameplay.

We call an ABT a *strategy*, if it can be used as an AI-player, able to complete the game while competing with a player. To this end we developed an adaptive strategy, consisting of almost 300 nodes, for the considered RTS game. Strategy's main idea is its ability to change the playstyle over time, due to multiple sub-strategies, i.e., initially AI can be focused on the development and defence, to become more aggressive later.

3.4 Similarity Metric

For the purpose of measuring similarity between two gameplays, a context gameplay $\mathcal{G}_C$ and an evaluated gameplay $\mathcal{G}_E$, we developed a metric, which heuristically assesses similarity between two multivariate time series, representing a generalized views on corresponding gameplays, from the evaluated player perspective.

Let $S_{\mathcal{G}} = (S_{\mathcal{G}}^1, \ldots, S_{\mathcal{G}}^n)$ be a Snapshot Time-Line of a gameplay $\mathcal{G}$, where $S_{\mathcal{G}}^i = (S_{\mathcal{G}}^{i,1}, \ldots, S_{\mathcal{G}}^{i,k})$ (a game snapshot at time i) describes a state of player's resources $S_{\mathcal{G}}^{i,1}, \ldots, S_{\mathcal{G}}^{i,k}$ at time i, $i \in \{1, \ldots, n\}$ and n is a number of snapshots, assuming they are sampled with a constant interval throughout the gameplay. In the presented approach, we sampled the amount of gold, the total number of units, the total number of buildings and numbers of entities of the same type as a representation of a game-state.

Let $S_{\mathcal{G}_C}$ and $S_{\mathcal{G}_E}$ be min-max normalized to $[-1, 1]$ interval. Then, a similarity matrix d, inspired by Self-Similarity matrix, often used recently and in the past in sequential data processing research [7,8], is computed in such a way, that each element $d_{i,j}$ represents similarity between $S_{\mathcal{G}_C}^i$ and $S_{\mathcal{G}_E}^j$ computed as averaged Manhattan distance, i.e.,

$$d_{i,j} = \frac{1}{k} \|S_{\mathcal{G}_C}^i, S_{\mathcal{G}_E}^j\|_1 = \frac{1}{k} \sum_{l=1}^{k} |S_{\mathcal{G}_C}^{i,l} - S_{\mathcal{G}_E}^{j,l}|.$$

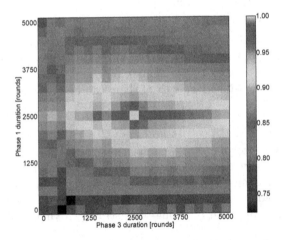

Fig. 1. Similarity metric values for different parameters of ABT.

The main idea of the presented approach is to promote similarities, in which pairs of non-distant snapshots are appearing after each other in the matrix d. To this end, let a trend t be a series of diagonal elements of matrix d, i.e., a trend t starting on $d_{i,j}$ of length $\|t\| = l$ is defined as

$$t = (d_{i,j}, d_{i+1,j+1}, \ldots, d_{i+l,j+l}).$$

Let $v(t)$, an evaluation of t, equal

$$v(t) = \left(\sum_{x \in t} x \right) * \|t\|^2.$$

We defined a trend-series $\mathcal{T} = \{t_1, \ldots, t_z\}$ as a set of trends, for which the following conditions hold:

- there is exactly one cell in each column of d belonging to some trend in $\mathcal{T}$,
- no two trends from $\mathcal{T}$ can be merged to form a longer trend.

Let an evaluation of $\mathcal{T}$, denoted as $V(\mathcal{T})$, equal

$$V(\mathcal{T}) = \frac{\sum_{t \in \mathcal{T}} v(t)}{(\sum_{t \in \mathcal{T}} \|t\|)^3}.$$

Then, a similarity metric $\mathcal{S}\left(\mathcal{G}_C, \mathcal{G}_E\right)$ between $\mathcal{G}_C$ and $\mathcal{G}_E$ gameplays is equal to a maximum value of trend-series over similarity matrix d obtained from $\mathcal{G}_C$ and $\mathcal{G}_E$. Corresponding value can be computed by a simple algorithm, utilizing prefix-sums and dynamic programming, of time complexity $O(\max(m, n)^3)$, where m and n are length of two Snapshot Time-Lines.

An example application of developed metric is presented in Fig. 1, showing values of similarity between a context game of T_A and T_B with fixed parameters, and gameplays in which values of two T_A parameters were changed, i.e., for the context gameplay the lengths of the first and third phases in the time-dependent selector node was set to 2500 rounds, while for evaluated gameplays their value was changed from 0 to 5000 rounds.

A distinct gradient can be seen, with peak reached at the point $(2500, 2500)$, reflecting comparison with target match. Similarity greatly differs in the vertical axis, the length of first phase - similarity metric appropriately distinguishes games varying in this parameter, because length of this phase strongly affects later course of the game. The metric has more difficulty in differentiating games varying in length of the third phase, that is justified by its similarity with the fourth phase, which complements the remaining game, up to 15 min (10000 rounds).

A similarity landscape, generated by developed metric, exhibits properties desirable by gradient-based optimization techniques. It properly reflects continuous changes in parameters values in the neighborhood of the global optimum - around values of context gameplay. On the other hand, discontinuities and small local extrema are to be dealt with design of metaheuristic optimization algorithm.

3.5 Optimization Problem

Given an ABT $T(p)$, its domain P and a context ATLs A and B, the goal is to find such $p^* \in P$, that p^* is feasible and similarity metric $\mathcal{S}$ between $\mathcal{G}(A, B)$ and $\mathcal{G}(T(p^*), B)$ is maximized, i.e.,

$$p^* = \arg\max_{p \in P'} \mathcal{S}\left(\mathcal{G}(A, B), \mathcal{G}(T(p), B)\right),$$

where $P' = \{p \in P | T(p) \text{ is feasible}\}$ ($\mathcal{G}(T(p), B)$ does not end with a *failure* status).

4   Experimental Evaluation

The methodology developed in Sect. 3 allows to cast a problem of automatic construction of AI-agents, that mimic and generalize given human gameplay, as an optimization problem. Let P, a domain of parameters of a given ABT, be a solution space to be searched for a solution $p^* \in P$ solving (Sect. 3.5). Such an optimization problem is characterized by a nontrivial objective function, requiring simulation of a gameplay for each solution $p \in P$.

To solve formulated optimization problem, in this section a hybrid metaheuristic [1] based on Memetic Search [15] is developed, to verify applicability and performance of the presented approach. Note, the presented algorithm is not constructed with time efficiency in mind. It is a proof of concept and a hint of promising optimization techniques, since in a solution to be deployed, the optimization problem is solved by a cloud-based parallel algorithm, tackling many Behaviour Trees at once with hierarchical optimization techniques, and using a scalable pool of game simulators.

4.1   Metaheuristic Optimization Algorithm

As a solution search environment we constructed an algorithm based on the Memetic Search hybrid-metaheuristic, which combines the strength of evolutionary algorithms in diversification of the search space exploration with complementary search intensification property of driven trajectory-method.

Criterion. The solution search process is obviously driven by the similarity metric S between context and evaluated gameplays, defined in Sect. 3.4. Nevertheless, not all solutions from P are feasible. In this case, we introduced a *penalty* for infeasible solutions:

$$penalty = \left| 1 - \frac{\|\mathcal{G}_E\|}{\|\mathcal{G}_C\|} \right|,$$

where $\mathcal{G}_C$ is a context gameplay, $\mathcal{G}_E$ is an evaluated gameplay and $\|\mathcal{G}\|$ denotes the length of a gameplay $\mathcal{G}$ (the number of rounds). Note, the less rounds were performed in an evaluated gameplay the greater a *penalty* value.

Finally, the criterion value $f(p)$ of a solution p used in the presented algorithm equals

$$f(p) = S\Big(\mathcal{G}(A, B), \mathcal{G}(T(p), B)\Big) -$$
$$\left| 1 - \frac{\|\mathcal{G}(T(p), B)\|}{\|\mathcal{G}(A, B)\|} \right|, \tag{1}$$

where A and B are ATLs of a context gameplay and $T(p)$ is evaluated ABT.

Since computation of a similarity measure between context and evaluated gameplays is a computationally demanding task, we implemented the algorithm in such a way, that once a solution is evaluated, its corresponding criterion value is cached for future usage.

Memetic Search. The Memetic Algorithm [15] in its basic form is a classic evolutionary algorithm, iteratively managing a set of solutions (a population) by applying crossover and mutation operators, additionally using an improvement algorithm as an intensification strategy.

To model and implement the algorithm we developed a *Heuristics Composition Engine*, in which algorithms are modeled as directed acyclic graphs. Nodes of such graphs represent some tasks performed on populations of solutions, and edges represent control and population flows. Nodes are executed in a topological order. An executed node forms its input population by merging output populations of its predecessors in a graph, and then computes its output population by performing its task.

Consider an initial population consisting of m solutions. The first node has an empty task - it only transfers initial population to subsequent nodes. In node 4 an input population, created from the best 5 solutions from initial population (node 2) and from $2m - 5$ solutions selected using *roulette wheel* rule (node 3), is shuffled. In node 5 a crossover operator is applied to consecutive pairs of solutions. Recombined solutions are mutated (with a small probability) in node 6. Node 7 applies an improvement algorithm to each solution from the input population, and finally node 8 selects the best m solutions from initial and improved populations. The procedure is repeated for n iterations.

Intensification Strategy. As an intensification strategy we used a Simulated Annealing (SA) [10]. The SA starts the search process from a given initial solution p. Then, at each iteration a new solution p' is randomly sampled from a neighborhood. The new solution p' replaces the old solution p either with a probability computed following the Boltzmann distribution $e^{-\frac{f(p)-f(p')}{\tau}}$ if $f(p') \leq f(p)$, or without a draw if $f(p') > f(p)$. The so-called temperature τ is decreased after each iteration $i = 1, 2, \ldots, i_{max}$ by a geometric cooling schedule, i.e., $\tau_i = \alpha\tau_{i-1}$, $\alpha \in (0,1)$.

4.2   Numerical Experiment

Taking into account limits induced by utilized delay-manager BT node, to show applicability of the presented approach, we designed a synthetic experiment setup as follows.

A gameplay $\mathcal{G}(T_A(p_A), T_B(p_B))$ of two ABTs $T_A(p)$ and $T_B(p)$ with fixed parameters $p_A \in P_A$ and $p_B \in P_B$ was recorded as a context gameplay. The ABTs represented two different strategies, described in Sect. 3.3.

The goal of the Memetic Search algorithm is to (re)discover of the parameters of a BHT-player. In the experiment, we use three ABT-driven players:

- Player 1 - $T_A(p)$, $p \in P_A$ - the ABT from the context gameplay,
- Player 2 - $T_A(p)$, $p \in P'_A$ - the domain P'_A of $T_A(p)$ was limited for one parameter in such a way, that $p_A \notin P'_A$,
- Player 3 - $T_B(p)$, $p \in P_B$ - the ABT of the opponent in the context gameplay.

As a by-product of such a setup, we eliminated the need for estimating delays
between actions performed by human players, as in both context and evaluated
gameplays delays are managed in the same way. On the other hand, since the
constructed ABTs limit the rate of performed actions to at most one per second,
the frequency of changes observed in the game-world is about 1 Hz. Therefore,
based on the Sampling Theorem [12], we set the sampling frequency of game
snapshots to 2 Hz[1].

Algorithms were coded in C# and simulations were run on a PC with CPU
Intel Core i7-3610QM 2.30 GHz and 16 GB RAM. Parameters of algorithms were
chosen empirically as follows: for the Memetic Search algorithm we set $m = 12$
and $n = 20$, and for subordinate SA we set $\tau_0 = 50$, $\alpha = 0.998$ and $i_{max} = 5$.
The initial population was drawn randomly from the domains of corresponding
ABTs. Due to non-deterministic nature of developed algorithms, each was ran
100 times, yielding 1 h per run on the average.

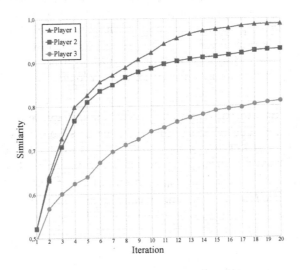

Fig. 2. Results of numerical experiment.

Figure 2 presents average values of similarity metric for the best found solution in subsequent iterations of the Memetic Search algorithm. It can be seen,
that developed heuristic similarity metric properly assess likelihood and differences between gameplays - parameters of Player 1 was determined correctly,
reaching almost perfect similarity, for Player 2 the similarity saturated at the
smaller value (0.93) - but reached limitations of its domain, and the value
obtained for Player 3 clearly renders his playstyle dissimilar.

---

[1] In a performed experiment (not showed here) we obtain 4 Hz sampling rate had no
effect on the convergence of the presented algorithm, while it dramatically increased
its run-time, due to $O(n^3)$ complexity of similarity metric. On the other hand,
with 1 Hz sampling frequency the algorithm was not able to find a good solution.

5   Conclusions

In this paper, we presented a novel approach to automatic construction of AI-agents, that mimic and generalize given human gameplays by adapting and tuning of ABTs - parameterized BTs, characterized by varying behaviors, diverse strategies and a range of skills and capabilities. To this end, we formulated mixed discrete-continuous optimization problem, in which topological and functional changes of the BT are reflected in numerical variables, and constructed a dedicated hybrid-metaheuristic, driven by developed similarity metric between source and BT gameplays. The performance of presented approach was confirmed experimentally on a prototype RTS game - ABT can be tuned so that it mimics human gameplay, given that it covers his playstyle.

The future work will be concentrated on mathematical models of delays in human gameplays [5]. In the presented approach, evaluated trees share the same simple model, therefore, it does not impact achievable similarity between parameterized BTs gameplays. On the other hand, adequate human-delay model, together with growing set of ABTs, covering a wide repertoire of playstyles, will enable on-demand generation of "ghost-players" - AI-agents mimicking requested human opponents, and in turn, monetization of such a feature.

Acknowledgments. The work was financially supported by the National Centre of Research and Development in Poland within GameINN programme under grant no. POIR.01.02.00-00-0108/16.

References

1. Blum, C., Roli, A., Sampels, M.: Hybrid Metaheuristics: An Emerging Approach to Optimization, vol. 114. Springer, Cham (2008)
2. Buckland, M., Collins, M.: AI Techniques for Game Programming. Premier Press (2002)
3. Colledanchise, M., Ögren, P.: Behavior Trees in Robotics and AI: An Introduction. CRC Press (2018)
4. Cowling, P.I., Devlin, S., Powley, E.J., Whitehouse, D., Rollason, J.: Player preference and style in a leading mobile card game. IEEE Trans. Comput. Intell. AI Games **7**(3), 233–242 (2014)
5. Gow, J., Baumgarten, R., Cairns, P., Colton, S., Miller, P.: Unsupervised modeling of player style with LDA. IEEE Trans. Comput. Intell. AI Games **4**(3), 152–166 (2012)
6. Harmer, J., et al.: Imitation learning with concurrent actions in 3D games. In: IEEE Conference on Computational Intelligence and Games (CIG), pp. 1–8 (2018)
7. Hirai, T., Sawada, S.: Melody2Vec: distributed representations of melodic phrases based on melody segmentation. J. Inf. Process. **27**, 278–286 (2019). https://doi.org/10.2197/ipsjjip.27.278
8. Jun, S., Hwang, E.: Music segmentation and summarization based on self-similarity matrix. In: Proceedings of the 7th International Conference on Ubiquitous Information Management and Communication, pp. 1–4 (2013)

9. Justesen, N., Risi, S.: Learning macromanagement in StarCraft from replays using deep learning. In: 2017 IEEE Conference on Computational Intelligence and Games (CIG), pp. 162–169. IEEE (2017)

10. Kirkpatrick, S., Gelatt, C.D., Vecchi, M.P.: Optimization by simulated annealing. Science **220**(4598), 671–680 (1983)

11. Liu, S., Louis, S.J., Ballinger, C.A.: Evolving effective microbehaviors in real-time strategy games. IEEE Trans. Comput. Intell. AI Games **8**(4), 351–362 (2016)

12. Luke, H.D.: The origins of the sampling theorem. IEEE Commun. Mag. **37**(4), 106–108 (1999)

13. McPartland, M., Gallagher, M.: Reinforcement learning in first person shooter games. IEEE Trans. Comput. Intell. AI Games **3**(1), 43–56 (2010)

14. Millington, I.: AI for Games. CRC Press (2019)

15. Moscato, P.: Memetic Algorithms: A Short Introduction, pp. 219–234. McGraw-Hill Ltd. (1999)

16. Oh, I.S., Cho, H., Kim, K.J.: Playing real-time strategy games by imitating human players' micromanagement skills based on spatial analysis. Expert Syst. Appl. **71**, 192–205 (2017)

17. Ontañón, S., Synnaeve, G., Uriarte, A., Richoux, F., Churchill, D., Preuss, M.: RTS AI problems and techniques. In: Lee, N. (ed.) Encyclopedia of Computer Graphics and Games. Springer, Cham (2019). https://doi.org/10.1007/978-3-319-08234-9_17-1

18. Robertson, G., Watson, I.: Building behavior trees from observations in real-time strategy games. In: 2015 International Symposium on Innovations in Intelligent Systems and Applications (INISTA), pp. 1–7. IEEE (2015)

19. Sagredo-Olivenza, I., Gómez-Martín, P.P., Gómez-Martín, M.A., González-Calero, P.A.: Trained behavior trees: programming by demonstration to support AI game designers. IEEE Trans. Games **11**(1), 5–14 (2017)

20. Shoulson, A., Garcia, F.M., Jones, M., Mead, R., Badler, N.I.: Parameterizing behavior trees. In: Allbeck, J.M., Faloutsos, P. (eds.) Motion in Games, pp. 144–155. Springer, Heidelberg (2011). https://doi.org/10.1007/978-3-642-25090-3_13

21. Song, S., Weng, J., Su, H., Yan, D., Zou, H., Zhu, J.: Playing FPS games with environment-aware hierarchical reinforcement learning. In: Proceedings of the 28th International Joint Conference on Artificial Intelligence, pp. 3475–3482. AAAI Press (2019)

22. Tomai, E., Flores, R.: Adapting in-game agent behavior by observation of players using learning behavior trees. In: Mateas, M., Barnes, T., Bogost, I. (eds.) Proceedings of the 9th International Conference on the Foundations of Digital Games, FDG 2014, Liberty of the Seas, Caribbean, 3–7 April 2014. Society for the Advancement of the Science of Digital Games (2014)

23. Vinyals, O., Babuschkin, I., Czarnecki, M.W., et al.: Grandmaster level in StarCraft II using multi-agent reinforcement learning. Nature **575**, 350–354 (2019)

24. Yannakakis, G.N., Togelius, J.: Artificial Intelligence and Games. Springer, Cham (2018)

Using Answer Set Programming
to Improve Sensor Network Lifetime

Artur Mikitiuk[(✉)] [ID], Krzysztof Trojanowski [ID], and Jakub A. Grzeszczak [ID]

Cardinal Stefan Wyszyński University in Warsaw, Wóycickiego 1/3,
01-938 Warsaw, Poland
{a.mikitiuk,k.trojanowski,jakub.grzeszczak}@uksw.edu.pl

Abstract. Sensor network lifetime maximization can be solved using heuristic methods, but they produce only suboptimal sensor activity schedules. However, knowing the quality of these solutions, we can use methods for solving decision problems to find better solutions than these suboptimal ones. We apply an answer set programming (ASP) system to answer the question, "Is there a schedule of length k?" where k is at least one unit higher than the best schedule returned by the heuristic method. First, we convert the problem's constraints and a particular data instance into a high-level constraint language theory. Then we use a *grounder* for this language and a *solver* for the language of grounder's output to find a more extended schedule or determine that no such schedule exists. The paper presents the conversion rules and the experiments' results with one of the ASP tools for selected classes of the SCP1 benchmark.

Keywords: Target coverage problems · Sensor network scheduling · Maximum lifetime optimization · Answer set programming

1 Introduction

Lifetime maximization of wireless sensor networks remains a subject of research interest. There are many variants of the problem depending on the real-world applications like, for example, border surveillance, monitoring of wildlife activity in hard-to-reach regions, or traffic control. Among them, we focus on the case where immobile battery-powered devices deployed within the target area gather and transfer information from the monitored field. Sensors have uniform limited battery capacity and sensing range, and the area contains a set of points of interest (POIs) to cover. The minimum level of coverage, that is, the percentage of POIs, located in the range of at least one working sensor, represents a necessary condition of feasible monitoring.

In the sensor network scheduling, we first identify coverage sets from the sensors, which satisfy the requested level of coverage, and then arrange them in a sequence, building a schedule for the sensor network. Because the maximum coverage sets scheduling problem is NP-hard [4], the application of heuristic

© The Author(s), under exclusive license to Springer Nature Switzerland AG 2023
L. Rutkowski et al. (Eds.): ICAISC 2022, LNAI 13588, pp. 399–410, 2023.
https://doi.org/10.1007/978-3-031-23492-7_34

methods is justified. Numerous heuristic methods have been proposed, which successfully find suboptimal solutions for the problem.

In this paper, we start where earlier methods end their work. The question is whether it is possible to improve the heuristic results. We want to identify classes of problems for which longer schedules can be found at a reasonable computational cost. To this end, we apply methods for solving decision problems.

The main contribution of this paper is encoding the problem in the language of one of the answer set programming (ASP) systems and using this ASP system to improve the results returned by heuristic methods. Our experiments show that this improvement in schedule length can be obtained when caring about the search space's size. This space expands rapidly as the numbers of sensors and POIs, and battery capacities grow. One could ask whether the improvement is worth additional computational efforts. The answer to this question depends on a particular application.

The paper consists of five sections. The Maximum Lifetime Coverage Problem is described in Sect. 2. Section 3 presents our representation of a sensor network, and application of an ASP system *clingo* to improve solutions obtained from heuristic methods. The experimental part of the research is described in Sect. 4. Section 5 concludes the paper.

2   Maximum Lifetime Coverage Problem (MLCP)

As a problem instance, we assume N_S immobile sensors randomly distributed over the network area to monitor N_P Points of Interest (POI). Every sensor of this network has the same sensing range and battery capacity T_{batt}. Battery usage is described with the discrete-time model of the sensor's activity. A sensor consumes one energy unit for every time slot it spends in its active state. For simplicity, the tiny energy consumption of a dormant sensor is ignored. For each time slot, an active sensor monitors all POIs within its sensing range. Effective monitoring of the network does not require every POI to be monitored at the same time. Usually, coverage (*cov*) of 80 to 90% is sufficient. In this case, we consider a single POI as covered if it is in the range of at least one active sensor.

In the model of the problem, we introduced some simplifications concerning energy consumption and communication. We assume that the sensors' activity control update is computed outside the network. A central computational unit has contact with all the sensors through mobile sinks moving around in the sensor field to collect data via short-distance radio communication. Therefore, we assume that the communication between the nodes is never disturbed. The use of mobile sinks and the exclusion of sinks from the scheduled set of sensors also allows assuming that communication evenly consumes sensors' energy. Hence, this consumption may be negligible from the point of view of the schedule optimization. Another reason for uneven energy consumption may be varying ambient temperature and frequent switching between states off and on. We decided to ignore their influence and allow sensors to switch between states without any additional cost, as the topic in question is not the focus of our studies.

Our goal is to find a sensor activity schedule that ensures a sufficient level of coverage every time step as long as possible for the model described above. This class of problems is referred to as the Maximum Lifetime Coverage Problem (MLCP) [1,2,7,8].

3   The ASP Approach

3.1   Sensor Activity Schedule Representation

We represent a schedule as a 0-1 matrix H. Its rows define the activity of the corresponding sensors over time. The columns define the state of all sensors during the corresponding time units (called slots). The H^j denotes the j-th column of H. $H^j[i] = 1$ ($H^j[i] = 0$, resp.) means that the i-th sensor is active (not active, resp.) at time j. For every column, all its active sensors cover the appropriate percentage of POIs, called the required level of coverage. Since every slot takes one unit of time, the number of slots is equal to the network's lifetime. The number of ones in a row of a schedule represents the working time of a sensor. Thus, it should not be higher than T_{batt}.

3.2   Adaptation to ASP Systems

Heuristic methods based on the matrix schedule representation produce a schedule where its length k represents its score. However, such schedules are suboptimal; that is, there is no guarantee that they are the longest possible schedules. Thus, the following question arises: "Is there a schedule of length $k + 1$?". To get an answer, we use a method for solving decision problems. In the case of a positive answer, we increment the value of k by one and repeat the question. As long as the answer is positive, we continue this process. This approach makes it theoretically possible to find an optimal solution to MLCP. However, there is a practical question of how long it takes to find a solution for a given problem instance and a specific value of k. The answer depends on the size of data and the efficiency of the software and hardware used for computations. If the data size is too big, the problem cannot be solved in a reasonable time. What is a reasonable time? It depends on a specific application.

Finding for a given network a schedule of a specific length can be solved using ASP systems. To do this, one needs to represent the constraints of a search problem as a theory P in a high-level constraint language. A specific instance of the problem has to be encoded separately as data D. Next, a specialized program called *grounder* compiles the pair (D, P) into a theory $T_{P,D}$ in some propositional target logic. The solutions to the problem instance must correspond to models of $T_{P,D}$ and be obtained from them quickly. Finally, a *solver* for this propositional target logic finds a model of $T_{P,D}$ or determines that no models exist.

If the target logic is classical propositional logic, one can use an off-the-shelf standard SAT solver to find solutions to the grounded theory. Another possibility is to use a logic allowing pseudo-boolean (PB) constraints and a SAT(PB) solver.

PB constraints such as cardinality constraints allow a more concise encoding of the problem. Moreover, using solvers dealing directly with PB constraints may be more effective than rewriting a ground theory to eliminate such constraints (which usually makes the theory larger) and using standard SAT solvers on the rewritten theory. The choice between a standard SAT solver and a SAT(PB) solver is arbitrary. Both types of solvers are the subject of intensive research and become more and more efficient. In the future, one could repeat all computations using a better solver.

3.3 Hypergraph Model Approach

The hypergraph model of the sensor network and the set of POI, as well as the matrix representation of a schedule are the basis for encoding the problem of finding a sensor network schedule as a logic program. In this model, sensors form the node set of the hypergraph while POIs correspond to its hyperedges. A hyperedge joins the set of nodes (i.e., sensors), which can monitor the corresponding POI. Figure 1 shows a small sensor network and its hypergraph model represented as a set of pairs $c(node, hyperedge)$ ($node$ belongs to $hyperedge$).

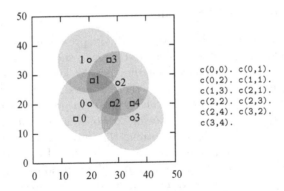

Fig. 1. Example of an operating sensor network: POIs 0, 1, 2, 3, and 4 (squares) and sensors 0, 1, 2, and 3 (circles) with their monitoring regions (gray disks around sensors)—on the left, and its representation as a set of data ready for *clingo*—on the right

Using this model, we can reduce the problem of finding a set of sensors covering a given set of POIs to finding a vertex cover set of a hypergraph, i.e., a set of nodes covering all the hyperedges.

3.4 Programming in *Clingo*

In our research, we decided to use a tool called *clingo* [3] (version 5.3.0), as it combines both (grounder and solver) functionalities for theories in a language of logic programs. This language allows aggregates, including cardinality constraints. As mentioned in Sect. 3.2, to use such a tool, we need to encode problem constraints and a particular problem instance separately.

Encoding Problem Constraints as a Logic Program. Finding a schedule for a sensor network was encoded by the logic program *prog.lp* presented in Algorithm 1. Constants t, ns, np, bat, and $mincov$ must be provided as the command line arguments when the program is called from a terminal, for example:

```
$ clingo prog.lp data.txt -c t=10 -c ns=9 -c np=4 -c bat=2
-c mincov=4
```

Data instance is encoded in file *data.txt* and the constants denote respectively: t the length of the schedule searched for, ns the maximal id of a sensor, equal to $N_S - 1$, np the maximal id of a POI, equal to $N_P - 1$, bat the initial battery load T_{batt}, and $mincov$ the minimal number of POIs which must be monitored in every time slot, equal to $N_P \times cov$.

The program begins with definitions of domains of data predicates *time*, *sensor*, *poi*, and *load* (lines 1–4). For example, line 2 means that valid sensor numbers are from 0 to ns. After these definitions, there are rules defining program predicates. The predicate $battery(S, T, L)$ means that sensor S's battery load at the beginning of time slot T is L. This predicate is defined in lines 6, 8, and 10. For every sensor, its battery load is initialized to bat (line 6). When a sensor is active in slot T, its battery load decreases (line 8); otherwise, its battery load does not change (line 10). The predicate $cvrd(P, T)$ means that POI P is covered during time slot T. P is covered if it is in the range of an active sensor (line 12). The predicate $on(S, T)$ ($off(S, T)$, resp.) means that sensor S is in an active (inactive, resp.) state during time slot T. A sensor with an empty battery is never active (line 14). At any time T, every sensor must be in an active or inactive state (line 16) but cannot be in both states at the same time (line 18).

Rules without a head enforce problem constraints. Their purpose is to eliminate unintended solution candidates. The first such rule appears in line 18. Its purpose is to filter out schedules in which a sensor is both active and inactive during the same time slot. Another rule without a head is in line 20. Its purpose is to eliminate candidates in which the number of monitored POI is below the required level of coverage at any time. The line contains the cardinality atom $\{cvrd(P, T) : poi(P)\}$, which represents the set of POIs covered in slot T. The inequality $\{cvrd(P, T) : poi(P)\} < mincov$ means that the cardinality of this set is smaller than $mincov$.

The last line in the program instructs *clingo* to output all atoms of the form $on(S, T)$ in the solution.

c(0,0). c(0,1). c(1,1). c(1,2). c(2,2). c(2,3). c(3,3). c(3,4). c(4,4). c(4,0).
c(5,0). c(5,2). c(6,1). c(6,3). c(7,2). c(7,4). c(8,0). c(8,3). c(9,1). c(9,4).

Fig. 2. File *data.txt* with a particular problem instance

Algorithm 1. A program for *clingo*.

```
1: time(0..t).
2: sensor(0..ns).
3: poi(0..np).
4: load(0..bat).
5: % initially all sensors have fully loaded batteries
6: battery(S, 0, bat) :- sensor(S).
7: % if a sensor is on, its battery load decreases
8: battery(S, T+1, L-1) :- battery(S, T, L), on(S, T), load(L), sensor(S), time(T),
   T<t.
9: % if a sensor is off, its battery load does not change
10: battery(S, T+1, L) :- battery(S, T, L), off(S, T), load(L), sensor(S), time(T), T<t.
11: % if S is on at time T and S covers P then P is covered at time T
12: cvrd(P, T) :- on(S, T), c(S, P), poi(P), sensor(S), time(T).
13: % a sensor with an empty battery cannot be on
14: off(S, T) :- battery(S, T, 0), sensor(S), time(T).
15: % a sensor can be on or off
16: on(S,T) ; off(S,T) :- sensor(S), time(T).
17: % but not both
18: :- on(S, T), off(S, T), sensor(S), time(T).
19: % the number of poi covered at time T cannot be less than mincov
20: :- {cvrd(P, T) : poi(P)} < mincov, time(T), T<t.
21: % Display
22: #show on/2.
```

Encoding a Problem Instance as Data. File *data.txt* (Fig. 2) contains an example problem instance encoded by statements of the form $c(s, p)$ meaning that sensor s covers POI p. We have a set of 10 sensors $\{0, 1, 2, 3, 4, 5, 6, 7, 8, 9\}$ monitoring 5 POIs $\{0, 1, 2, 3, 4\}$. In this network each sensor can monitor exactly two POIs and every POI is monitored by four sensors. For $i = 0, \ldots, 4$, sensor i monitors POIs i and $i+1$. For $i = 5, \ldots, 9$, sensor i monitors POIs $(i-5)$ *mod* 5 and $(i-3)$ *mod* 5.

Output Produced by Clingo. Once we encoded general problem constraints as a logic program and a particular problem instance as a data set, we are ready to execute *clingo* in the way shown above.

For the program *prog.lp* from Algorithm 1 and the data set *data.txt* from Fig. 2, we invoked *clingo* with the following constants: the schedule length $t = 10$, the maximal sensor id $ns = 9$, the maximal POI id $np = 4$, the battery capacity $bat = 2$, and the number of POI that must be covered during every time slot $mincov = 4$ (i.e. $cov = 0.8$). We obtained the output schedule presented in Fig. 3. This schedule is represented by the extension of the predicate $on(S, T)$, meaning that sensor S is active at time slot T.

In addition to printing positive atoms as directed by the instruction #*show* in the last line of *prog.lp*, *clingo* also displays additional information, including CPU time used during computations.

on(0,5) on(0,9) on(1,2) on(1,3) on(2,0) on(2,1) on(3,5) on(3,9) on(4,1) on(4,2)
on(5,7) on(5,8) on(6,6) on(6,8) on(7,4) on(7,6) on(8,3) on(8,4) on(9,0) on(9,7)

Fig. 3. A solution found by *clingo*.

As mentioned in Sect. 3.2, when we find a schedule of length k, we increase k by one and repeat the exercise. In this case, we increased t to 11 and executed *clingo* with the remaining command-line arguments as before. This time we got a message that the problem is unsatisfiable. Thus, we found out that for the instance of the problem in question, the maximum length of a schedule is 10— we have an optimal solution.

If a search problem has no solution, we need a much longer time to get the answer. The solver has to look through the whole search space to confirm this. In the opposite case, when at least one solution exists, the solver stops earlier—when the first solution is found. In our case, the order of the slots does not influence the schedule feasibility. Therefore, we get families of schedules consisting of the slots permutations. Chances to find any of those schedules are relatively high.

4   Experiments

4.1   Benchmark

To test the viability of the given method, we used the SCP1 dataset [5]. A single instance from this dataset is described as a rectangular or triangular grid of POIs with a constraint of a square area with sides of 13, 16, 19, 22, 25, or 28 abstract units. To introduce irregularities between instances, each grid node inside the square has an 80% chance to act as a POI for the network. As a result, triangular grids consisted of 199 to 240 POIs. And rectangular of 166 to 221 POIs. For each network, exactly 2000 sensors were placed inside the network area. Their placement was determined with the help of a random or a Halton generator. Exactly eight classes were specified as a combination of the area side, grid shape, and a generator used to define the network. For each of them, 40 instances have been generated. In our experiments we assumed the minimal satisfiable level of network coverage *cov* as 80%.

4.2   Methodology and Plan of Experiments

The computational complexity of the SAT problem is an issue when we use a solver for MLCP. In the first experimental tests, we observed that application of *clingo* to SCP1 problem instances with battery sizes as in experiments with heuristic methods described in [5] never found solutions in a reasonable time. Therefore, we decided to reduce the size of the search space by decreasing the sensor battery capacity. A new set of schedules was constructed for batteries of 3–10 units.

This reduction has no particular impact on the adjustment of this model to real-world circumstances. Since the battery capacity is expressed in abstract units, we can use for experiments the same battery as was used by [5] but assume longer time slots. It means that if the working time is, for example, 150 min, and we assume a battery load of 30 abstract units, one unit has to last 5 min. If we assume a battery load of three abstract units, one unit has to last 50 min.

As the method explained in previous sections can be considered a brute force approach to judge the quality of the heuristic results, we decided to set an additional constraint. *Clingo* was given precisely three days to solve each problem instance. Such cut-off was an arbitrary decision, but it was motivated by the increase of uncertainty in weather forecasts for extended time-frames. We assume that the weather conditions impact the network performance, even if our model does not take them into account.

For each instance of the problem, we have the length k of the best–found schedule returned by the heuristic method LS_{HMA} from [6]. This method was selected because it usually gives results better than many other heuristic methods (see [5]). *Clingo* was first tasked to find a schedule of the length $k + 1$. If it was able to find a solution, the instance was fed back with a task to find a schedule longer by one more unit (and so on). Otherwise, the previous result was saved. This way, we kept track of just the last successful execution of each problem. The timer was reset for each additional iteration.

The obtained results are grouped in regards to used battery capacity and input SCP1 class in three tables. Table 1 depicts the average length of schedules obtained by LS_{HMA} (the column "base") and the average increase of the schedule length (the column "incr."). The averages are calculated only for the instances where *clingo* answered the question in a specified time at least once.

The overall completion rate of each class is presented in Table 2, which contains the number of both successful *clingo* executions (the column "ans.") and the time constraint-based terminations of the process (the column "no ans."). In no case *clingo* returned the result that the theory is unsatisfiable. All failures were due to exceeding the time limit. Using a more powerful computer or allowing a longer time for computation could increase the success rate.

Table 3 presents the average completion time (CPU time spent on the process) of the final successful *clingo* iteration for each instance. Figure 5 shows the same data in a graphical form. All experiments were conducted on a machine equipped with Intel® Xeon® Processor E5-2660 v3.

4.3   The Results

SCP1 consists of eight classes of problems. However *clingo* was able to find a solution within the given time limit for only four of them, namely Classes 4, 5, 7, and 8, according to the numbering given in [5].

Class 4 consists of instances generated over a square area of 19 × 19 distance units. Sensors were placed within that area using a random generator while POIs formed a rectangular grid. Classes 7 and 8 both consist of a triangular grid of POIs, but the square areas have different side lengths: 25 distance units for Class

7, and 28—for Class 8. Additionally, for sensor distribution in Class 8, a Halton generator was used instead.

In this paper, the aforementioned classes are referred to as sq_r_19 (Class 4), tr_r_25 (Class 7), and tr_h_28 (Class 8), respectively. Class 5, labeled as tr_r_19 (triangle grid of POIs, random generator for sensor placement and the area of size 19×19), proved to be too difficult for *clingo*. It was able to produce a result for only a single instance of this class, for the lowest considered battery capacity. Therefore, we decided to omit this class along with the other four from the following summary.

Table 1. Schedule length gain

Battery	3		4		5		6	
	base	incr	base	incr	base	incr	base	incr
tr_h_28	26.38	1.00	35.35	1.29	44.38	1.38	53.30	1.44
tr_r_25	30.00	1.57	40.15	1.77	50.08	2.11	60.15	2.09
sq_r_19	47.79	1.58	63.28	1.78	78.85	1.77	92.60	1.80
Battery	7		8		9		10	
	base	incr	base	incr	base	incr	base	incr
tr_h_28	62.21	1.50	71.13	1.39	80.10	1.00	89.00	1.00
tr_r_25	70.00	2.12	80.00	1.80	89.64	1.64	99.37	1.47
sq_r_19	–	–	–	–	–	–	–	–

Table 1 shows absolute schedule length gains for various battery capacities for classes tr_h_28, tr_r_25, and sq_r_19. Figure 4 illustrates percentage length gains for these classes. One can see that an average improvement found in each case is relatively low, usually by one or two time slots, that is, 1–5% of the original result.

Table 2. Scheduled tasks' success rates

Battery	3		4		5		6	
	ans.	no ans.	ans.	no ans.	ans.	no ans.	ans.	no ans.
tr_h_28	32	8	31	9	32	8	27	13
tr_r_25	40	0	39	1	38	2	34	6
sq_r_19	24	16	18	22	13	27	5	35
Battery	7		8		9		10	
	ans.	no ans.	ans.	no ans.	ans.	no ans.	ans.	no ans.
tr_h_28	24	16	23	17	21	19	13	27
tr_r_25	33	7	30	10	28	12	19	21
sq_r_19	–	–	–	–	–	–	–	–

Since there were only five positive results in class *sq_r_19* for the battery capacity of 6 units (see Table 2), we did not perform computations for higher battery capacities.

Even if we can see in Table 1 that in some cases, for larger battery capacity, the gain grows, the base schedule length grows in these cases even more. Thus, the percentage length gain continuously decreases as the battery capacity grows, as shown in Fig. 4.

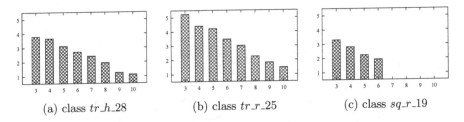

(a) class *tr_h_28*   (b) class *tr_r_25*   (c) class *sq_r_19*

Fig. 4. Average schedule length increase (percentage) for the sensor battery capacity from 3 to 10

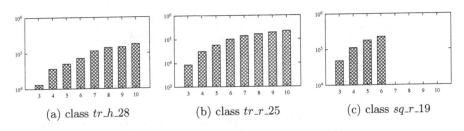

(a) class *tr_h_28*   (b) class *tr_r_25*   (c) class *sq_r_19*

Fig. 5. Average CPU time (in seconds) spent on the final successfull *clingo* execution for the sensor battery capacity from 3 to 10

Table 2 shows that when the battery capacity grows, the completion rate decreases (there is one exception to this rule). The reason is obvious—the larger battery capacity, the longer base schedule. Consequently, the grounded theory for *clingo* becomes larger and requires more time for processing. Therefore, the computations for more instances exceed given time limit.

However, it is interesting that for the class *tr_r_25* we got a higher completion rate than for the class *tr_h_28*. It is surprising because the former one has longer base schedules and larger grounded theories for *clingo*. We have a similar observation for Classes 4 and 6 (according to the numbering from [5], Class 4 is *sq_r_19*; Class 6 has a triangular grid of POIs, the square area with side size 22, and a Halton generator was used for sensor placement). For Class 4, we obtained many positive results. In contrast, Class 6 has got zero completion rate even if base schedules are shorter and grounded theories smaller than in Class 4. Maybe these counter–intuitive results are related to the application of different generators for sensor placement (random generators for Classes 4 and 7, Halton generators for Classes 6 and 8).

Table 3. Average CPU time

Battery	Time [s]			
	3	4	5	6
tr_h_28	13193.35	37011.04	50564.68	72068.46
tr_r_25	8663.52	31153.81	56766.13	99389.61
sq_r_19	47934.54	110696.50	178439.49	223236.20
Battery	7	8	9	10
tr_h_28	115988.15	142643.89	147235.82	177781.64
tr_r_25	138407.59	164017.78	189885.76	219544.08
sq_r_19	–	–	–	–

As seen in Table 3 and Fig. 5, the average completion time varies a lot between the data categories. It is higher for larger battery capacities.

Moreover, for the classes with smaller monitored area, that is, with the higher coverage redundancy and thus longer base schedules, we have a longer average completion time.

However, there is an exception. When we compare results for classes tr_r_25 and tr_h_28, we notice that for lower battery capacities the time is shorter for the instances of the former class.

With the presented results, it is safe to assume that the more time-consuming categories could be improved even further if given more time.

5   Conclusions

In this research, we apply the ASP system *clingo* to sensor network lifetime maximization. The network consists of immobile sensors monitoring a set of POIs in a given area. Sensors have a homogeneous nature in terms of sensing and communication capabilities, and the sensors' connectivity is not a part of the solved problem. Time is discrete. Therefore every schedule defines sensors' activity for time units of the same length called slots. Previous research used heuristic methods, which returned suboptimal schedules. The presented research shows that the representation of a schedule as a collection of fixed-interval slots containing sensors control allows defining the problem in the language of ASP tools. The novelty also lies in showing that ASP systems can improve results obtained by heuristic methods, and in many cases, the system finds schedules more than one slot longer.

A disadvantage of the ASP approach could be the computational cost necessary to get the answer. We use a threshold for the acceptable computational time. In every case, when the time is longer than the threshold, the computations break, and the question remains unanswered. As a result of this procedure, the obtained tables show the range of improvements in the worst-case scenario and leave chances that they may be even better.

It is worth noting that in all cases *clingo* either found a solution or the execution was interrupted due to exceeding the time limit. In other words, for none of the requested schedule lengths, the problem was proven to be unsatisfiable. Therefore, there is no proof that the optimal solutions have already been found—further improvement is still possible.

The experiments conducted for three out of eight classes of the SCP1 benchmark show that the schedule length returned by heuristic methods from [5] can be improved by one to three slots or by 1–5%. For one more class, we found a solution just once for the lowest battery capacity, while for the remaining four classes, all computations were terminated due to exceeded time limit.

References

1. Cardei, M.: Coverage problems in sensor networks. In: Pardalos, P.M., Du, D.-Z., Graham, R.L. (eds.) Handbook of Combinatorial Optimization, pp. 899–927. Springer, New York (2013). https://doi.org/10.1007/978-1-4419-7997-1_72
2. Dargie, W., Poellabauer, C.: Fundamentals of Wireless Sensor Networks: Theory and Practice. Wiley Series on Wireless Communications and Mobile Computing. Wiley (2010). https://doi.org/10.1002/9780470666388
3. Gebser, M., Kaminski, R., Kaufmann, B., Schaub, T.: Clingo = ASP + control: preliminary report. CoRR abs/1405.3694 (2014)
4. Luo, C., Hong, Y., Li, D., Wang, Y., Chen, W., Hu, Q.: Maximizing network lifetime using coverage sets scheduling in wireless sensor networks. Ad Hoc Netw. **98**, 102037 (2020). https://doi.org/10.1016/j.adhoc.2019.102037
5. Mikitiuk, A., Trojanowski, K.: Maximization of the sensor network lifetime by activity schedule heuristic optimization. Ad Hoc Netw. **96**, 101994 (2020). https://doi.org/10.1016/j.adhoc.2019.101994
6. Trojanowski, K., Mikitiuk, A., Kowalczyk, M.: Sensor network coverage problem: a hypergraph model approach. In: Nguyen, N.T., Papadopoulos, G.A., Jędrzejowicz, P., Trawiński, B., Vossen, G. (eds.) ICCCI 2017. LNCS (LNAI), vol. 10448, pp. 411–421. Springer, Cham (2017). https://doi.org/10.1007/978-3-319-67074-4_40
7. Wang, B.: Coverage Control in Sensor Networks. Computer Communications and Networks, Springer, Cham (2010). https://doi.org/10.1007/978-1-84800-328-6
8. Yetgin, H., Cheung, K.T.K., El-Hajjar, M., Hanzo, L.H.: A survey of network lifetime maximization techniques in wireless sensor networks. IEEE Commun. Surv. Tutor. **19**(2), 828–854 (2017). https://doi.org/10.1109/COMST.2017.2650979

Spectroscopy-Based Prediction of In Vitro Dissolution Profile Using Random Decision Forests

Mohamed Azouz Mrad$^{(\boxtimes)}$, Kristóf Csorba, Dorián László Galata,
Zsombor Kristóf Nagy, and Brigitta Nagy

Budapest University of Technology and Economics, Műegyetem rkp. 3,
Budapest 1111, Hungary
mmrad@edu.bme.hu

Abstract. In the pharmaceutical industry, dissolution testing is part of the target product quality that is essential in the approval of new products. The prediction of the dissolution profile based on spectroscopic data is an alternative to the current destructive and time-consuming method. Raman and near-infrared (NIR) spectroscopies are two complementary methods, that provide information on the physical and chemical properties of the tablets and can help in predicting their dissolution profiles. This work aims to use the information collected by these methods by creating partial least squares models to predict the content of the pills. The predicted values are then used along with the measured compression force as input data to Random Decision Forests in order to predict the dissolution profiles of the scanned tablets. It was found that Random Decision Forests models were able to predict the dissolution profile within the acceptance limit of the f_2 factor.

Keywords: Random Decision Forests · Partial least squares · PLS · Dissolution prediction · Raman spectroscopy · NIR spectroscopy

1 Introduction

In the pharmaceutical industry, a target product quality profile is a term used for the quality characteristics that a drug product should process to satisfy the promised benefit from the usage and are essential in the approval of new products or the post-approval changes. A target product quality profile would include different important characteristics, very often one of these is the in vitro (taking place outside of the body) dissolution profile [1]. A dissolution profile represents the concentration rate at which capsules and tablets emit their drugs into the bloodstream over time. It is especially important in the case of tablets that yield a controlled release into the bloodstream over several hours. That offers many advantages over immediate release drugs like reducing the side effects due to the reduced peak dosage and better therapeutic results due to the balanced drug release [2]. In vitro dissolution testing has been a subject of scientific research

© The Author(s), under exclusive license to Springer Nature Switzerland AG 2023
L. Rutkowski et al. (Eds.): ICAISC 2022, LNAI 13588, pp. 411–422, 2023.
https://doi.org/10.1007/978-3-031-23492-7_35

for several years and became a vital tool for accessing product quality performance [3]. However, this method is destructive since it requires immersing the tablets in a solution simulating the human body and is time-consuming as the measurements usually take several hours. As a result, the tablets measured represent only a small amount of the tablets produced, also called a batch. Therefore, there is a need to find different methods that do not have the limitations of the in vitro dissolution testing. The prediction of the dissolution profile based on spectroscopic data is an alternative on which many articles have been published and showed promising results. Raman and near-infrared (NIR) spectroscopies are two complementary methods that are applied in the pharmaceutical industry. They offer the opportunity to obtain information on the physical and chemical properties of the tablets that can help predict their dissolution profiles in a few minutes without destroying them. Hence, Raman and NIR are recognized as straightforward, cost-effective alternatives and non-destructive tools in the quality control process [4,5]. However, these spectroscopies produce a large amount of data as they consist of measurements of hundreds of wavelengths. This data can be filtered out or maintained depending on how much useful information can be extracted from it. This can be achieved using multivariate data analysis techniques such as Principal Component Analysis (PCA). Several researchers have used the spectroscopies data along with the multivariate data analysis techniques in order to predict the dissolution profiles. Zan-nikos et al. worked on a model that permits hundreds of NIR wavelengths to be used in the determination of the dissolution rate [6]. Donoso et al. used the NIR reflectance spectroscopy to measure the percentage of drug dissolution from a series of tablets compacted at different compressional forces using linear regression, nonlinear regression, and partial least square (PLS) models [7]. Freitas et al. created a PLS calibration model to predict drug dissolution profiles at different time intervals and for media with different pH using NIR reflectance spectra [8]. Hernandez et al. used PCA to study the sources of variation in NIR spectra and a PLS-2 model to predict the dissolution of tablets subjected to different levels of strain [9]. Galata et al. developed a PLS model to predict the contained drotaverine (DR) and the hydroxypropyl methylcellulose (HPMC) content of the tablets which are respectively the drug itself and a jelling material that slows down the dissolution, based on both Raman and NIR Spectra, and used the predicted values along with the measured compression force as input to an ANN model in order to predict the dissolution profiles of the tablets defined in 53-time points [10]. Mrad et al. used NIR and RAMAN spectroscopy data reduced using PCA along with compression force to predict dissolution profiles using Artificial neural network models [11]. Random decision forests are suitable for complex problems and have been used in the pharmaceutical industry in many aspects, such as predicting the drug activity against cancer cells based on minimal genomic information and chemical properties [12], Identifying predictive markers of chemosensitivity of breast cancer [13]. Random forests have been also used in the pharmaceutical industry for learning drug functions from chemical structures [14] and also to predict drug vehicles that are most suited to reduce a drug's toxicity [15]. In all these papers, Random decision forests showed great and promising results. Using NIR and Raman spectra to predict the DR and HPMC content of the tablets then along with the concentration force predicting the dissolution profile is a fast method that requires a minimal amount of human labor and which makes it easier to evaluate a larger amount of the batch. Our goal was to extract the useful information directly from the NIR and RAMAN spectra using a multivariate data analysis technique and use it as an input for PLS models to predict the DR and HPMC content of the tablets, then use the predicted values along with the measured compression force as an input for the Random Decision Forests model in order to predict the dissolution profiles of the tablets.

2   Data and Methods

In this section, the data used will be described, and the methods used for the data pre-processing will be presented. The PLS and the Random Forests models created will be presented and finally the error measurement methods adopted to evaluate the results.

2.1   Data Description

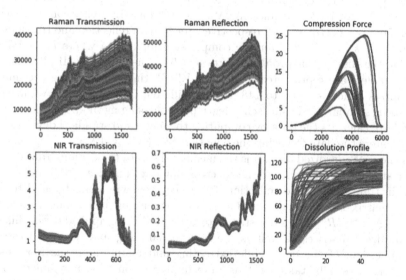

Fig. 1. Dataset

We have been provided with the measurements of the NIR and RAMAN spectroscopy, along with the pressure curves extracted during the compression of the tablets. The data consists of the NIR reflection and transmission, Raman reflection and transmission spectra, the compression force-time curve, and the dissolution profile of 148 tablets. The tablets were produced with a total of 37 different settings. Three parameters were varied: drotaverine content, HPMC content, and the compression force. From each setting, four tablets were selected for analysis ($37 * 4$). The spectral range for NIR reflection spectra was 4000–$10{,}000\ \mathrm{cm}^{-1}$, with a resolution of $8\ \mathrm{cm}^{-1}$, which represents 1556 wavelength points. NIR transmission spectra were collected in the 4000–$15{,}000\ \mathrm{cm}^{-1}$ wavenumber range with $32\ \mathrm{cm}^{-1}$ spectral resolution, which represents 714 wavelength points. Raman spectra were recorded in the range of 200–$1890\ \mathrm{cm}^{-1}$ with $4\ \mathrm{cm}^{-1}$ spectral resolution for both transmission and reflection measurements which represents 1691 points. Two spectra were recorded for each tablet in both NIR and Raman. The pressure during the compression of the pill was recorded in 6037 time points. The dissolution profiles of the tablets were recorded using an in vitro dissolution tester. The length of the dissolution run was 24 h. During this period, samples were taken at 53 time points (at 2, 5, 10, 15, 30, 45 and 60 min, after that once in every 30 min until 1440 min) (Fig. 1).

2.2   Data Analysis

The collected data were visualized and analyzed using MATLAB and Excel in order to detect and fix missed and wrong values: Setting first point of the dissolution curves to zero, detecting missed values, and fixing negative values found due to error of calibration, etc. Specifically, the data is represented in matrices N_i^n for NIR transmission data and M_j^n for NIR Reflection data, where $i = 1556$, $j = 714$. R_k^n and Q_k^n respectively for Raman reflection and transmission data where $k = 1691$. C_l^n for the compression force data where $l = 6037$ and P_s^n for the dissolution profiles where $s = 54$. The DR and HPMC contents of the tablets are represented in matrice V_2^n. With n representing the number of samples which is equal to 148. All the different NIR and RAMAN matrices have been standardized using scikit-learn preprocessing method: StandardScaler. StandardScaler fits the data by computing the mean and standard deviation and then centers the data following the equation $Stdr(NS) = (NS - u)/s$, where NS is the non-standardized data, u is the mean of the data to be standardized, and s is the standard deviation. All the spectroscopy data matrices (NIR and Raman both in transmission and reflection) have been row-wise concatenated to form a new matrix D_m^n where $n = 148$ and m is the sum of their columns as follow: $D_m^n = (N_i^n|M_j^n|R_k^n|Q_k^n)$. After standardization, PCA was applied to the different standardized matrices as well as the merged data D_m^n and in order to reduce the dimension of the data while extracting and maintaining the most useful variations. Basically, taking D_m^n as an example we construct a symmetric $m*m$ dimensional covariance matrix Σ that stores the pairwise covariances between the different features calculated as follow:

$$\sigma_{j,k} = \frac{1}{n}\sum_{i=1}^{n}(x_j^{(i)} - \mu_j)(x_k^{(i)} - \mu_k) \tag{1}$$

With μ_j and μ_k are the sample means of features j and k. The eigenvectors of Σ represent the principal components, while the corresponding eigenvalues define their magnitude. The eigenvalues were sorted by decreasing magnitude in order to find the eigenpairs that contains most of the variances. Variance explained ratios represents the variances explained by every principal components (eigenvectors), it is the fraction of an eigenvalue λ_j and the sum of all the eigenvalues. The following plot Fig. 2 shows the variance explained rations and the cumulative sum of explained variances. It indicates that the first principal components alone accounts for 63% of the variance. The second component account for approximately 18% of the variance. The plot indicates that the seven five principal components combined explain more than 96% of the variance in D. These components are used to create a projection matrix W which we can use to map D to a lower dimensional PCA subspace D' consisting of less features:

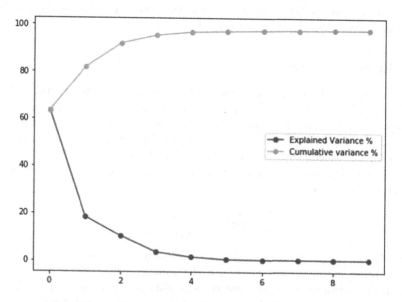

Fig. 2. PCA explained and cumulative variances (x:n components, y:% explained)

$$D = [d_1, d_2, d_3, \ldots d_m], d \in R^m \rightarrow D' = DW, W \in R^{m*v} \tag{2}$$

$$D' = [d_1, d_2, d_3, \ldots d_m], d \in R^m \tag{3}$$

However, for the compression force the maximum value for each row was identified as it is the most important feature to be extracted.

2.3 PLS and Random Decision Forests

PLS model was used to predict the DR and HPMC contents of the pills. The model was created using the python library sklearn. The input for the model was the extracted information which are described later in this paper and the target was the DR and HPMC contents of the tablets (matrix V_2^n). The predicted values along with the maximum values of the compression force curves were used for Random decision forests (RF) created also using the python library sklearn (RandomForestRegressor). The target for the RF models were the measured dissolution profiles described in 54 dissolution curve points and represented in P_s^n. The number of estimators used for the RF model was 100, the criterion was MSE (mean squared error) and the maximum depth of the tree was not specified. While fitting both PLS and RF models, 111 samples have been used for the fitting corresponding to three pills out of every setting the pill was created with, while the remaining pill from each setting was used for validation corresponding to 37 samples. The accuracy of the PLS model was represented by Percentage error between the predicted content and the real one. The Accuracy of the RF model prediction was calculated by evaluating the similarity of the predicted and measured dissolution profiles using the f_2 value.

2.4   Error Measurement

Two mathematical methods are described in the literature to compare dissolution profiles [16]. A difference factor f_1 which is the sum of the absolute values of the vertical distances between the test and reference mean values at each dissolution time point, expressed as a percentage of the sum of the mean fractions released from the reference at each time point. This difference factor f_1 is zero when the mean profiles are identical and increases as the difference between the mean profiles increases.

$$f_1 = \frac{\sum_{t=1}^{n} |R_t - T_t|}{\sum_{t=1}^{n} |R_t|} * 100 \tag{4}$$

where R_t and T_t are the reference and test dissolution values at time t.

The other mathematical method is the similarity function known as the f_2 measure, it performs a logarithmic transformation of the squared vertical distances between the measured and the predicted values at each time point. The value of f_2 is 100 when the test and reference mean profiles are identical and decreases as the similarity decreases.

$$f_2 = 50 log_{10}[(1 + \frac{1}{n}\sum_{t=1}^{n}(R_t - T_t)^2]^{-0.5}) * 100 \tag{5}$$

Values of f_1 between zero and 15 and of f_2 between 50 and 100 ensure the equivalence of the two dissolution profiles. The two methods are accepted by the FDA (U.S. Food and Drug Administration) for dissolution profile comparison, however the f_2 equation is preferred, thus in this paper maximizing the f_2 will be used.

3   Results and Discussions

In this section the results after the PCA decompositions will be discussed. The results of the PLS Prediction and the performance of the Random Decision Forests models created will also be presented in this part.

3.1   Dimensionality Reduction Using PCA

Principal component analysis transformation was applied in a first step to the training samples of the standardized NIR and Raman spectra recorded in reflection and transmission mode (N_i^n, M_j^n, R_k^n, Q_k^n matrices), and in a second step on all training samples of the spectroscopy data merged in matrix D_m^n in order to investigate the effect of the transformation on the merged and the separated data. The resulting PCA decompositions, showed that in the case of NIR reflection, three principal components explaining 82.46%, 9.34% and 7.29% of the total variance in the data, respectively, leading to a cumulative explained variance of more than 99%. Five principal components explained around 75% of the total variances of the NIR transmission data. However, for Raman transmission and Raman Reflection, only the first principal components explain 99.85% and 98.85% of the variance in the data, respectively. For the training samples in matrix D_m^n, 80 principal components explain more than 99% of the merged standardized data. The extracted information from the PCA dimension reduction of all the spectroscopy data merged (D_m^n) were used as the input for fitting the PLS model created to predict the DR and HPMC contents of the tablets (Fig. 3).

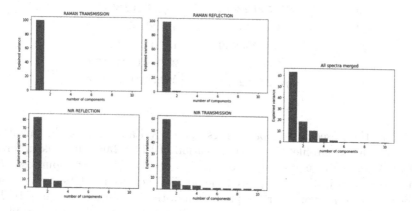

Fig. 3. Explained variance of each spectral data and all of them merged

3.2 Predicting the Dissolution Profile Using PLS and Random Forests

The PLS model was able to fit the training spectroscopy data and to predict the DR contents with an accuracy of 92.40% and to predict the HPMC contents with an accuracy of 88.51%. On the other side, the Random Decision Forests model was able to predict the dissolution profiles of the different tablets represented in 54 dissolution points using the predicted DR and HPMC contents along with the measured maximum value of the compression force curves as an input. The average f_2 value out of 1000 fittings was $f_2 = 68.99$ and the best performing model had an $f_2 = 69.86$. These f_2 results ensure the equivalence of the predicted and measured dissolution profiles (between 50 and 100) and thus are accepted in the pharmaceutical industry. As a result, using PLS models along with Random Decision Forests model is a valuable method for the prediction of dissolution profiles (Table 1 and Fig. 4).

Table 1. Example of some PLS Predictions of DR and HPMC contents

Formulation	DR Real/**Predicted**	HPMC Real/**Predicted**
8	8/**8.24602334**	30/**29.75093989**
11	8/**8.13028671**	10/**11.18509565**
13	6/**5.95800699**	20/**17.56516425**
18	10/**10.30174923**	30/**28.7375516**
28	7/**7.54732992**	20/**22.86372352**
29	7.5/**7.20805507**	20/**21.99106006**
30	8.5/**7.86280624**	20/**22.56387168**
32	8/**8.44795391**	5/**4.82566509**
33	8/**7.32268229**	15/**16.556136**
34	8/**7.33666094**	25/**25.16244172**
35	8/**8.24012551**	35/**35.81103737**

The method used in this paper presents a novel approach in the prediction of the dissolution profiles as the combination of PLS, Random decision forests and PCA for data analysis was not applied previously for this purpose. The results were promising and improved the previous achieved results of when using Artificial neural networks for the prediction of the dissolution profiles [11]. The results during this experiment are reproducible, as the models created could be reused or trained if the used features are remeasured.

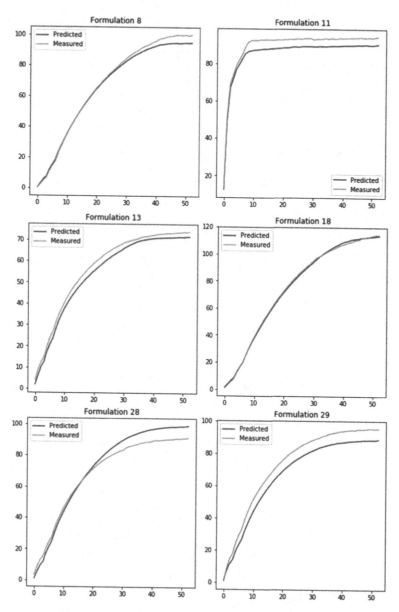

Fig. 4. Sample predicted dissolution curves using Random Decision Forests (x: dissolution points, y: dissolution rate)

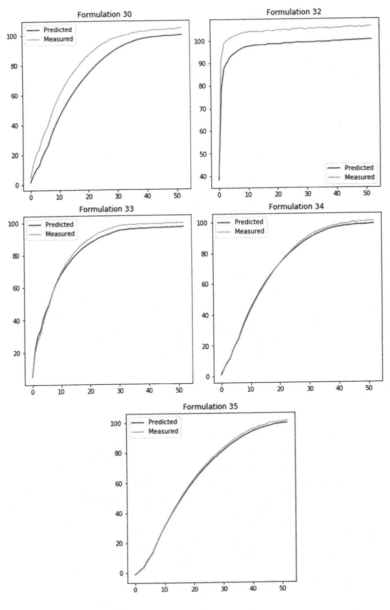

Fig. 4. (*continued*)

4   Conclusion

The current work aimed to utilize the recorded NIR and Raman spectroscopy data along with the compression force to predict the dissolution profiles of tablets produced with 37 different settings. The spectroscopy data was merged together, standardized then dimensionality of the data was reduced using PCA. The resulted data was then fitted to a PLS model to predict the DR and HPMC contents of the tablets. The PLS model was able to predict the content of the pills using the extracted data from the NIR and RAMAN spectroscopy data. Using the predicted values along with the maximum value of the compression force curve. Random Decision forests model was able to predict the dissolution profiles withing the acceptance range of the f_2 factor. The results show that the in vitro dissolution testing can be replaced by more advanced methods such as Random Decision Forests that use similar data providing a large amount of information about the tablets.

Acknowledgments. Project no. FIEK_16-1-2016-0007 has been implemented with the support provided from the National Research, Development and Innovation Fund of Hungary, financed under the Centre for Higher Education and Industrial Cooperation Research infrastructure development (FIEK_16) funding scheme.

References

1. Lawrence, X.Y.: Pharmaceutical quality by design: product and process development, understanding, and control. Pharm. Res. **25**(4), 781–791 (2008)
2. Susto, G.A., McLoone, S.: Slow release drug dissolution profile prediction in pharmaceutical manufacturing: a multivariate and machine learning approach. In: 2015 IEEE International Conference on Automation Science and Engineering (CASE), pp. 1218–1223. IEEE (2015)
3. Patadia, R., Vora, C., Mittal, K., Mashru, R.: Dissolution criticality in developing solid oral formulations: from inception to perception. Crit. Rev. Therap. Drug Carrier Syst. **30**(6) (2013)
4. Hédoux, A.: Recent developments in the Raman and infrared investigations of amorphous pharmaceuticals and protein formulations: a review. Adv. Drug Deliv. Rev. **100**, 133–146 (2016)
5. Porep, J.U., Kammerer, D.R., Carle, R.: On-line application of near infrared (NIR) spectroscopy in food production. Trends Food Sci. Technol. **46**(2), 211–230 (2015)
6. Zannikos, P.N., Li, W.-I., Drennen, J.K., Lodder, R.A.: Spectrophotometric prediction of the dissolution rate of carbamazepine tablets. Pharm. Res. **8**(8), 974–978 (1991)
7. Donoso, M., Ghaly, E.S.: Prediction of drug dissolution from tablets using near-infrared diffuse reflectance spectroscopy as a nondestructive method. Pharm. Dev. Technol. **9**(3), 247–263 (2005)
8. Freitas, M.P., et al.: Prediction of drug dissolution profiles from tablets using NIR diffuse reflectance spectroscopy: a rapid and nondestructive method. J. Pharmac. Biomed. Anal. **39**(1–2), 17–21 (2005)
9. Hernandez, E., et al.: Prediction of dissolution profiles by non-destructive near infrared spectroscopy in tablets subjected to different levels of strain. J. Pharm. Biomed. Anal. **117**, 568–576 (2016)
10. Galata, D.L., et al.: Fast, spectroscopy-based prediction of in vitro dissolution profile of extended release tablets using artificial neural networks. Pharmaceutics **11**(8), 400 (2019)

11. Mrad, M.A., Csorba, K., Galata, D.L., Nagy, Z.K., Nagy, B.: Spectroscopy-based prediction of in vitro dissolution profile using artificial neural networks. In: Rutkowski, L., Scherer, R., Korytkowski, M., Pedrycz, W., Tadeusiewicz, R., Zurada, J.M. (eds.) ICAISC 2021. LNCS (LNAI), vol. 12854, pp. 145–155. Springer, Cham (2021). https://doi.org/10.1007/978-3-030-87986-0_13

12. Lind, A.P., Anderson, P.C.: Predicting drug activity against cancer cells by random forest models based on minimal genomic information and chemical properties. PLoS One **14**(7), e0219774 (2019)

13. Hu, W., et al.: Identifying predictive markers of chemosensitivity of breast cancer with random forests. J. Biomed. Sci. Eng. **3**(01), 59 (2010)

14. Meyer, J.G., Liu, S., Miller, I.J., Coon, J.J., Gitter, A.: Learning drug functions from chemical structures with convolutional neural networks and random forests. J. Chem. Inf. Model. **59**(10), 4438–4449 (2019)

15. Mistry, P., Neagu, D., Trundle, P.R., Vessey, J.D.: Using random forest and decision tree models for a new vehicle prediction approach in computational toxicology. Soft. Comput. **20**(8), 2967–2979 (2016)

16. Moore, J., Flanner, H.: Mathematical comparison of dissolution profiles. Pharm. Technol. **20**(6), 64–74 (1996)

Validation of Labelling Algorithms for Abstract Argumentation Frameworks: The Case of Listing Stable Extensions

Samer Nofal$^{(\boxtimes)}$, Amani Abu Jabal, Abdullah Alfarrarjeh, and Ismail Hababeh

Department of Computer Science, German Jordanian University, Amman, Jordan
{samer.nofal,amani.abujabal,abdullah.alfarrarjeh,
ismail.hababeh}@gju.edu.jo

Abstract. An *abstract argumentation framework* (AF for short) is a pair (A, R) where A is a set of *abstract arguments* and $R \subseteq A \times A$ is the *attack* relation. Let $H = (A, R)$ be an AF, $S \subseteq A$ be a set of arguments and $S^+ = \{y \mid \exists x \in S \text{ with } (x, y) \in R\}$. Then, S is a *stable extension* in H if and only if $S^+ = A \backslash S$. In this paper, we present a thorough, formal validation of a known labelling algorithm for listing all stable extensions in a given AF.

Keywords: Nonmonotonic reasoning · Abstract argumentation · Stable semantics · Labelling algorithm · Backtracking algorithm · NP-hard

1 Introduction

An *abstract argumentation framework* (AF for short) is a pair (A, R) where A is a set of *abstract arguments* and $R \subseteq A \times A$ is the *attack* relation between them. Let $H = (A, R)$ be an AF, $S \subseteq A$ be a subset of arguments and $S^+ = \{y \mid \exists x \in S \text{ with } (x, y) \in R\}$. Then, S is a *stable extension* in H if and only if $S^+ = A \backslash S$.

Since introduced in [8] as a formalism for nonmonotonic reasoning, AFs have attracted a substantial body of research (see e.g. [1,2,13]) due to their promising applications in different domains such as medicine, agriculture, law, and e-government. Stable extension enumeration is a fundamental computational problem within the context of AFs. Listing all stable extensions in an AF is an NP-hard problem, see for example the complexity results presented in [9,10]. In the literature one can find different proposed methods for listing all stable extensions in a given AF, such as dynamic programming and reduction-based methods, see for example [4,5] for a fuller review. However, this paper is centered around backtracking algorithms for listing stable extensions in an AF.

In related work, backtracking algorithms are often called *labelling* algorithms. From now on, we may use "backtracking" and "labelling" interchageably. The work of [6] proposed a backtracking algorithm that can be used for generating stable extensions of AFs. Later, the work of [14] enhanced the algorithm of [6]

L. Rutkowski et al. (Eds.): ICAISC 2022, LNAI 13588, pp. 423–435, 2023.
https://doi.org/10.1007/978-3-031-23492-7_36

by a look-ahead mechanism. Most recently, the system of [12] implemented a backtracking algorithm that is similar to the essence of the algorithm of [14]; however, [12] made use of heuristics to examine their effects on the practical efficiency of listing stable extensions. All these works (i.e. [6,12,14]) presented *experimental* backtracking algorithms.

Generally, there are several empirical studies on algorithmic methods in abstract argumentation research, see e.g. [3,11,15]. However, we do not see in the literature a comprehensive, formal demonstration of a backtracking algorithm for listing stable extensions in a given AF. Therefore, in Sect. 2, we give a rigorous, formal validation of a backtracking algorithm (which is comparable to the core structure of the algorithm of [14]) for listing all stable extensions in an AF. We close the paper with some concluding remarks in Sect. 3.

2   Our Validation

Let $H = (A, R)$ be an AF and $T \subseteq A$ be a subset of arguments. Then,

$$T^+ \stackrel{\text{def}}{=} \{y \mid \exists x \in T \text{ with } (x, y) \in R\},$$
$$T^- \stackrel{\text{def}}{=} \{y \mid \exists x \in T \text{ with } (y, x) \in R\}.$$

For the purpose of enumerating all stable extensions in H, let S denote an under-construction stable extension of H. Thus, we start with $S = \emptyset$ and then let S grow to a stable extension (if any exists) incrementally by choosing arguments from A to join S. For this, we denote by *choice* a set of arguments eligible to join S. More precisely, take $S \subseteq A$ such that $S \cap S^+ = \emptyset$, then *choice* $\subseteq A \backslash (S \cup S^- \cup S^+)$. Additionally, we denote by *tabu* the arguments that do not belong to $S \cup S^+ \cup choice$. More specifically, take $S \subseteq A$ such that $S \cap S^+ = \emptyset$, and *choice* $\subseteq A \backslash (S \cup S^- \cup S^+)$. Then, *tabu* $= A \backslash (S \cup S^+ \cup choice)$. The next example shows these structures in action.

Example 1. To list the stable extensions in the AF H_1 depicted in Fig. 1, apply the following steps:

1. Initially, $S = \emptyset$, $S^+ = \emptyset$, *choice* $= \{a, b, c, d, e, f\}$, and *tabu* $= \emptyset$.
2. Select an argument, say a, to join S. Then, $S = \{a\}$, $S^+ = \{b\}$, *choice* $= \{c, d\}$ and *tabu* $= \{e, f\}$.
3. As $\{c, d\}^- \subseteq S^+ \cup tabu$, $\{c, d\}$ must join S. Otherwise, $S = \{a\}$ without $\{c, d\}$ will never expand to a stable extension since c and d never join S^+. Thus, $S = \{a, c, d\}$, $S^+ = \{b, e, f\}$, *choice* $= \emptyset$ and *tabu* $= \emptyset$. As $S^+ = A \backslash S$, S is stable.
4. To find another stable extension, backtrack to the state of step 1 and then try to build a stable extension without a. Hence, $S = \emptyset$, $S^+ = \emptyset$, *choice* $= \{b, c, d, e, f\}$ and *tabu* $= \{a\}$.
5. Select an argument, say b, to join S. Then, $S = \{b\}$, $S^+ = \{c, d\}$, *choice* $= \{e, f\}$ and *tabu* $= \{a\}$.

6. As $\{e\}^- \subseteq S^+$, e must join S. Otherwise, $S = \{b\}$ without e will not grow to a stable extension since e never joins S^+. Hence, $S = \{b, e\}$, $S^+ = \{a, c, d, f\}$, *choice* $= \emptyset$ and *tabu* $= \emptyset$. Since $S^+ = A\backslash S$, S is stable.
7. Backtrack to the state of step 4 and then attempt to build a stable extension excluding b. Thus, $S = \emptyset$, $S^+ = \emptyset$, *choice* $= \{c, d, e, f\}$ and *tabu* $= \{a, b\}$.
8. Since $\{d\}^- \subseteq tabu$, d must join S. Otherwise, $S = \emptyset$ without d will never grow to a stable extension because d never joins S^+. Subsequently, $S = \{d\}$, $S^+ = \{b, e, f\}$, *choice* $= \{c\}$ and *tabu* $= \{a\}$.
9. As $\{a\}^- \subseteq S^+$ and $a \in tabu$, a will never join S^+ and so $S = \{d\}$ will never grow to a stable extension.
10. At this point, we confirm that there are no more stable extensions to find. This is because the state of step 7, being analyzed in steps 8 & 9, an assertion is concluded that trying to build a stable extension from *choice* $= \{c, d, e, f\}$ (i.e. excluding *tabu* $= \{a, b\}$) will never be successful. Note, we already tried to build a stable extension including a (step 2), and later (step 5) we tried to construct a stable extension including b but without a.

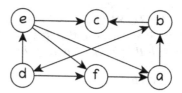

Fig. 1. Argumentation framework H_1.

Our first proposition captures stable extensions when *choice* $= tabu = \emptyset$.

Proposition 1. *Let* $H = (A, R)$ *be an* AF, $S \subseteq A$, $S \cap S^+ = \emptyset$, *choice* $\subseteq A\backslash(S \cup S^+ \cup S^-)$, *tabu* $= A\backslash(S \cup S^+ \cup choice)$. *Then*, S *is a stable extension in* H *if and only if choice* $= \emptyset$ *and tabu* $= \emptyset$.

Proof. If *tabu* $= \emptyset$ and *choice* $= \emptyset$, then $\emptyset = A\backslash(S \cup S^+ \cup \emptyset)$ because *tabu* $= A\backslash(S \cup S^+ \cup choice)$. Thus, $S \cup S^+ = A$. As $S \cap S^+ = \emptyset$, then $S^+ = A\backslash S$. On the other hand, if S is stable, then $S^+ = A\backslash S$ and so $S \cup S^+ = A$. Thus, *tabu* $= A\backslash(S \cup S^+ \cup choice) = A\backslash(A \cup choice) = \emptyset$. Further, *choice* $= \emptyset$ since *choice* $\subseteq A\backslash(S \cup S^+ \cup S^-)$ and $S \cup S^+ = A$. □

Now, we present three propositions that are essential for efficiently listing stable extensions. These propositions are inspired by the excellent work of [7], which presented a backtracking algorithm for solving a different computational problem in AFs.

Proposition 2. *Let* $H = (A, R)$ *be an* AF, $S \subseteq A$, $S \cap S^+ = \emptyset$, *choice* $\subseteq A\backslash(S \cup S^+ \cup S^-)$, *tabu* $= A\backslash(S \cup S^+ \cup choice)$, *and* $x \in choice$ *be an argument with* $\{x\}^- \subseteq S^+ \cup tabu$. *If there is a stable extension* $T \supseteq S$ *such that* $T\backslash S \subseteq choice$, *then* $x \in T$.

Proof. Suppose $x \notin T$. Then, $x \in T^+$ because T is stable. Thus,

$$\exists y \in \{x\}^- \text{ such that } y \in T. \tag{2.1}$$

As $\{x\}^- \subseteq S^+ \cup tabu$,

$$\forall y \in \{x\}^- \text{ it is the case that } y \in S^+ \cup tabu. \tag{2.2}$$

Hence, (2.1) together with (2.2) imply that

$$T \cap (S^+ \cup tabu) \neq \emptyset. \tag{2.3}$$

Observe, $T \subseteq S \cup choice$ since $T \supseteq S$ and $T \backslash S \subseteq choice$. As $S \cap (S^+ \cup tabu) = \emptyset$ and $choice \cap (S^+ \cup tabu) = \emptyset$, it holds that

$$T \cap (S^+ \cup tabu) = \emptyset. \tag{2.4}$$

Note the contradiction between (2.3) and (2.4). □

Proposition 3. *Let $H = (A, R)$ be an* AF, *$S \subseteq A$, $S \cap S^+ = \emptyset$, choice $\subseteq A \backslash (S \cup S^+ \cup S^-)$, tabu $= A \backslash (S \cup S^+ \cup choice)$, and $x \in choice$ be an argument such that for some $y \in tabu$ it is the case that $\{y\}^- \cap choice = \{x\}$. If there is a stable extension $T \supseteq S$ such that $T \backslash S \subseteq choice$, then $x \in T$.*

Proof. Suppose $x \notin T$. As $x \in choice$,

$$T \backslash S \subseteq choice \backslash \{x\}. \tag{3.1}$$

As $T \supseteq S$ and $T \backslash S \subseteq choice$, it is the case that $T \subseteq S \cup choice$. Because $tabu \cap (S \cup choice) = \emptyset$, it holds that

$$T \cap tabu = \emptyset. \tag{3.2}$$

Referring to the premise of this proposition, as $y \in tabu$, $y \notin T$. Subsequently, $y \in T^+$ since T is stable. Thus,

$$\{y\}^- \cap T \neq \emptyset. \tag{3.3}$$

Since $y \in tabu$ and $tabu \cap S^+ = \emptyset$, $y \notin S^+$. Thus,

$$\{y\}^- \cap S = \emptyset. \tag{3.4}$$

Due to (3.1), (3.3), and (3.4), it holds that $\{y\}^- \cap (choice \backslash \{x\}) \neq \emptyset$, which is a contradiction with $\{y\}^- \cap choice = \{x\}$, see the premise of this proposition. □

Proposition 4. *Let $H = (A, R)$ be an* AF, *$S \subseteq A$, $S \cap S^+ = \emptyset$, choice $\subseteq A \backslash (S \cup S^+ \cup S^-)$, tabu $= A \backslash (S \cup S^+ \cup choice)$, and $x \in tabu$ be an argument with $\{x\}^- \subseteq S^+ \cup tabu$. Then, there does not exist a stable extension $T \supseteq S$ such that $T \backslash S \subseteq choice$.*

Proof. Assume there exists a stable extension $T \supseteq S$ with $T \backslash S \subseteq choice$. Since $(S \cup choice) \cap (S^+ \cup tabu) = \emptyset$,

$$T \cap (S^+ \cup tabu) = \emptyset. \tag{4.1}$$

As $x \in tabu$ and due to (4.1), $x \notin T$. Thus, $x \in T^+$ because T is stable. Subsequently,

$$T \cap \{x\}^- \neq \emptyset. \tag{4.2}$$

Because $\{x\}^- \subseteq S^+ \cup tabu$,

$$T \cap (S^+ \cup tabu) \neq \emptyset \tag{4.3}$$

Note the contradiction between (4.1) and (4.3).   □

We now give Algorithm 1 for listing all stable extensions in a given AF. Before presenting its proof, we demonstrate the execution of Algorithm 1.

Algorithm 1: stb(S, *choice*, *tabu*)

---

1 **repeat**
2   **if** there is $x \in tabu$ with $\{x\}^- \subseteq S^+ \cup tabu$ **then** return;
3      $\alpha \leftarrow \{x \in choice \mid \{x\}^- \subseteq S^+ \cup tabu\}$;
4      $S \leftarrow S \cup \alpha$;
5      $choice \leftarrow choice \backslash (\alpha \cup \alpha^+ \cup \alpha^-)$;
6      $tabu \leftarrow (tabu \cup \alpha^-) \backslash S^+$;
7   **if** $\neg(\exists y \in choice, \exists x \in tabu, \{x\}^- \cap choice = \{y\})$ **then** $\beta \leftarrow \emptyset$ **else**
8          $\beta \leftarrow \{y\}$;
9          $S \leftarrow S \cup \beta$;
10         $choice \leftarrow choice \backslash (\beta \cup \beta^+ \cup \beta^-)$;
11         $tabu \leftarrow (tabu \cup \beta^-) \backslash S^+$;
12 **until** $\alpha = \emptyset$ and $\beta = \emptyset$;
13 **if** $choice = \emptyset$ **then**
14   **if** $tabu = \emptyset$ **then** S is stable;
15   return;
    // For some $x \in choice$.
16 stb($S \cup \{x\}$, $choice \backslash (\{x\} \cup \{x\}^+ \cup \{x\}^-)$, $(tabu \cup \{x\}^-) \backslash (S^+ \cup \{x\}^+)$);
17 stb(S, $choice \backslash \{x\}$, $tabu \cup \{x\}$);

Example 2. Apply Algorithm 1 to list the stable extensions in H_1:

1. Initially, call $stb(\emptyset, \{a, b, c, d, e, f\}, \emptyset)$.
2. Perform the repeat-until block of Algorithm 1 to get $S = \emptyset$, $choice = \{a, b, c, d, e, f\}$, and $tabu = \emptyset$.
3. Call $stb(\{a\}, \{c, d\}, \{e, f\})$, see line 16 in the algorithm.
4. Now, execute the repeat-until block to get $S = \{a, c, d\}$, $choice = \emptyset$, and $tabu = \emptyset$. Applying line 14, S is stable. Now, apply line 15 and return to the state: $S = \emptyset$, $choice = \{a, b, c, d, e, f\}$, and $tabu = \emptyset$.

5. Call $stb(\emptyset, \{b, c, d, e, f\}, \{a\})$, see line 17 in the algorithm.
6. After performing the repeat-until block, $S = \emptyset$, $choice = \{b, c, d, e, f\}$, and $tabu = \{a\}$.
7. Call $stb(\{b\}, \{e, f\}, \{a\})$, see line 16.
8. Apply the repeat-until block to get $S = \{b, e\}$, $choice = \emptyset$, and $tabu = \emptyset$. Performing line 14, we find S stable. Now apply line 15, and so return to the state: $S = \emptyset$, $choice = \{b, c, d, e, f\}$, and $tabu = \{a\}$.
9. Call $stb(\emptyset, \{c, d, e, f\}, \{a, b\})$, see line 17.
10. Perform a first round of the repeat-until block. Thus, $S = \{d\}$, $choice = \{c\}$, and $tabu = \{a\}$. In a second round of the repeat-until block, the algorithm returns (see line 2), and eventually the algorithm halts.

In what follows we give another set of propositions that together with the previous ones will establish the validity of Algorithm 1. Thus, we denote by T_i the elements of a set T at the algorithm's state i. The algorithm enters a new state whenever S is updated. In other words, the algorithm enters a new state whenever lines 4, 9, 16, or 17 are executed. Focusing on the under-construction stable extension S and consistently with the algorithm's actions, in the initial state of the algorithm we let

$$S_1 = \emptyset, \tag{1}$$

and for every state i it is the case that

$$\begin{aligned} &S_{i+1} = S_i, \text{ or} \\ &S_{i+1} = S_i \cup \delta_i \text{ with } \delta_i \in \{\alpha_i, \beta_i, \{x\}_i\}; \end{aligned} \tag{2}$$

see respectively lines 17, 4, 9, and 16 in the algorithm.

Proposition 5. *Let $H = (A, R)$ be an* AF *and Algorithm 1 be started with*

$$stb(\emptyset, A \backslash \{x \mid (x, x) \in R\}, \{x \mid (x, x) \in R\}).$$

For every state i, $S_i \cap S_i^+ = \emptyset$.

Proof. As $S_1 = \emptyset$, $S_1 \cap S_1^+ = \emptyset$. We now show that for every state i

$$S_i \cap S_i^+ = \emptyset \implies S_{i+1} \cap S_{i+1}^+ = \emptyset. \tag{5.1}$$

Suppose $S_i \cap S_i^+ = \emptyset$. Using (2) we write

$$\begin{aligned} S_{i+1} \cap S_{i+1}^+ &= (S_i \cup \delta_i) \cap (S_i^+ \cup \delta_i^+) \\ &= (S_i \cap S_i^+) \cup (S_i \cap \delta_i^+) \cup (S_i^+ \cap \delta_i) \cup (\delta_i \cap \delta_i^+). \end{aligned} \tag{5.2}$$

In fact, $\delta_i \subseteq choice_i$, recall (2) and lines 3, 7, and 16. Considering (5.2), we proceed the proof by showing that

$$(S_i \cap S_i^+) \cup (S_i \cap choice_i^+) \cup (S_i^+ \cap choice_i) \cup (\delta_i \cap \delta_i^+) = \emptyset. \tag{5.3}$$

In other words, we need to prove that for every state i

$$S_i \cap choice_i^+ = \emptyset, \tag{5.4}$$

$$S_i^+ \cap choice_i = \emptyset, \tag{5.5}$$

$$\delta_i \cap \delta_i^+ = \emptyset. \tag{5.6}$$

Now we prove (5.4). For $i = 1$, $S_1 \cap choice_1^+ = \emptyset$ since $S_1 = \emptyset$. Then, we will show that

$$\forall i \ (S_i \cap choice_i^+ = \emptyset \implies S_{i+1} \cap choice_{i+1}^+ = \emptyset). \tag{5.7}$$

Suppose $S_i \cap choice_i^+ = \emptyset$. Referring to lines 5, 10 and 16, it holds that $choice_{i+1} = choice_i \backslash (\delta_i \cup \delta_i^- \cup \delta_i^+)$. And using (2), we note that

$$S_{i+1} \cap choice_{i+1}^+ = (S_i \cup \delta_i) \cap (choice_i \backslash (\delta_i \cup \delta_i^- \cup \delta_i^+))^+. \tag{5.8}$$

Since $S_i \cap choice_i^+ = \emptyset$, it holds that $S_i \cap (choice_i \backslash (\delta_i \cup \delta_i^- \cup \delta_i^+))^+ = \emptyset$. Additionally, we note that

$$\delta_i \cap (choice_i \backslash (\delta_i \cup \delta_i^- \cup \delta_i^+))^+ = \emptyset \iff \delta_i^- \cap (choice_i \backslash (\delta_i \cup \delta_i^- \cup \delta_i^+)) = \emptyset. \tag{5.9}$$

As $\delta_i^- \cap (choice_i \backslash (\delta_i \cup \delta_i^- \cup \delta_i^+)) = \emptyset$, (5.8) holds. For the case of line 17 in the algorithm, where $S_{i+1} = S_i$, note that

$$S_{i+1} \cap choice_{i+1}^+ = S_i \cap (choice_i \backslash \{x\}_i)^+ = \emptyset \tag{5.10}$$

since $S_i \cap choice_i^+ = \emptyset$. That concludes the proof of (5.4). Now we prove (5.5). For $i = 1$, it holds that $choice_1 \cap S_1^+ = \emptyset$ since $S_1 = \emptyset$. Then, we need to show that for every state i

$$choice_i \cap S_i^+ = \emptyset \implies choice_{i+1} \cap S_{i+1}^+ = \emptyset. \tag{5.11}$$

Referring to line 5, 10, and 16, it is the case that $choice_{i+1} = choice_i \backslash (\delta_i \cup \delta_i^+ \cup \delta_i^-)$. Considering (2), observe that

$$\begin{aligned} choice_{i+1} \cap S_{i+1}^+ &= (choice_i \backslash (\delta_i \cup \delta_i^+ \cup \delta_i^-)) \cap (S_i \cup \delta_i)^+ \\ &= (choice_i \backslash (\delta_i \cup \delta_i^+ \cup \delta_i^-)) \cap (S_i^+ \cup \delta_i^+). \end{aligned} \tag{5.12}$$

Thus, (5.11) holds. Likewise, for the case of line 17 in the algorithm,

$$choice_{i+1} \cap S_{i+1}^+ = (choice_i \backslash \{x\}_i) \cap S_i^+ = \emptyset \tag{5.13}$$

since $choice_i \cap S_i^+ = \emptyset$, see the premise of (5.11). That completes the proof of (5.5). Now we prove (5.6). From (2), for all i, if $S_{i+1} = S_i \cup \delta_i$, then it holds that $\delta_i \in \{\alpha_i, \beta_i, \{x\}_i\}$. Considering the algorithm's actions (lines 3, 7 and 16), $\delta_i \subseteq choice_i$. If $\delta_i \in \{\beta_i, \{x\}_i\}$, then $|\delta_i| = 1$, and so $\delta_i \cap \delta_i^+ = \emptyset$ holds since $\{x \mid (x, x) \in R\} \not\subseteq choice_i$. If $\delta_i = \alpha_i$, then from line 3 in the algorithm we note that

$$\delta_i = \{x \in choice_i \mid \{x\}^- \subseteq tabu_i \cup S_i^+\}. \tag{5.14}$$

Therefore,

$$\delta_i^- \subseteq tabu_i \cup S_i^+. \tag{5.15}$$

To establish $\delta_i \cap \delta_i^+ = \emptyset$ it suffices to show that

$$\delta_i \cap \delta_i^- = \emptyset. \tag{5.16}$$

Thus, given (5.15) with the fact that $\delta_i \subseteq choice_i$, we will prove (5.16) by showing that

$$choice_i \cap (tabu_i \cup S_i^+) = \emptyset. \tag{5.17}$$

Because we already showed that $choice_i \cap S_i^+ = \emptyset$, recall (5.5), the focus now is on demonstrating that for all i

$$choice_i \cap tabu_i = \emptyset. \tag{5.18}$$

For $i = 1$, (5.18) holds since $choice_1 = A\backslash\{x \mid (x,x) \in R\}$ and $tabu_1 = \{x \mid (x,x) \in R\}$. Now we will show that

$$\forall i \ (choice_i \cap tabu_i = \emptyset \implies choice_{i+1} \cap tabu_{i+1} = \emptyset). \tag{5.19}$$

Let $choice_i \cap tabu_i = \emptyset$. Observe that

$$\begin{aligned} choice_{i+1} &= choice_i\backslash(\delta_i \cup \delta_i^- \cup \delta_i^+) \quad \text{(lines 5, 10, and 16)}\\ tabu_{i+1} &= (tabu_i \cup \delta_i^-)\backslash(S_i^+ \cup \delta_i^+) \quad \text{(lines 6, 11, and 16)} \end{aligned} \tag{5.20}$$

Thus,

$$choice_{i+1} \cap tabu_{i+1} = (choice_i\backslash(\delta_i \cup \delta_i^- \cup \delta_i^+)) \cap ((tabu_i \cup \delta_i^-)\backslash(S_i^+ \cup \delta_i^+)) = \emptyset. \tag{5.21}$$

For the case of line 17, observe that

$$choice_{i+1} \cap tabu_{i+1} = (choice_i\backslash\{x\}_i) \cap (tabu_i \cup \{x\}_i) = \emptyset. \tag{5.22}$$

That concludes the demonstration of (5.6), and so the proof of (5.1) is now complete. $\quad\square$

Proposition 6. *Let $H = (A, R)$ be an* AF *and Algorithm 1 be started with*

$$stb(\emptyset, A\backslash\{x \mid (x,x) \in R\}, \{x \mid (x,x) \in R\}).$$

For every state i, $choice_i \subseteq A\backslash(S_i \cup S_i^+ \cup S_i^-)$.

Proof. We note that $choice_1 \subseteq A\backslash(S_1 \cup S_1^+ \cup S_1^-)$. This is because $choice_1 = A\backslash\{x \mid (x,x) \in R\}$ and, $S_1 = \emptyset$. Now, we show that for every state i

$$choice_i \subseteq A\backslash(S_i \cup S_i^+ \cup S_i^-) \implies choice_{i+1} \subseteq A\backslash(S_{i+1} \cup S_{i+1}^+ \cup S_{i+1}^-). \tag{6.1}$$

Given the premise of (6.1), we write

$$choice_i\backslash(\delta_i \cup \delta_i^- \cup \delta_i^+) \subseteq A\backslash(S_i \cup S_i^+ \cup S_i^- \cup \delta_i \cup \delta_i^- \cup \delta_i^+). \tag{6.2}$$

According to lines 5, 10 and 16 in the algorithm we observe that

$$choice_{i+1} = choice_i\backslash(\delta_i \cup \delta_i^+ \cup \delta_i^-). \tag{6.3}$$

Hence, using (6.3) along with (2) we rewrite (6.2) as

$$choice_{i+1} \subseteq A\backslash(S_{i+1} \cup S_{i+1}^+ \cup S_{i+1}^-), \qquad (6.4)$$

which is the consequence of (6.1). But to complete the proof of (6.1) we need to illustrate the case of line 17 in the algorithm where for some $x \in choice_i$ it holds that

$$choice_{i+1} = choice_i\backslash\{x\}_i, \qquad (6.5)$$

and

$$S_{i+1} = S_i. \qquad (6.6)$$

Given the premise of (6.1), observe that

$$choice_i\backslash\{x\}_i \subseteq A\backslash(S_i \cup S_i^+ \cup S_i^-). \qquad (6.7)$$

Using (6.5) with (6.6), we note that $choice_{i+1} \subseteq A\backslash(S_{i+1} \cup S_{i+1}^+ \cup S_{i+1}^-)$. □

Proposition 7. *Let $H = (A, R)$ be an* AF *and Algorithm 1 be started with*

$$stb(\emptyset, A\backslash\{x \mid (x, x) \in R\}, \{x \mid (x, x) \in R\}).$$

For every state i, $tabu_i = A\backslash(S_i \cup S_i^+ \cup choice_i)$.

Proof. We note that $tabu_1 = A\backslash(S_1 \cup S_1^+ \cup choice_1)$. Observe, $S_1 = \emptyset$, $choice_1 = A\backslash\{x \mid (x, x) \in R\}$ and $tabu_1 = \{x \mid (x, x) \in R\}$. Now we prove that for every state i,

$$tabu_i = A\backslash(S_i \cup S_i^+ \cup choice_i) \implies tabu_{i+1} = A\backslash(S_{i+1} \cup S_{i+1}^+ \cup choice_{i+1}). \quad (7.1)$$

Considering lines 6, 11, and 16 we note that

$$tabu_{i+1} = (tabu_i \cup \delta_i^-)\backslash(S_i^+ \cup \delta_i^+). \qquad (7.2)$$

Given the premise of (7.1), rewrite (7.2) as

$$tabu_{i+1} = ((A\backslash(S_i \cup S_i^+ \cup choice_i)) \cup \delta_i^-)\backslash(S_i^+ \cup \delta_i^+). \qquad (7.3)$$

According to lines 5, 10 and 16 in the algorithm we observe that

$$choice_{i+1} = choice_i\backslash(\delta_i \cup \delta_i^+ \cup \delta_i^-). \qquad (7.4)$$

Note that $\delta_i \subseteq choice_i$, see (2). For the case where $\delta_i^+ \cup \delta_i^- \subseteq choice_i$, using (7.4) and (2), we rewrite (7.3) as

$$\begin{aligned}
tabu_{i+1} &= ((A\backslash(S_i \cup S_i^+ \cup choice_{i+1} \cup \delta_i \cup \delta_i^+ \cup \delta_i^-)) \cup \delta_i^-)\backslash(S_i^+ \cup \delta_i^+) \\
&= ((A\backslash(S_{i+1} \cup S_{i+1}^+ \cup choice_{i+1} \cup \delta_i^-))\backslash(S_i^+ \cup \delta_i^+)) \cup (\delta_i^- \backslash(S_i^+ \cup \delta_i^+)) \\
&= ((A\backslash(S_{i+1} \cup S_{i+1}^+ \cup choice_{i+1} \cup \delta_i^-))\backslash S_{i+1}^+) \cup (\delta_i^- \backslash S_{i+1}^+) \\
&= (A\backslash(S_{i+1} \cup S_{i+1}^+ \cup choice_{i+1} \cup \delta_i^-)) \cup (\delta_i^- \backslash S_{i+1}^+)
\end{aligned}$$

$$(7.5)$$

Note, it is always the case that $\delta_i^- \cap (choice_{i+1} \cup S_{i+1}) = \emptyset$, see Proposition 5 and (7.4). Thus, (7.5) can be rewritten as

$$tabu_{i+1} = (A\backslash(S_{i+1} \cup S_{i+1}^+ \cup choice_{i+1} \cup (\delta_i^- \backslash S_{i+1}^+))) \cup (\delta_i^- \backslash S_{i+1}^+) \quad (7.6)$$
$$= A\backslash(S_{i+1} \cup S_{i+1}^+ \cup choice_{i+1}).$$

For the case where $(\delta_i^+ \cup \delta_i^-) \cap choice_i = \emptyset$ with $\delta_i^+ \cup \delta_i^- \subseteq S_i^+ \cup tabu_i$, we write (7.2) using the premise of (7.1) as

$$tabu_{i+1} = (tabu_i \cup \delta_i^-)\backslash(S_i^+ \cup \delta_i^+) = tabu_i\backslash\delta_i^+ = (A\backslash(S_i \cup S_i^+ \cup choice_i))\backslash\delta_i^+. \quad (7.7)$$

Using (7.4) we rewrite (7.7) as

$$tabu_{i+1} = (A\backslash(S_i \cup S_i^+ \cup choice_{i+1} \cup \delta_i))\backslash\delta_i^+ = A\backslash(S_i \cup \delta_i \cup S_i^+ \cup \delta_i^+ \cup choice_{i+1}). \quad (7.8)$$

Using (2), (7.8) is equivalent to

$$tabu_{i+1} = A\backslash(S_{i+1} \cup S_{i+1}^+ \cup choice_{i+1}). \quad (7.9)$$

For the case where $\delta_i^+ \cup \delta_i^- \subseteq S_i^+ \cup tabu_i \cup choice_i$, let $\delta_i^+ = \delta_{it}^+ \cup \delta_{ic}^+ \cup \delta_{is}^+$ and $\delta_i^- = \delta_{it}^- \cup \delta_{ic}^- \cup \delta_{is}^-$ such that $\delta_{it}^+ \cup \delta_{it}^- \subseteq tabu_i$, $\delta_{ic}^+ \cup \delta_{ic}^- \subseteq choice_i$, and $\delta_{is}^+ \cup \delta_{is}^- \subseteq S_i^+$. Then, we rewrite (7.2) as

$$
\begin{aligned}
tabu_{i+1} &= (tabu_i \cup \delta_{ic}^- \cup \delta_{is}^- \cup \delta_{it}^-)\backslash(S_i^+ \cup \delta_{ic}^+ \cup \delta_{it}^+ \cup \delta_{is}^+)\\
&= (tabu_i \cup \delta_{ic}^- \cup \delta_{is}^-)\backslash(S_i^+ \cup \delta_{ic}^+ \cup \delta_{it}^+)\\
&= ((A\backslash(S_i \cup S_i^+ \cup choice_i)) \cup \delta_{ic}^- \cup \delta_{is}^-)\backslash(S_i^+ \cup \delta_{ic}^+ \cup \delta_{it}^+)\\
&= ((A\backslash(S_i \cup S_i^+ \cup choice_i)) \cup \delta_{ic}^-)\backslash(S_i^+ \cup \delta_{ic}^+ \cup \delta_{it}^+)\\
&= ((A\backslash(S_i \cup S_i^+ \cup choice_i)) \cup \delta_{ic}^-)\backslash(\delta_{ic}^+ \cup \delta_{it}^+)\\
&= ((A\backslash(S_i \cup S_i^+ \cup choice_{i+1} \cup \delta_i \cup \delta_{ic}^+ \cup \delta_{ic}^-)) \cup \delta_{ic}^-)\backslash(\delta_{ic}^+ \cup \delta_{it}^+) \quad (7.10)\\
&= (A\backslash(S_i \cup S_i^+ \cup choice_{i+1} \cup \delta_i \cup \delta_{ic}^+))\backslash(\delta_{ic}^+ \cup \delta_{it}^+)\\
&= (A\backslash(S_i \cup S_i^+ \cup choice_{i+1} \cup \delta_i \cup \delta_{ic}^+))\backslash\delta_{it}^+\\
&= A\backslash(S_i \cup S_i^+ \cup choice_{i+1} \cup \delta_i \cup \delta_{ic}^+ \cup \delta_{it}^+)\\
&= A\backslash(S_i \cup S_i^+ \cup choice_{i+1} \cup \delta_i \cup \delta_i^+)\\
&= A\backslash(S_{i+1} \cup S_{i+1}^+ \cup choice_{i+1}).
\end{aligned}
$$

For the special case of line 17 in the algorithm, which indicates that $tabu_{i+1} = tabu_i \cup \{x\}_i$, $S_{i+1} = S_i$, and $choice_{i+1} = choice_i\backslash\{x\}_i$, we write

$$
\begin{aligned}
tabu_{i+1} &= tabu_i \cup \{x\}_i\\
&= (A\backslash(S_i \cup S_i^+ \cup choice_i)) \cup \{x\}_i\\
&= (A\backslash(S_{i+1} \cup S_{i+1}^+ \cup choice_i)) \cup \{x\}_i \quad (7.11)\\
&= (A\backslash(S_{i+1} \cup S_{i+1}^+ \cup choice_{i+1} \cup \{x\}_i)) \cup \{x\}_i\\
&= A\backslash(S_{i+1} \cup S_{i+1}^+ \cup choice_{i+1}).
\end{aligned}
$$

This concludes our proof of Proposition 7. $\qquad\square$

Proposition 8. *Let $H = (A, R)$ be an* AF *and Algorithm 1 be started with*

$$stb(\emptyset, A\backslash\{x \mid (x,x) \in R\}, \{x \mid (x,x) \in R\}).$$

The algorithm computes exactly the stable extensions of H.

Proof. We will show that the following two statements (*P1* & *P2*) hold.

P1. For every set, S_i, reported by Algorithm 1 at line 14 at some state i, S_i is a stable extension in H.

P2. For all Q, if Q is a stable extension in H and Algorithm 1 is sound (i.e. *P1* is established), then there is a stable extension S_i, reported by the algorithm at line 14 at some state i, such that $S_i = Q$.

Regarding *P1*, to show that $S_i^+ = A \backslash S_i$ we need to prove that

$$S_i \cap S_i^+ = \emptyset, \tag{8.1}$$

and

$$S_i \cup S_i^+ = A. \tag{8.2}$$

Note, (8.1) is proved in Proposition 5. For (8.2), it can be easily established by using Proposition 1 if we prove that for every state i

$$choice_i \subseteq A \backslash (S_i \cup S_i^+ \cup S_i^-), \tag{8.3}$$

and

$$tabu_i = A \backslash (S_i \cup S_i^+ \cup choice_i). \tag{8.4}$$

However, (8.3) and (8.4) are proved in Proposition 6 and 7 respectively. Thus, referring to line 14 in the algorithm, we note that S_i is reported stable if and only if $tabu_i = choice_i = \emptyset$. Considering (8.4), we note that $\emptyset = A \backslash (S_i \cup S_i^+ \cup \emptyset)$, and hence (8.2) holds. The proof of *P1* is complete.

Regarding *P2*, we rewrite *P2* (by modifying the consequence) into *Ṕ2*.

Ṕ2: For all Q, if Q is a stable extension in H and Algorithm 1 is sound, then there is a stable extension S_i, reported by the algorithm at line 14 at some state i, such that for all $a \in Q$ it holds that $a \in S_i$.

We establish *Ṕ2* by contradiction.. Later, we show that the consequence of *Ṕ2* is equivalent to the consequence of *P2*. Now, assume that *Ṕ2* is false.

Negation of Ṕ2: There is Q such that Q is a stable extension in H, Algorithm 1 is sound, and for every S_i reported by the algorithm at line 14 at some state i, there is $a \in Q$ such that $a \notin S_i$.

We identify four cases.

Case 1. For $a \in choice_1$, if the algorithm terminates at line 2 during the very first execution of the repeat-until block (but not necessarily from the first round), then, since the algorithm is sound, H has no stable extensions. This contradicts the assumption that $Q \supseteq \{a\}$ is a stable extension in H. Hence, *Ṕ2* holds.

Case 2. If $(a, a) \in R$, then this is a contradiction with the assumption that $Q \supseteq \{a\}$ is a stable extension in H. Hence, *Ṕ2* holds.

Case 3. With $a \in choice_1$, assume that after the very first execution of the repeat-until block (i.e. including one or more rounds), $a \in choice_k$ for some state $k \geq 1$. Then, for a state $i \geq k$, let $x = a$ (see line 16 in Algorithm 1). If for all subsequent states $j > i$, the set $S_j \supseteq \{a\}$ is not reported stable by the algorithm, then, since the algorithm is sound, a does not belong to any stable extension. This contradicts the assumption that $Q \supseteq \{a\}$ is a stable extension in H. Hence, $\acute{P}2$ holds.

Case 4. With $a \in choice_1$, assume that after the very first execution of the repeat-until block (i.e. including one or more rounds), $a \notin choice_i$ for some state $i > 1$. This implies, according to the repeat-until block's actions, that $\{a\} \subseteq S_i \cup S_i^+ \cup tabu_i$. For $\{a\} \subseteq S_i$, if for all subsequent states $j > i$, the set $S_j \supseteq \{a\}$ is not reported stable by the algorithm, then, since the algorithm is sound, a does not belong to any stable extension. This contradicts the assumption that $Q \supseteq \{a\}$ is a stable extension in H. Likewise, for $\{a\} \subseteq S_i^+ \cup tabu_i$, since the algorithm's actions are sound, this implies that a does not belong to any stable extension. This contradicts the assumption that $Q \supseteq \{a\}$ is a stable extension in H. Hence, $\acute{P}2$ holds.

Now we rewrite the consequence of $\acute{P}2$.

The Consequence of $\acute{P}2$: There is a stable extension S_i, reported by the algorithm at line 14 at some state i, such that $Q \subseteq S_i$.

Q being a proper subset of S_i is impossible because otherwise it contradicts that the algorithm is sound or that Q is stable. Therefore, the consequence of $\acute{P}2$ can be rewritten as next.

The Consequence of $\acute{P}2$: There is a stable extension S_i, reported by the algorithm at line 14 at some state i, such that $Q = S_i$.

This is exactly the consequence of *P2*. □

3 Conclusion

We presented formal validation of a known backtracking algorithm for listing all stable extensions in a given abstract argumentation framework. Our validation process given in this paper may encourage more investigations in the research arena of abstract argumentation. Despite being experimentally verified, several existing backtracking algorithms for abstract argumentation can be reinforced with formal validation, which might be done in the spirit of this article. A natural extension to this work is to apply formal validation to a finer-implementation level of Algorithm 1 where set operations are implemented using characteristic functions (or *labellings* as referred to in the literature). More broadly, we note that since NP-hard problems can be solved using backtracking procedures, this paper might stimulate further work to validate backtracking procedures for solving NP-hard problems that occasionally arise in the field of artificial intelligence.

References

1. Atkinson, K., et al.: Towards artificial argumentation. AI Mag. **38**(3), 25–36 (2017)
2. Baroni, P., Caminada, M., Giacomin, M.: An introduction to argumentation semantics. Knowl. Eng. Rev. **26**(4), 365–410 (2011)
3. Bistarelli, S., Rossi, F., Santini, F.: Not only size, but also shape counts: abstract argumentation solvers are benchmark-sensitive. J. Log. Comput. **28**(1), 85–117 (2018)
4. Cerutti, F., Gaggl, S.A., Thimm, M., Wallner, J.P.: Foundations of implementations for formal argumentation. FLAP **4**(8), 2623–2705 (2017)
5. Charwat, G., Dvořák, W., Gaggl, S.A., Wallner, J.P., Woltran, S.: Methods for solving reasoning problems in abstract argumentation - a survey. Artif. Intell. **220**, 28–63 (2015)
6. Dimopoulos, Y., Magirou, V., Papadimitriou, C.H.: On kernels, defaults and even graphs. Ann. Math. Artif. Intell. **20**(1–4), 1–12 (1997)
7. Doutre, S., Mengin, J.: Preferred extensions of argumentation frameworks: query, answering, and computation. In: Goré, R., Leitsch, A., Nipkow, T. (eds.) IJCAR 2001. LNCS, vol. 2083, pp. 272–288. Springer, Heidelberg (2001). https://doi.org/10.1007/3-540-45744-5_20
8. Dung, P.M.: On the acceptability of arguments and its fundamental role in non-monotonic reasoning, logic programming and n-person games. Artif. Intell. **77**(2), 321–358 (1995)
9. Dunne, P.E.: Computational properties of argument systems satisfying graph-theoretic constraints. Artif. Intell. **171**(10), 701–729 (2007)
10. Dvořák, W., Dunne, P.E.: Computational problems in formal argumentation and their complexity. FLAP **4**(8), 2557–2622 (2017)
11. Gaggl, S.A., Linsbichler, T., Maratea, M., Woltran, S.: Summary report of the second international competition on computational models of argumentation. AI Mag. **39**(4), 77–79 (2018)
12. Geilen, N., Thimm, M.: Heureka: a general heuristic backtracking solver for abstract argumentation. In: Black, E., Modgil, S., Oren, N. (eds.) TAFA 2017. LNCS (LNAI), vol. 10757, pp. 143–149. Springer, Cham (2018). https://doi.org/10.1007/978-3-319-75553-3_10
13. Modgil, S., et al.: The added value of argumentation. In: Ossowski, S. (ed.) Agreement Technologies, Law, Governance and Technology Series, vol. 8, pp. 357–403. Springer, Dordrecht (2013). https://doi.org/10.1007/978-94-007-5583-3_21
14. Nofal, S., Atkinson, K., Dunne, P.E.: Looking-ahead in backtracking algorithms for abstract argumentation. Int. J. Approx. Reasoning **78**, 265–282 (2016)
15. Thimm, M., Villata, S., Cerutti, F., Oren, N., Strass, H., Vallati, M.: Summary report of the first international competition on computational models of argumentation. AI Mag. **37**(1), 102 (2016)

Author Index

Printed in the United States
by Baker & Taylor Publisher Services